# The Artificial and the Natural

**Dibner Institute in the History of Science and Technology**

**Dibner Institute Studies in the History of Science and Technology**
George Smith, general editor

Bernadette Bensaude-Vincent and William R. Newman, editors, *The Artificial and the Natural: An Evolving Polarity*

Jed Z. Buchwald and I. Bernard Cohen, editors, *Isaac Newton's Natural Philosophy*

Jed Z. Buchwald and Andrew Warwick, editors, *Histories of the Electron: The Birth of Microphysics*

Geoffrey Cantor and Sally Shuttleworth, editors, *Science Serialized: Representations of the Sciences in Nineteenth-Century Periodicals*

Michael Friedman and Alfred Nordmann, editors, *The Kantian Legacy in Nineteenth-Century Science*

Anthony Grafton and Nancy Siraisi, editors, *Natural Particulars: Nature and the Disciplines in Renaissance Europe*

J. P. Hogendijk and A. I. Sabra, editors, *The Enterprise of Science in Islam: New Perspectives*

Frederic L. Holmes and Trevor H. Levere, editors, *Instruments and Experimentation in the History of Chemistry*

Agatha C. Hughes and Thomas P. Hughes, editors, *Systems, Experts, and Computers: The Systems Approach in Management and Engineering, World War II and After*

Manfred D. Laubichler and Jane Maienschein, editors, *From Embryology to Evo-Devo*

Brett D. Steele and Tamera Dorland, editors, *The Heirs of Archimedes: Science and the Art of War through the Age of Enlightenment*

N. L. Swerdlow, editor, *Ancient Astronomy and Celestial Divination*

# The Artificial and the Natural

## An Evolving Polarity

edited by Bernadette Bensaude-Vincent and William R. Newman

The MIT Press
Cambridge, Massachusetts
London, England

For information about special quantity discounts, please e-mail special_sales@mitpress.mit.edu

This book was set in Bembo on 3B2 by Asco Typesetters, Hong Kong.
Printed and bound in the United States of America.

Library of Congress Cataloging-in-Publication Data

The artificial and the natural : an evolving polarity / edited by Bernadette Bensaude-Vincent and William R. Newman.
p. cm. — (Dibner institute studies in the history of science and technology)
Includes bibliographical references and index.
ISBN 978-0-262-02620-8 (hardcover : alk. paper)
1. Science—Europe—History. 2. Science—Philosophy—History. 3. Philosophy, European—History. 4. Science, Medieval. I. Bensaude-Vincent, Bernadette. II. Newman, William R.
Q127 .E8A78 2007
501—dc22 2007001896

10 9 8 7 6 5 4 3 2 1

## Contents

# The Artificial and the Natural

# 1

## Introduction: The Artificial and the Natural: State of the Problem

Bernadette Bensaude-Vincent and William R. Newman

With each passing day the traditional boundary between the natural and the artificial becomes less distinct. Consider a few examples from the realm of biology. Bioengineering has proposed strawberries with genes taken from fish, "genetic artists" boast of having made a phosphorescent rabbit by implanting DNA from jellyfish, and in Mexico, genetically modified "Frankencorn" has possibly made its way into the wild, hybridizing with varieties of maize hitherto untouched by humans. Are such living entities rendered "artificial" by the human intervention that modified their genetic makeup? If so, does it not follow that hybrids produced by old-fashioned cross-breeding are human-made as well, and that every tomato or pear that we eat is an "artificial" product? Perhaps the reader will recoil at this suggestion, since it would imply that virtually every fruit, vegetable, meat, or drink that serves for our nutrition is factitious. So let us imagine for the moment that the products of hybridization and contemporary biotechnology are "natural." In that case, further problems arise. If we do not label the products of bioengineering as artificial, then what right does any human-made product have to the term? Why should a polymer or dyestuff made by tinkering with coal-tar molecules be any more artificial than a rabbit whose DNA has been altered so that it glows in the presence of a certain wavelength of light? If we turn to the realm of cold, hard silicon, similar questions emerge. Computer science has bridged the chasm between man and machine, giving us "Deep Blue," the IBM product that defeated Garry Kasparov at chess. Unsatisfied with this conquest, the computational laboratories of MIT are building robots that simulate human emotion, while researchers at Carnegie Mellon are devising ways humans may one day give up their biological bodies, allowing computers to become the recipients of their consciousness, digitized and uploaded into a suitable machine-readable matrix.[1] Assuming the eventual feasibility of this science fiction scenario, where then would the line be drawn between an artificial and a natural human being? The triumph of Ian Wilmut and his team in engineering a sheep cloned from a mammary cell just a decade or so ago now seems a mere memory—a distant prelude to the polyvalent symphony of human intervention that is sure to follow.

The recent products and future dreams of biotechnology and artificial intelligence present striking challenges to the commonsense distinction between the natural and the artificial. But in reality this dichotomy has always been confounded by human activities in the form of even the most primitive machines and technologies. All materials, whether natural or artificial, are first extracted from nature and then processed according to human purposes. Cotton, wool, or wine—items that we usually consider natural—are in fact the result of a long manufacturing process including many sophisticated chemical and mechanical operations. At the same time, artifacts are never really unnatural. As physical and chemical systems they belong to nature and generate a number of effects independent of the intentions of their designers. And of course our artifacts have such a profound effect on ecosystems that their mass production increasingly raises important environmental issues.

If we turn from individual "natural products" to nature in the wild, it is clear that we will fail to find the absolutely natural here either. Over centuries and millennia of agricultural and industrial activities nature has been deeply reconfigured by humans. "Native forest" in the sense of woodland absolutely untouched by humans exists nowhere but in the human imagination. No part of the earth has been completely unaffected by the effects of human technologies. This is by no means a recent discovery resulting from an increasing concern with environmental issues. As early as the eighteenth century, when the first artificial soda was synthesized and when gardens with sheep grazing in meadows became fashionable—supplanting the jardins à la française—Jean-Jacques Rousseau clearly realized that the state of nature was an intellectual construction, an indispensable fiction for ascertaining the foundations of the political order.[2] Given these and other considerations, we should reasonably conclude that there is no such thing as a great divide between nature and art. More precisely, instead of opting for an absolute distinction of quality between the artificial and the natural, one should accept only a gradual distinction of degree.

But the omnipresence of this divide in our culture and its persistence in contemporary debates cannot be overlooked. As Roald Hoffman pointed out in *The Same and Not the Same*, the "rational" arguments used by modern chemists in order to fight the popular prejudice against chemicals are largely useless, because they ignore the cultural aspects of the issue.[3] The concept of nature functions and has always been used as a cultural value, a social norm, and a moral authority. Debates over art and nature generally conceal the broad questions that undergird and drive them: is *technē* a continuation of nature's activity (tools being viewed as something like the prolongation of a person's hand), a rebellion against nature, or a challenge to nature? The nature of technology and its legitimacy, the situation of humans as technicians among other animals, and the status of artisans in society are among the broad issues at stake. Because of the importance of such philosophical implications and cultural roots in all the debates over the impact of technologies, we cannot simply dismiss the distinction between art and nature as a "popular prejudice" or as an "irrational nostalgia for the past." Rather we have to disentangle the cultural roots of current debates about new technologies.

How then can we expect to reflect on our current situation without having taken a census of the past? The present book originated out of precisely this perceived need, for the editors were struck by the absence of any combined attempt by specialists of different periods and various disciplines to consider discussions of art (in the broad sense) and nature in their respective fields of expertise. It is not too much to suggest that the resulting book is the first collective effort to devote itself specifically to the issue of the artificial and the natural over the *longue durée*, incorporating the perspectives of historians of science, art, and philosophy.

## Changing Boundaries

The task is difficult, since discussions about art and its power form a deep and perennial issue with roots in the civilization of classical antiquity. Given the paucity of research that scholars have devoted to this subject, we could hardly hope to engage the topic on a comprehensive scale. Rather than aiming at a grand history of the art-nature dichotomy in the Western world, we have therefore assembled a collection of disciplinary and chronological core samples in an effort to understand them in their specific contexts. For various periods, we try to bring the concepts of nature and art into confrontation with the scientific and technological practices of their times and with contemporaneous philosophical and religious debates. Even the few selective samples that we have brought to light reveal surprising patterns. Perhaps the most striking result that emerges from the following chapters is the fact that while the opposition between art and nature is itself a major leitmotif throughout the history of the West, the forms that this dichotomy takes are themselves far from stable or constant over the centuries. On the contrary, the divide is continuously challenged and reassessed. The pseudo-Aristotle of the *Mechanical Problems*, for example, was amazed by the power of circular wheels set in tandem to drive mechanical devices. In recognition of the circle's ability to make heavy things move against their natural tendencies, pseudo-Aristotle remarked that mechanics works "against nature" (*para physin*). Centuries later, the spell of machines had grown so much in power that Descartes and his followers would reject this now-hoary dictum, claiming that machines were not only not contrary to nature, but that nature itself was mechanical. If one could only extend his field of vision to the suitably small, he would see nothing but mechanical operations underlying our sensible universe. On the other hand, the historian can find many instances where the power of a given artificial product to challenge nature diminishes over time instead of becoming more compelling. The superficial patinas and colorations applied to metals by late antique alchemists were thought in some cases to yield products that were "superior to the natural." In the Middle Ages, however, such products came to be routinely disparaged by alchemists themselves as "sophistical," fraudulent, and artificial. Clearly the bar had been raised in the challenge between artifice and nature, partly as a result of technological progress in the art of alchemy.

The shifting boundaries between the artificial and the natural are revealed in another way as well that will appear throughout this book. We refer to the absence of any clear and unambiguous terminology for distinguishing artificial and natural products in the English language (or any language that we know of). This may seem a surprising claim, given the rich vocabulary at our disposal for distinguishing the genuine and natural from the inauthentic and artificial. But we have already pointed to some borderline cases where living creatures have been modified to the point that many would call them unnatural and even artificial. Indeed, living things such as plants, seeds, genetic materials, and transgenic animals are considered patentable in a number of countries, and fall under the same legal strictures as "inventions" on the assumption that they result from human art or intervention.[4] But are they artificial in the sense that an imitation leather belt or a plastic woodgrain desk bears that attribute? Surely not, one may say, since a bioluminescent rabbit is a new creation, not an imitation of anything else. The rabbit, moreover, has not been created out of whole cloth, but is merely a normal rabbit whose nature has been tweaked by science. At what point, then, does the natural object cease to be natural, and become artificial?

## A Spectrum of Relations

The distinction between the artificial and the natural has traditionally been addressed by considering the limiting cases supplied by examples of pure artisanship and unaided nature. A most influential instance of this approach can be seen in book 2, chapter 1 of Aristotle's *Physics*. The Stagirite argues there that a bed cannot be natural, since if a planted bed could grow and bloom, it would not sprout beds, but trees. The shape and structure of the bed are merely human impositions on the unchanged matter that remains a natural product. The *physis*—or "nature"—of the wood itself remains unaffected by the artificial form imposed on it by the carpenter. But in a later passage—often ignored by modern commentators—in the same book of the *Physics* (chapter 8) Aristotle undermines this clear distinction, pointing out that there are two sorts of arts—those that imitate nature and those that lead it to a greater state of perfection. The physician's art can be taken as an example of the second sort, since it brings the diseased body to health without changing the essential nature of the body itself. Well and good, one may say, but where does this perfecting process stop? To return to the example of alchemy, the scholastic practitioners of that discipline were often wont to say that they did not make artificial precious metals but genuine natural ones, since they merely perfected the base metals, achieving what nature could have done beneath the earth if there had been enough time and sufficiently pure materials. Pushing this line still further, they argued that their art could make metals and minerals better than those available in nature, and that such products could sometimes serve in turn as macrobiotic medicines for humans. Not only could the natural human lifetime be greatly extended by the ingestion of such products, some alchemists argued; it was possible even to refashion humans themselves by a process of artificial incubation in a flask. The result of this process, the homunculus, would have remarkable powers unshared by other human beings, such as the gift of preternatural intelligence. All of this discussion was couched in the language of the traditional debate between the artificial and the natural, and the alchemists almost invariably saw themselves as perfecting nature in the Aristotelian sense.

Despite the extravagant character of the homunculus discussion, one can see how it underscores the problems of the approach taken by

Aristotle in the famous example of the bed. The limiting cases of the purely natural tree and the utterly artificial bed allow for no intermediate gray area, and are inadequate in trying to determine whether something like a homunculus, or for that matter a transgenic rabbit, is a real human or a real rabbit. Or to use a less exotic example, it would be difficult to employ Aristotle's criteria in determining whether a product of grafting, such as the tree created when an apricot scion is inserted on a plum tree, remains a natural plum tree. Could one deny that the physis of the tree has been changed by human intervention, given that it now produces apricots rather than plums? How far can the nature of a thing be pushed before that thing ceases to belong to its original species in the natural world? The more general question is far from trivial, and points to the inherently relativistic character of the categories "natural" and "artificial."

The problem is only intensified for the Aristotelians by the position that the Stagirite takes in other works beyond the *Physics*. Book 4 of Aristotle's *Meteorology*, a work whose authenticity has been questioned by philologists (though not in the premodern era), argues that the "artificial" boiling and roasting carried out in a kitchen are analogous, and perhaps identical, to processes that occur naturally within the earth. After all, these arts and others originated from human attempts to mimic nature, as Aristotle points out. But this opens up an entirely distinct avenue for asserting the naturalness of human products. If we take the emphasis off of the product itself, and focus on its mode of production, then we can say that something as seemingly unnatural as glass is actually a product of nature. After all, by one interpretation of *Meteorology* 4 the heat employed in fusing sand and alkali together into a hard, clear substance is the same as the heat that melts stone in volcanoes. Since we use nature's own agencies in making glass, the product is itself natural by this line of reasoning. Hence one tradition in medieval scholasticism argues that manufactured glass is a "stone," and just as natural as any stone found in the world at large.[5]

Such ambiguities are hardly the exclusive province of our ancestors. We moderns often discuss the differences between the artificial and the natural without explicitly considering the kinds of action that art is said to exert on nature. Verbs matter here insofar as the substantives *nature* and *art* are defined by their mutual relation. Does art mimic nature? Represent nature? Simulate nature? Complete nature? Improve on nature? Counterfeit nature? Violate nature? In addition to the difficulties engendered by ignoring these verbal distinctions, a group of concepts and terms clustering around the troublesome idea of imitation provides particular problems. When we say that something is an artificial imitation of a natural product, do we mean that the former is necessarily different from the latter in some respect other than the mere fact that it is human-made? Let us consider the nineteenth-century examples of celluloid and ivory, which clearly bore the same relationship to one another as fake fur and real fur—one was viewed as the genuine thing, the other as a poor substitute, a sort of counterfeit. But what about the artificial vitamin C manufactured by pharmaceutical supply houses and the natural vitamin C extracted from rose hips? The ascorbic acid that makes each of these substances capable of being called "vitamin C" is the same in both cases. We cannot simply say that one is fake and the other real.

For this reason, chemists speak of "synthesizing" natural products —that is, reproducing the very molecules that nature employs—when they want to express a relation of identity between the natural and the manufactured product. But even chemists acknowledge that while they can produce pure substances by synthetic processes, they cannot introduce all of the impurities that are typically found in a natural product. As Roald Hoffmann has pointed out, the vitamin C from rose hips will contain a host of other molecules in varying proportions that the chemist does not try to reproduce. Although the impurity of natural substances is not a problem for someone making ascorbic acid, it does present huge difficulties for those trying to reproduce natural odors and tastes. Anyone who has tasted synthetic strawberry or watermelon flavoring can vouch for the reality of the artificial-natural dichotomy, even if the main active ingredients of the manufactured flavoring are identical to the preponderant molecules in the natural substance.

In a certain sense, then, even the products of synthetic organic chemistry can be viewed as ersatz—they often replace (*ersetzen*) a natural substance, but are not always equivalent to it in every respect. Similar problems arise in the laboratory manufacture of precious stones, natural pigments, and medical products. The synthetic product is often too pure to do the job of the natural one and hence the former is artificial in at least two senses—first, by the brute fact that it is a product of human intervention, and second, because it is chemically or physically different from its natural exemplar. Still, one has to admit that a synthetic pure substance bears a closer relationship to its natural model than does an outright counterfeit, like margarine, polystyrene pearls, or simulated leather. The latter are mere substitutes for a natural product that work by deluding the senses. Just as we may be fooled by a trompe l'oeil painting when looking at it from a distance, but recognize its illusory character on approaching, so taste, touch, or smell immediately reveals the fraudulent character of imitation butter, pearls, or hide (at least in their incarnations as of 2007).

## Conflicting Perspectives

As we have now seen, the terms *natural* and *artificial* mean quite different things, depending on the context and point of view in which they occur. Indeed, one can reach quite opposite conclusions when starting from different standpoints. A thing's "naturalness" or "artificiality" has one meaning when we are talking about its origin (extracted from nature versus human-made) and quite another when we are discussing its inherent qualities. For instance, metals are "natural" since they are extracted from nature whereas plastics are "artificial" since they have to be synthesized. But in the language of ordinary life and commerce, plastics are often more "natural" than metals because they are more flexible and soft, and less conductive of heat and electricity, making them more like biological tissue. Similar conflicts occur when we are talking of the end product itself or of its mode of manufacture. Let us first consider product rather than process—in this case, an important distinction hinges on whether the producer has consciously imitated a natural product or rather altered a naturally occurring thing in some way that yields a result not found in nature. Assuming that the product is in some sense an imitation, however, the measure of its artificiality is radically conditioned by the extent of variation between the product and its exemplar—fake leather is not artificial in the same sense as synthetic strawberry flavoring, even though both may be obviously different from their natural models. In the second case, where process is the determining factor, the range of artificiality is itself a historically conditioned artifact of the manufacturing techniques present in a given time. To give but one example of this fact, the Italian painter of the fifteenth century Cennino Cennini labeled such pigments as vermilion produced by subliming mercury with sulfur and minium made by calcining lead as "artificial." Few of us today (with the possible exception of those who specialize in the manufacture and use of "historical pigments") would think of these chemicals in such terms, any more than we call metallic silver or copper "artificial" merely because they have been reduced from a sulfide ore at some point in their existence in order for us to arrive at a more useful product. As the manufacturing process for a particular yield becomes more commonplace, the distinction between the artificial and the natural loses its force.

## Conceptual and Material Practices

Needless to say, the full range of nuance in the antithetical terms *artificial* and *natural* cannot be addressed in a single introduction. Rather than dwelling on generalities, the subsequent chapters invite the reader to consider the interplay between the conceptual dichotomy "art-nature" and specific practices such as medicine, painting, collecting, building machines, or performing chemical syntheses. To what extent can the materials worked on and the actions of the technician transforming the raw materials shape and reshape our concepts of nature, or of life itself? Or to invert the question, how much have the cultural patterns of representation of nature and art influenced specific technological changes? Fierce debates over the artificial and the natural have been raised by very diverse practices over the course of time, including medicine, alchemy and metallurgy, mechanics and the making of automata, agriculture and gardening (grafting), breeding (hybridization and selection), painting and sculpture, chemical synthesis, materials technologies, cybernetics and artificial intelligence, genetics, and even patenting (since the distinction between art and nature, invention and discovery, is the basis of most patent legislations).

This book cannot cover such a wide range of fields. We have been forced to leave out many interesting subjects, such as the eighteenth-century enthusiasm for artificial flowers or Darwin's comparison between natural selection and breeders' selection.[6] However, in order to give a sense of the diversity of practices that formed the core of the debates over time, we have not limited the discussion to the artificial-natural divide in the sciences alone. Until very recent times, the term *art* referred both to the field now called "fine arts"—mainly painting, the plastic arts, and literature—and to what is now called "technology." One major advantage of the historical perspective developed in this book is to go beyond the recent divide between "the two cultures" and to restore *art* to something like its full set of traditional meanings. Without an understanding of the original domain of *art*, one cannot appreciate the development of the art-nature dichotomy. Classical civilization already had problems with the issue of trompe l'oeil technique in visual art and persuasive rhetoric in poetry, both of which were thought by

Plato and various others to be morally questionable. Over the course of time, similar arguments have arisen in fields as diverse as alchemy and bioengineering. We therefore invite the reader to consider the similarities and differences that emerge from these discussions in their respective disciplines as they develop over the period of emerging Greek civilization up to the present.

## Overview of the Book

Our book begins with Heinrich von Staden's magisterial contribution "Physis and Technē in Greek Medicine" (chapter 2). Von Staden opens his chapter with the observation that physis—nature—had multiple meanings and uses in the Greece of the Hippocratic writers, beginning in the fifth century BCE. Not only was there nature at large, but also the natures of individual beings, the natures of their parts, and the natures of the poisons and remedies that acted on those parts. In contrast to these multiple natures was the technē—the art of medicine—which employed the *dynameis* or powers of drugs, regimens, and of the body itself, to combat disease. At times, Von Staden points out, the adversarial relationship between the medical art and disease expanded into a surprisingly general account of the relationship between art and nature as a whole. Hence, in a seemingly Baconian fashion, the Hippocratic work *On the Technē* speaks of art as violating nature by means of "forcible constraints." Using the metaphor of judicial torture, the Hippocratic author recommends that drugs be employed to make the patient evacuate humors that will reveal the inner state of his body—hence nature is forced to inform on itself in the same way that a slave would be forced to reveal incriminating information. Although this strikingly agonistic opposition between art and nature was not developed further by Greek physicians, Von Staden presents additional material to show that the possibly artifactual character of dissection and vivisection led to a reluctance on the part of the ancients to develop anatomy to its fullest potential. Finally, he considers the remarkable work of the Hellenistic physician Erasistratus, who not only rejected the Greek taboo on vivisection, but who explicitly viewed the body in mechanical terms. In a way that will bring to mind the millennia-later discoveries of Harvey and Descartes, Erasistratus seems to have borrowed the image of the recently invented double-action pump from the engineer Ctesibius and used it to explain the workings of the human heart.

Chapter 3, Francis Wolff's "The Three Pleasures of Mimesis According to Aristotle's *Poetics*," carries the discussion of technē and physis beyond the realm of medicine and natural philosophy into the arena of aesthetics. As Wolff points out, early modern writers on the fine arts tended to employ arguments taken from Platonic and Aristotelian discussion of the arts as a whole (including all branches of technology) when writing about painting, sculpture, and poetry. In doing so, however, they often failed to note that Aristotle himself had carefully separated off the fine arts and created a separate category for them in his *Poetics*. There Aristotle designates poetry, drama, and the visual arts as *technai mimētikai*—mimetic arts, because their raison d'être lies in the realm of mimicry. Wolff argues, nonetheless, that Aristotle saw the technai mimētikai as being analogous to the other arts in an important respect. Most of us are familiar with Aristotle's theory of four causes, developed in book 2, chapter 3 of his *Physics* and elsewhere. In the famous example of a statue, for example, Aristotle argues that the material cause is the bronze out of which the effigy is made, the efficient cause is the sculptor or his hands, the formal cause is the idea of the statue in his mind, and the final cause is the purpose for which the statue is made. Wolff argues that a parallel system of causation implicitly operates in Aristotle's discussion of the mimetic arts. By this line of reasoning, the material cause is the medium that the artist employs—shapes and colors in painting, words for literature, rhythm and melody for music, and so forth. The formal cause is the idea in the artist's mind of the thing represented, the efficient cause is the agency of representation, for example the narrator or the actors in epic poetry and drama, and the final cause for all the technai mimētikai is pleasure—the pleasure of mimicking and of observing mimicry. Wolff thereby provides an important new element to our understanding of the concept of "art" in antiquity.

The book's fourth chapter, "Art and Nature in Ancient Mechanics," by Mark J. Schiefsky, provides a nuanced and original reading of the complex relationship between art and nature in the writings of ancient mechanical engineers. Beginning with the foundational *Mechanical Problems* of pseudo-Aristotle, Schiefsky shows that this text cannot be used to support the mistaken idea that Aristotelian science ruled out knowledge of nature arrived at by employing "artificial" techniques and interventionist processes (this widespread view has elsewhere been referred to as the "noninterventionist fallacy").[7] Schiefsky is particularly concerned with discrediting the oft-cited position of the historian Fritz

Krafft that Greek mechanicians viewed their discipline as "tricking" or subverting nature itself, with the implication that this made mechanics an invalid way of learning about nature. Indeed, Schiefsky points out that Aristotle himself often draws elaborate analogies between art and nature, and at times even seems to blur the distinction between the two. In support of his position, Schiefsky offers an important and controversial new reading of *Physics* II 8 199a15–17, where Aristotle distinguishes between arts that mimic nature and those that complete nature or bring it to a state that it could not otherwise attain. In Schiefsky's novel reading, mechanics itself could be viewed as an art that completes nature, thus throwing further doubt on the thesis of Krafft and others that the ancient mechanists viewed their discipline as a field necessarily occupying an antithetical relationship to nature and its products. Schiefsky further argues that mechanics operated in a way similar to an Aristotelian subordinate or "middle" science in the Stagirite's view, occupying much the same relationship to physics as did harmonics, astronomy, and optics. Schiefsky's sustained and detailed defense of this fresh viewpoint is bound to open up new questions for scholars of ancient, medieval, and early modern mechanics, with ramifications leading up to Galileo and Descartes.

William R. Newman's contribution, "Art, Nature, Alchemy, and Demons: The Case of the *Malleus maleficarum* and Its Medieval Sources" (chapter 5), carries the Aristotelian analysis of art and nature into a highly unexpected venue, namely, the use that scholastic theologians and inquisitors made of alchemy in determining the power of demons and witches in the Middle Ages and early modern period. From the time of Albertus Magnus in the mid-thirteenth century, scholastic authors used alchemy as a test case within their highly Aristotelian thought world for determining the limits of human and demonic power. Demons were typically thought to be restricted to the use of technology—they could not create by mere will alone as God was said to do, but had to join active natural substances to passive ones in order to achieve their ends. Alchemy, on the other hand, was the one art that offered par excellence to transmute species by imposing new substantial forms on matter. If humans could really convert one metal into another by transmuting its species, then demons should be able to do the same. And if such radical changes wrought by humans were admitted as a general principle, then demons and their servants, the witches, should be able to alter matter by imposing new forms that would also make it possible for them to work horrific effects on their enemies, resulting in disease, deformation, and even death. Evidence of the widespread use of alchemy as a technological benchmark receives significant support from the fact that Heinrich Kramer and Jakob Sprenger employed the aurific art in precisely that fashion in the most influential witch-hunting manual of all time—the *Malleus maleficarum* of 1487, a work that would become symbolic of the great witch hunt that followed over the course of the next two centuries.

Dennis Des Chene's chapter, "Forms of Art in Jesuit Aristotelianism (with a Coda on Descartes)" (chapter 6), deals mainly with the changing fortunes of the art-nature relationship in Jesuit commentaries and textbooks of the sixteenth and seventeenth centuries. As Des Chene points out, Jesuit authors tended to argue that human art was secondary in that it could only imitate nature, superficial in that it could only employ local motion and outward figure, and subordinate in that its artificial products lack the innate activity of natural ones. Nonetheless, certain arts, such as the making of automata, the art of magic, and alchemy, seemed to challenge this devaluation. Of these three arts, the Jesuits managed to make short work of automata and magic, reducing them to either fraud or manipulation of outward figure. That left alchemy, which the Coimbrans—relying on the same scholastic tradition outlined in chapter 5—admit as an area where art can possibly challenge or exceed the products of nature. Nonetheless, it remained for Descartes to restrict nature to the same status as art by reducing matter to extension and eliminating the powers and virtues that characterized natural substances for the Aristotelian. The result of this elision appears most fully in Descartes's *Dioptrique*, where the philosopher prescribes a cyborglike combination of man and machine in the form of a water tube implanted in the eye to improve vision. Despite the novelty of Descartes's idea, one cannot help but be reminded of Erasistratus's fusion of mechanics and medicine as described by Von Staden. The combination of a mechanistic matter theory in both authors, along with a physiology that blurs the distinction between art and nature, suggests that these ideas have an innate contextuality rather than acquiring their juxtaposition merely from coincidence.

The explicit contest between art and nature that we see in Des Chene's Jesuit commentators—where art is generally the loser—appears in a very different light in the paintings of Giuseppe Arcimboldo, as analyzed by Thomas DaCosta Kaufmann in chapter 7, "The Artificial and the Natural: Arcimboldo and the Origins of Still Life." Arcimboldo's style is typically viewed as the height of artifice and chimerical fantasy in painting, as we find it described by his admirer Gregorio Comanini. Despite the fame of Arcimboldo's playful "composed heads," compositions made up of sea life, combustibles, and so forth, Kaufmann argues convincingly that we should also situate the Milanese painter at the origins of the early modern still life. Arcimboldo compiled careful studies of individual animals before integrating them into his illusionistic compositions, displaying a keen desire to imitate nature by means of his art. The skill with which he carried out such nature studies reveals itself in a particular type of painting in which he excelled, where the upright painting shows a head, but when inverted, it reveals a still life. Kaufmann dwells on one recently discovered Arcimboldo painting of this sort—a head that shows a fruit basket when inverted. Here as elsewhere in Arcimboldo's art, the painter was seen by his contemporaries as being engaged in a contest with a personified Nature. Whereas she could only make humans from human members, Arcimboldo was able to weave plants and their parts together to compose his humans—hence he had not only challenged Nature but surpassed her. Art's use of figure and surface derided by Jesuit authors now became the very means of outdoing Nature in the game of producing very different compositions from the same pictorial elements.

The sources used by Anthony Grafton for his "Renaissance Histories of Art and Nature" (chapter 8) display many of the characteristics found in Kaufmann's. Grafton's chapter focuses on the many Renaissance discussions of human invention and its relationship to nature that preceded the technological optimism of Francis Bacon and Tommaso Campanella. Locating this positive view of technē's progress partly in the princely tradition of *Kunst- und Wunderkammern*, Grafton sees a similarity between Samuel Quiccheberg's 1565 "theater" of "artificial and miraculous things" and the inventories of human artifice described by Campanella and Bacon. But the interaction of art and nature was not always seen as one of linear progress, as Grafton also points out. The jurisconsultant Guido Pancirolli and his commentator Heinrich Salmuth both stressed that the ancients had possessed arts now lost, though perhaps made up for by the "modern" inventions of Greek fire and the compass. The same emphasis on art's ability to evolve, and in so doing to surpass nature, is also seen in another of Grafton's cases, the famous humanist and writer on the visual arts, Leon Battista Alberti. As in Kaufmann's treatment of Arcimboldo and Comanini, Grafton stresses the claim of Alberti that art can even exceed the creative powers of nature, in this case by creating the composite likeness of a perfect female. A similar expression of art's power entered into Alberti's discussion of engineering feats like the building of the Florentine duomo, and this rhetoric made its way, via the unlikely source of Heinrich Cornelius Agrippa's *De occulta philosophia,* into the revived discipline of Renaissance magic. Agrippa and his heirs, such as the Jesuit writer Gaspar Schott, emphasized that their "mathematical magic" could surpass and even overpower nature by means of marvelous machines. As Grafton concludes, then, such early modern genres as the Kunst- und Wunderkammer literature, the histories of the arts, and the extensive writings on learned magic all emphasized a growing fascination with the topos of art progressing beyond nature.

Horst Bredekamp's contribution, "Leibniz's Theater of Nature and Art and the Idea of a Universal Picture Atlas" (chapter 9), picks up chronologically where Grafton's leaves off. Beginning with a discussion of the current vogue of recreated Kunst- und Wunderkammern, Bredekamp points to the interesting contrast between the highly visual character of these protomuseums and the abstract character of our increasingly digital world. As Bredekamp makes clear, this interesting antithesis finds a prototype in the work of the brilliant codiscoverer of the calculus, Gottfried Wilhelm Leibniz, who was also an enthusiast of Kunstkammern. In a way that will be surprising to those who know Leibniz only as a mathematician or philosopher, Bredekamp manages to link his interest in the visual organization of knowledge to the great Renaissance pictorial mnemotechnics of writers such as Campanella, Johann Valentin Andreae, and above all Jan Amos Comenius. Perhaps it should come as no surprise that Leibniz the librarian and maven of universal languages would also have an interest in imagisticially organized knowledge, and yet the degree of his enthusiasm for this subject is indeed impressive. For a quarter of a century he promoted a scheme for an *Atlas universalis*, a pictorial compendium of knowledge that would inculcate the arts and sciences into the tender brains of young students. Bredekamp concludes by suggesting that Leibniz's emphasis on the visual as a means of learning coupled with his love of mathematical abstraction may contain clues to our own bipolar culture, with its seemingly antithetical love of the image and the algorithm.

The case of Spinoza presented by Alan Gabbey—in chapter 10, "Spinoza on the Natural and the Artificial"—is of special interest for our topic because the Dutch philosopher refused any "artificialization" of nature. This was rather exceptional in a period when the mechanical explanations that prevailed in natural philosophy led to descriptions of nature as a machine and God as its supreme engineer or as a clockmaker. Moreover, Spinoza's metaphysics, as Gabbey emphasizes, denied any possible distinction between nature and human art. All objects whether human-made or extracted from nature should be the products of necessity. Even the artists' intentions derive from a natural necessity rather than from free will. Gabbey argues, however, that Spinoza used a dual language for art. As an artisan, a lens grinder, he could not realistically consider lenses as a simple product of nature's necessity. As a political thinker he insisted on the artificial nature of states and governments. How could he sustain the contradiction? Here is a splendid illustration of the interplay between practices and concepts. As a metaphysician, aiming at understanding the eternal truth, Spinoza blurred the distinction between art and nature. But as a practioner of one art concerned with human welfare, Spinoza assumed a difference between nature and art. These conflicting views suggest that whatever the rational arguments against the dichotomy between art and nature, it remains implicit in all human actions and indispensable for understanding them.

Creating artificial life is an old interest dating back to antiquity and currently pursued in many laboratories. It does not mean that this project is the expression of one and the same long-standing project. Rather, as Jessica Riskin shows in chapter 11, "Eighteenth-Century Wetware," various attempts to create lifelike artificial creatures mirror the changing views of life and matter, or of humans, animals, and machines. In the long story of artifical life, Riskin singularizes the second half of the eighteenth century as a crucial moment when artisans and engineers designed automata for testing the mechanistic understanding of life shaped by materialist philosophers. The spectacular automata built by Jacques Vaucanson or by the Jacquet-Droz family were much more than clock mechanisms performing rigid motions. Their attempts went so far as to simulate the soft and wet texture of living matter. They performed not only locomotion, writing, or music playing but also inner physiological processes such as digestion and defecation. According to Riskin, it would be unrealistic to consider the theatrical performances of such automata as counterfeits intended to fool the spectators, even though they provided—and can still provide when they work in museums—the kind of pleasure that Aristotle conferred on the mimetic arts. These machines, Riskin argues, were experimental models used in the same manner as modern simulations. They worked in two ways: they illustrated the mechanization of life since they were animal-like machines and at the same time they animated the machinery. Finally,

Riskin emphasizes the similarity between late eighteenth-century automata and late twentieth-century artificial life developed as simulations of life.

Nineteenth-century attempts at reproducing the products of life are the subject of John Hedley Brooke's chapter, "Overtaking Nature? The Changing Scope of Organic Chemistry in the Nineteenth Century" (chapter 12). Brooke revisits a landmark episode often presented as a triumphal step in art matching nature. Friedrich Wöhler's production of urea in 1828 was the first in vitro synthesis of a substance up to then exclusively produced by living organisms. Nineteenth-century chemistry textbooks assumed that this synthesis proved that chemists had the power to reproduce organic compounds artificially and that their art destroyed the metaphysical belief in the existence of a vital force. In fact Wöhler's synthesis could not—and did not—challenge vitalism or the theological view of nature. First, it was not a complete synthesis since it started from organic products such as horn. Second, the replication of a natural product did not imitate nature's process. Reconsidering this famous synthesis in the long perspective of the debates over art and nature as well as in the context of nineteenth-century culture, Brooke develops an alternative view of its impact. Far from securing the triumph of materialism, the Faustian ambitions of the synthetic chemists prompted objections that favored further distinctions between products and processes and between various types of syntheses. Moreover, Brooke emphasizes the contrast between the triumphalist rhetoric of the champions of organic synthesis and its dramatic effect on the discipline of chemistry, whose theoretical framework was consequently split between inorganic and organic chemistry. Despite the vaunting verbiage used by the promoters of organic syntheses, these syntheses in reality challenged the identity and consistence of chemistry as a field.

The contrast between the chemical optimism raised by nineteenth-century attempts at synthesis and the cultural perception of chemical synthesis toward the end of the twentieth century is striking. The antithesis between "chemical" and "natural" did not work to the benefit of chemistry. To what extent do the deep changes that affect the cultural image of chemistry result from changes in the aims and practices of chemical synthesis? In chapter 13, "Reconfiguring Nature through Syntheses: From Plastics to Biomimetics," Bernadette Bensaude-Vincent addresses this question through a review of three different strategies of synthesis: polymerization, combinatorial chemistry, and biomimetic syntheses. She argues that synthetic polymers favored a view of nature as a rigid set of limited resources opposed to the plasticity and profusion of synthetic artifacts. By contrast, the more recent strategy of making drugs through combinatorial chemistry favored the view of nature as a huge library of resources gathered through random processes of combination. Mimicking nature in this case means mimicking the stupid and blind process of natural selection. Art thus loses one of its major distinctive features, intentionality. An alternative view of nature as an unrivaled engineer underlies the attempts at making artificial materials with characteristics analogous to the variety of properties offered by natural materials such as muscle, blood, or spider silk. Hence we arrive at the conclusion that the notions of nature and art are mutually constructed. Nature and art are continuously and mutually redefined in coordination with the intellectual and materials strategies used for designing artifacts.

Thus the dance that Roald Hoffmann imagines at the end of our book—in chapter 14, "Concluding Comments"—goes on over the centuries. Art and nature are two inseparable partners whose movements continuously shape and reshape the map of those cultures that have inherited the ancient yet modern distinction between technē and physis.

## Notes

1. For the obtaining of antifreeze protein from winter flounder, see http://www.actahort.org/books/484/484_99.htm. For Eduardo Kac and the bioluminescent rabbit, see http://www.ekac.org/. For the argument that transgenic maize has made its way into the wild in Mexico, see David Quist and Ignacio H. Chapela, "Transgenic DNA Introgressed into Traditional Maize Landraces in Oaxaca, Mexico, *Nature* 414 (2001): 541–543. For robotics and artificial intelligence, see Rodney A. Brooks, *Flesh and Machines: How Robots Will Change Us* (New York: Pantheon Books, 2002), and Hans Moravec, *Mind Children: the Future of Robot and Human Intelligence* (Cambridge, MA: Harvard University Press, 1988).

2. Monique Mosser and Georges Teyssot, *Histoire des jardins de la Renaissance à nos jours* (Paris: Flammarion, 1991): Jean-Jacques Rousseau, *Discours sur l'origine de l'inégalité parmi les hommes* (Paris, 1755). Bernadette Bensaude-Vincent, "L'alchemie du jardinage," in Georges Farhat, ed., *André Lenôtre. Fragments d'un paysage culturel* (Paris: Musée de l'Ile du France, 2006) 152–161.

3. Roald Hoffman, *The Same and Not the Same* (New York: Columbia University Press, 1995), 87–125.

4. Jean-Pierre Clavier, *Les catégories de la propriété intellectuelle à l'épreuve des créations génétiques* (Paris: Lharmattan, 1998); J. Rifkin, *Le siècle Biotech: Le commerce des gènes* (Paris: La découverte, 1998); American College of Medical Genetics 1999, *Position Statement on Gene Patents and Accessibility to Gene Testing*; Maurice Cassier and Jean-Pierre Gaudillière, "Recherche, médecine et marché: la génétique du cancer du sein,

*Sciences sociales et Santé* 18 (2000): 29–51; UNESCO, Division of Human Sciences, "Ethics, Intellectual Property and Genomics, International Symposium, Paris, January 30–February 1, 2001.

5. William R. Newman, *Promethean Ambitions: Alchemy and the Quest to Perfect Nature* (Chicago: University of Chicago Press, 2004), 247, 266.

6. Christine Velut, *La rose et l'orchidée: Les usages sociaux et symboliques des fleurs à Paris au XVIIIe siècle* (Paris: Larousse, 1993), 111–134.

7. For the expression "noninterventionist fallacy," see Newman, *Promethean Ambitions*, chapter 5.

# 2

## Physis and Technē in Greek Medicine

Heinrich von Staden

Natures differ greatly from natures.

—Hippocrates, *On Joints*

### Introduction: The "Art" and the Many "Natures"

The recurrent Hippocratic distinction between a thing's visible, external form or appearance (its idéa or eîdos), its invisible but inferentially knowable capacities and susceptibilities (its dynameis), and its invisible "nature" (physis) is of considerable significance for some Hippocratics' understanding of the relation between nature and technē.[1] Such distinctions reflect a new conceptual armature—in large measure appropriated and expropriated from ordinary language—that was central to the well-known efforts of a number of Greeks, starting in the fifth century BCE, to explore innovative, nonmythic ways of giving an account of the human body, of health, of illness, and of its therapy. Some of these Greeks were philosophers, others medical practitioners, but they tended to have in common that they tried to develop accounts of the body, in illness and in health, that would leave no room for unpredictable interventions by divine caprice or for magic (even if most of them, including some Hippocratics, did not abandon the categories "divinity" and "the divine"). Despite the fact that they had notoriously disparate doctrinal, stylistic, rhetorical, and methodological commitments, many of the innovators seem to have shared the belief that there are unbreachable regularities in the natural world (i.e., also in the body)—regularities that not even the gods violated. Furthermore, many medical authors believed that familiarity with these regularities was a cornerstone of the professional expertise that would allow one to practice medicine responsibly and efficaciously.

Some medical writers made two central moves in their development of a new conceptual armature. The first was the use of *physis* (normally translated "nature") to refer to natural regularity, to normativity, to the intelligibility of things by reason, or to nondivine agency in natural entities. The second was the use of *technē* (often translated "art") to refer to a result-oriented professional expertise, based on a knowledge of such regularities, and to a rule-based practice in accordance with this expertise.

Neither *physis* nor *technē* had to be invented by these often audacious innovators. Both words are attested as early as Homeric epic.[2] In ancient Greek the semantic range of *physis* extends from "origin," "appearance," and "natural constitution" to "genitalia," not to mention numerous other shades of meaning. But in certain cultural niches, *physis* became emblematic of nontraditional ways of investigating or conceptualizing the physical and biological worlds. It often is used, for example, in the articulation of the view that biological or physical entities or, for that matter, the universe as a whole and all its parts, are characterized by rational design or by recurrent, rationally communicable processes.

In the earliest extant Greek medical writings—the collection of texts known as the Hippocratic Corpus, most of which were written in the fifth and fourth centuries BCE—*physis* frequently is used to refer to the "nature" of things in the sense of the regularly recurring cluster of characteristics by which one can always recognize a thing as what it is. But there are so many different things a Hippocratic healer has to be able to recognize, that there are numerous "natures." Human beings have a physis by which we identify them as human beings, but each human being also has her or his own physis. Furthermore, the physis of a woman differs from that of a man, the physis of a child is unlike that of an adult or an elderly person, and the physis of a strong person differs from a weak person's. Moreover, every part and constituent of the body has its own physis, by which, under favorable conditions, it can be recognized directly or indirectly, and by which the limits and nature of its behavior can reliably be predicted; the heart, liver, and brain each has its own physis, and so do the eyes, each limb, and so on. Each humor likewise has a different, distinctive physis, as does each type of bodily secretion. Every kind of food and drink also displays its own physis, and the physician not only must be able to recognize this physis but also must know the physis of every drug—compound or simple—and of every drug ingredient. Furthermore, every disease has its distinctive physis, as does every locality in which a patient resides. Various airs, waters, and places all have their distinctive "natures." In the view of those Hippocratic writers who resort to the concept of physis—and not all do—a physician therefore has to know an enormous range of internal and external "natures," many of which interact, if he aspires to possess and practice the "art" (i.e., the technē) that constitutes professional expertise.

On questions such as whether and how one can know the physis of anything, whether and to what extent one can manipulate it, and just what the physis of any given thing is, the Hippocratics, like many later medical writers, disagreed. Those who gave physis center stage tended, however, to link its knowability to another innovative concept of enormous consequence, namely, *dynamis*, which nowadays tends to be rendered by "faculty," "quality," or "property" (but which is derived from *dynamai*, "I can, I am able to"). *Dynamis* too is not a new word; it appears as early as the *Iliad* and the *Odyssey*, usually to refer to "power," "might," and especially to "bodily strength," a meaning that remained common in medical writings too. Some Hippocratics seem to have pioneered the use of *dynamis* to refer to the distinctive power or capacity of a given thing to have or to experience specifiable effects. Through its characteristic effects or susceptibilities the dynamis of a thing can lead us beyond the surface, beyond the external appearances and visible forms or shapes of a thing to its physis or "nature."

In part for this reason, the pairing of dynamis and "nature" (physis) is conspicuously present in some Hippocratic works, notably in *On Ancient Medicine*, *On the Nature of a Human Being*, *On Sacred Disease*, and *On Regimen*.[3] These and many later medical writings, including Galen's, tend to explain a thing's "nature" in terms of the capacities, powers, and susceptibilities imparted by its various constituent parts.[4] They recognized that the various dynameis (plural) of a given thing are an essential part of that cluster of recurrent, stable characteristics by which—as Gregory Vlastos remarked with reference to physis—"we can recognize that thing and can anticipate the limits within which it can act upon other things and be acted upon by them."[5] In other words, the dynameis are an essential feature of the cluster of recurrent characteristics that allow us to get at, and to speak of, a thing's "nature." Because the distinctive effects or experiences enabled by the dynameis of a given natural entity tend to become observable, these "capacities" can help unveil the distinctive but invisible physis of that entity.

In some Hippocratic writings, as in later philosophical and medical works, the dynameis of a thing therefore not only allow one to move from the outward appearance or visible form (eîdos, idéa) of the thing toward its nature (physis) but also permit one to draw distinctions (*diorízesthai*) between the "natures" of things. Distinction and differentiation are, of course, as fundamental to "scientific" medicine as they are to

Greek philosophy. In this context an influential expression, *dynamis idia* (*idiē* in the Hippocratics' Ionic Greek), a thing's "own, distinctive capacity," first becomes prominent in Hippocratic writings. Constitutive parts, natural qualities, and composite natural entities are described by some early Hippocratics as each having its own (idia) distinctive dynamis (or, in the plural, dynameis), which, through its effects, reveals the "nature" of the part or the quality or the entity.[6] The use of *dynamis* as a means of both detecting and differentiating "natures" is given a variety of expressions (not all resorting to the qualifier "own"), often in the context of what the practitioner of the medical technē must know in order to attain therapeutic success.

The Hippocratic author of *On Ancient Medicine* (c. 420–410 BCE[7]), for example, more than once emphasizes that, because different people have different "natures" (some, for instance, have a weaker physis, others a stronger), the practitioner must precisely "fashion" or "mold" each patient's ingestion of food and drink in relation to that person's dynameis and thus to her or his invisible yet detectable "nature(s)."[8] As will be shown in the next section, it was precisely the invisibility or inaccessibility of the many "natures" in and of the human body that led to notorious ancient controversies about the lengths to which technē can and should go in order to gain knowledge of any given bodily "nature." The move from visible appearances and effects to invisible natures later assumed divergent forms, but physis and dynamis often remained a significant feature of this move.

As the Hippocratic author of *On the Technē* (probably late fifth century BCE[9]) suggests, the intelligence of technē can gain power even over the invisible interior parts and processes of the body, if their "natures" are indirectly—that is, especially through the observable effects of their dynameis—available for scrutiny (provided, however, that the "natures" of the investigators themselves are suited to such an investigation).[10] And the author of the Hippocratic treatise *On Regimen* (late fifth or early fourth century BCE) argues that the practitioner of the technē should know not only the physis of every human being but also the dynamis of each of the foods and drinks. Furthermore, this writer distinguishes between the dynamis of things "by nature" and their dynamis "through forcible constraint by human technē." An expert knowledge of the "capacities" of things and hence, through inference, of their "natures," will enable the practitioner to use his "art" to reduce the "capacity" or "power" of substances that are by their nature (*physei*) too strong for a given patient's nature. Conversely, his "art" will allow him to strengthen things that are weak by nature.[11]

More generally, *On Ancient Medicine* offers an elaborate account of what the practitioner of the technē should know about "nature." A few excerpts will illustrate its tenor and, in particular, its view of the relation between action and knowledge of "nature":

> It therefore seems necessary to me that a healer have the following *knowledge concerning nature* and that he make every effort to know this if he intends to do any of the required things: what a person *is* in relation to things both eaten and drunk, what he is in relation to the rest of his habits of life, and what happens to each person as a consequence of each thing, and not simply like this: "cheese is a bad food; for it causes troubles to a person who overindulged in it"; rather, [one must know] exactly which trouble [it causes], and for what reason, and to which of the things in a human being it is unsuitable. There are, after all, many other bad foods and drinks. . . . The natures (*physies*, plural) of humans differ from one another, and they differ with respect to exactly what thing in them is hostile to cheese and is aroused by it and set in motion by it.[12]

Closely related to these views is the recurrent Hippocratic theme of the differences in the "natures" of things, even when things belong to the same class or species or type. Some Hippocratics believed that the enormous variety and variability in nature, and in particular the variety and variability of human bodies and of their parts, pose a daunting challenge to the technē.[13] In a discussion of dislocations, the author of the Hippocratic surgical treatise *On Joints* remarks, for example, with reference to joint sockets and to the attachment of ligaments, that "natures (*physies*) differ greatly from natures (*physiōn*)."[14] Later he again emphasizes that the differences in the natures (physies) in humans is great, also with reference to the degree of difficulty of reducing dislocations in any given patient. The practitioner, he concludes, therefore not only must be familiar with the most powerful tools in the arsenal of the technē for each type of case but must also know how to use them on each particular occasion where they seem appropriate.[15]

Not unrelated is the tension between generalization and particularization that becomes visible in many ancient medical and biological works, beginning with the Hippocratics. On the one hand, generalizations—often expressed through recourse to physis (i.e., to the "nature" of a given entity)—seemed necessary inasmuch as they provided norms and the stable, communicable, pedagogically transmissible foundation and framework of expert professional practice. The alternative would be to face a bewildering chaos of countless seemingly disconnected, unrelated particulars in one's daily practice. On the other hand, the need for the individualization of treatment and for acute attentiveness to distinctive, idiosyncratic particulars was impressed on some Hippocratic physicians by their observation of variability and by their recognition that many of their prized generalizations did not hold in every particular case. The physician's technē therefore constantly had to bear in mind not only a multiplicity of things with generalizable "natures" and "capacities" pertinent to every individual case of illness, but also the pervasive presence of particular swerves from the various norms expressed by "nature" (physis). This tension between the general and the particular, between universalizing and individualizing, at times also is reflected in some Hippocratics' pioneering uses of an expression that later became central to Aristotle's philosophy of science,[16] namely, "for the most part" (*hōs epì tó polý* and related phrases). The Hippocratic introduction of "for the most part" to characterize generalizations or rules that do not hold universally but nevertheless are too useful to abandon, because they account better for a majority of the observed cases that fall under them than do any other generalizations or rules, is a reflection of the struggle by the technē to come to terms with exceptions, that is, with the frequent observation or inference of particulars that represent deviations from the norms expressed by *physis* and by related terms.[17]

Before I turn to two examples of a direct, adversarial confrontation between "nature" and technē, a few more general observations on Hippocratic views of technē might be helpful. First, a number of Hippocratic works depict the technē as having a bad reputation.[18] The author of *On the Technē* even complains that there are those who have made a technē out of bad-mouthing the technai (plural) and who attempt to discredit the medical technē through a technē of ignoble discourses.[19] As has often been suggested, the preoccupation of some Greek medical writers with the means of securing a good reputation for the healer, also through prediction, perhaps reflects the insecurity of the medical technē within the social order.[20] Indeed, some Greeks even seem to have questioned the very existence of a medical technē. A number of Hippocratic treatises responded by offering arguments for its existence.[21] Some of these arguments were based on the efficacy of the therapies prescribed by a consistently reliable technē, notably on the efficacy of therapeutic regimens, while others pointed out that even those who do not believe that the medical technē exists at times end up being saved precisely by this "art." Still others cited the ability of the technē persuasively to define the boundaries between correct and incorrect treatment as proof of its existence.

While the legitimacy of claims that there is in fact a medical technē had to be defended, the existence of "nature" never seems to have been in doubt, other than in traditions of skepticism. Some medical writers admittedly thought that "nature" cannot be known (see below), and others questioned the pragmatic utility of the concept, but it does not seem to have become the target of the same kind of radical ontological challenge as did "art."

If the power of technē was widely acknowledged, so were its limitations. In *On the Technē* medicine is defined as knowing how to do three things: to remove completely the troubles of the ill; to blunt the vehemence of disease; and *not* to try treating those who have been overpowered by diseases.[22] The famous principle of nonintervention in incurable cases, or in cases where intervention may aggravate the patient's condition, is articulated in varying modulations in Hippocratic medicine. *Prognostic* (c. 440–410 BCE), for example, emphasizes at the outset that it is impossible for the technē to make all patients healthy again, and that some always will die,[23] while *On the Technē* argues that the technē would be justified if it refrained from trying to treat diseases that cannot be rectified easily.[24] Indeed, according to the latter, it belongs to the delirious ignorance of insanity to demand that a technē be capable of things for which it is not by nature a technē or, for that matter, to demand that a given physis be capable of things for which it is not by nature a physis. The author of *On the Technē* adds that we can work effectively—"we can be craftsmen" (*dēmiourgoí*)—only in cases where we can gain control through the tools (*órgana*) of the "natures" or the tools of the technai. Severe affections that are too strong for the tools of the medical technē simply cannot be conquered by technē.[25] More concretely, the Hippocratic author of *Prorrhetic* II (perhaps mid-fourth century BCE) remarks that older patients suffering from gout or hard accretions around the joints cannot possibly become healthy by any human technē, and likewise that certain patients with a rupture of the eye and a protrusion of the iris cannot be helped—neither by time nor by the technē.[26] A famous principle announced in the first book of the Hippocratic *Epidemics* (late fifth century BCE) probably also belongs to this context: "to help or not to harm" (which only much later appears in the distorting but much better known refraction *primum non nocere*).[27]

The limitations of the technē stem, however, not only from the boundaries of its own capacities and from the intractable "natures" of certain diseases, but also from human error: from the mistakes made even by expert practitioners and from the failure of some patients to follow the prescriptions of professional experts in the technē. Some Hippocratic writers in fact became famous already in antiquity for having confessed and recorded their own therapeutic errors, including those that caused the death of a patient. The technē itself and the cognitive limitations of human beings (of doctors and patients alike), not to mention the innumerable, interactive, external, and internal invisible "natures," therefore all contribute to the limitations displayed by the technē and acknowledged by the good healer. Yet there never was any doubt among medical authors that their technē had significant powers. In a statement with which many of them would have agreed, the Hippocratic author of *On Regimen in Acute Diseases* (end of the fifth century BCE) summarizes its power as twofold: to bring those who are ill into health, and to bring those who are healthy into enduring bodily safety.[28]

A further noteworthy feature of Hippocratic characterizations of technē is its figuration in agonal terms. The physician by means of the "art" enters into an agon with each disease[29]—a struggle that it does not always win but to which it brings a formidable and, on the whole, efficacious arsenal of resources. The agon of the technē is, however, not only with diseases; at times it also is with nature (physis).

## "Art" against "Nature": Concealing and Revealing, a Coercive Semiotics

Two Hippocratic characterizations of the relation between technē and nature as harmonious and nonconflictual have received much attention in modern scholarship. One is the depiction of technē as having its origin in the mimesis of nature or, more specifically, of natural human life or of the human body.[30] A second is the portrayal of technē as the servant of physis (and the healer in turn as the servant of technē). No later than the late fifth or early fourth century BCE, however, an agonal representation of the relation between "art" and "nature" begins to appear in Greek medical writings. Here the language of violence and force is used to suggest the possibility—and, at times, the necessity—of a technē that will violate physis, that will transgress natural boundaries, and that enters into a fierce agon with "nature."

A remarkable passage in the Hippocratic work *On the Téchnē* appears to initiate this strand in Greek medical thought. The author comments that the medical *téchnē*, though deprived of the possibility of "seeing" (*ideîn*) any of the affections in the thoracic or abdominal cavity—such as ailments of the liver and the kidneys—"with eyesight" (*ópsei*), nonetheless has discovered (*heûre*) some auxiliary resources that allow it to recognize such affections.[31] These tools include the signs (*sēmeîa*) offered to the expert practioner's five senses by patients' bodies, for example, the signs detectable in the voices, breathing, habitual excretions, odor, color, and slenderness or fatness of a given patient.[32] He then proceeds to the more difficult cases where nature completely conceals things without yielding any such semiotic clues:

> But whenever nature (*physis*) itself does not willingly yield the [signs] that are informers [or: denouncers], the medical *téchnē* has discovered forms of duress [*anánkai*, forcible constraints] by which nature, when it has been overpowered by force (*biastheîsa*), with impunity surrenders [the denunciating/informing signs]. And, once released [from duress], nature makes clear, to those who know the things belonging to the *téchnē*, what should be done.[33]

After stating the general principle that the technē at times should violate an unwilling, recalcitrant nature in order to force it to yield signs of the naturally invisible, the author offers four examples of such "technical" uses of "force" or "compulsion" against nature:

> (1) And by means of the harshness of [prescribed] foods and drinks, [the medical technē] forcibly compels (*biázetai*) the congenital [interior, invisible] phlegm to spill some [visible] pus, so that, on the basis of an observed (*ophthén*) sign, one will make conjectures concerning those things whose [direct] observation by the *téchnē* lies in the realm of the impracticable. (2) And likewise, by means of steep roads and by runs, it [the technē] forces (*ekbiâtai*) the pneuma [breath] to inform against (*katēgoreîn*, to denounce) the things of which pneuma is a betrayer (*katēgoros*, denouncer, accuser). (3) By inducing perspiration through the above-mentioned means, and through exhalations of hot liquids, the *téchnē* also judges [the invisible]. (4) There are also things that pass out through the bladder [as a consequence of a prescribed harsh regimen] and are more apt at revealing the disease than those exiting through the flesh. The *téchnē* therefore has discovered drinks and foods which, because they are hotter than the warming things [in the body], are of such a kind that they both melt those [things in the

> body] and make them flow through [along a way] where they would not have flowed, had they not been subjected to this action [on the part of technē].[34]

Four features of the language used in the formulation of the general principle and in the examples are noteworthy. First, it is a language of violence, of force, and of compulsion (even if the examples themselves are relatively unremarkable for their "violence"). Here there is no trace of the view that technē is a servant of nature, imitates nature, is a benign, co-operative extension of nature, or helps to restore nature when things in the body have gone awry. Rather, the representative of technē here argues in favor of the use of violent compulsion (*biastheîsa, biázetai, ekbiâtai*) by technē against nature. Furthermore, while the Greek word *anánkē* in the singular often means "necessity" in the Hippocratic writings,[35] here, quite significantly, it is used in the plural (*anánkai*): "The medical technē has discovered *anánkai*" by which it can overpower nature and force it to surrender normally hidden, inaccessible signs of "natures" and of their "powers." This plural form often is used to refer to torture, to instruments of torture, or to violent punishment, rather than to "necessity."[36] Theodor Gomperz here plausibly translated it with "Folterzwang," the "coercion by torture."[37]

Second, the means of torture and of forcible constraint are described as a discovery made by the technē itself. It is striking that the author thrice in this brief passage uses the verb "to discover" (*heurískein*) in the active sense, with technē as its subject,[38] to refer to the origin of the new tools of compulsion: they did not exist before the technē invented them, let alone before there was a technē. Without technē, nature never could have been forced to yield its secrets. Moreover, as pointed out above, the author of *On the Technē* elsewhere presents the discoveries made by the technē as one of the proofs of the existence of the "art." The "coercive torture instruments" invented and efficaciously applied by technē against a secretive, recalcitrant nature, in order to pry knowledge of invisible things from it, accordingly are among the proofs of the very existence of the technē. They thus also belong to the larger context of the debate about whether or not there is in fact a medical technē.

Third, much of the language is drawn from the legal sphere. Nature is imaged as an unwilling witness who possesses invaluable information but refuses to divulge it and therefore must be tortured, like those accused of treason or the slaves described by Antiphon and others in judicial contexts.[39] The silent, reluctant witness has to be compelled by force to testify about—or, in the Hippocratic imagery, "to inform against" or "denounce" or "betray"—what is hidden by it. Geoffrey Lloyd has contributed significantly to our understanding of the extent to which legal and political cross-examination and scrutiny or testing of evidence were seen by ancient Greeks as paradigmatic of scientific testing and examining.[40] But here the legal imagery is not introduced, at least not in the first instance, to evoke analogies between legal and scientific acts of testing evidence for its validity. Rather, it belongs to the forensic figuration of the agonistic, in part violent relation between "art" and "nature," with "art" as the prosecutor using torture, and "nature" as the silent, refractory witness subjected to physical force (and, by further implication, treated like a misbehaving slave).[41]

Fourth, a semiotic dimension is fundamental to the characterization of this agonal relation between technē and nature. It is a contest about difficult signs, about concealing and revealing signs of invisible, nonevident things. Under artificial constraint physis might surrender snitching signs to a well-trained practitioner of the technē, but the author of *On the Technē* recognized that such signs belong to a different, more difficult, semiotic system than the signs regularly provided by nature when it is not subjected to technē's tortures. The signs produced by means of violence against nature are depicted not only as invaluable informants and as interpreters or translators of hidden bodily facts, but also as speakers of a foreign language. Furthermore, the author says, there are many different signs, coming from different places: different signs exit through different parts of the body; different signs betray different things. Consequently, the author remarks,

> it is not astonishing that one's convictions about them [the meanings of the forced signs] take longer to form in these cases, while shorter periods of time [are left] for the [physician's] interventions, because they [the signs] are interpreted through foreign translations (*hermēneuoménōn allotríōn hermēneiôn*) for the understanding [intelligence] that tries to heal.[42]

The complex process of indirect diagnosis by means of the interpretation of signs forcibly wrung from nature takes longer, leaving less time for effective therapeutic intervention. The adjective *allótrios* is striking in this context: its basic meaning is "belonging to another," but it more often is used to refer to "foreign(er)," "stranger," or "alien," also in the hostile sense of "enemy." The multiple, varied, physical signifiers forced into the open by artificial means thus are indispensable yet alien, perhaps hostile interpreters of the body's interior, not actually showing its invisible "natures," but merely indirectly signaling aspects thereof. Artificially provoking these indispensable foreign interpreters of a secretive, concealing nature is necessary for the "therapeutic intelligence," even if its time-consuming complexity reduces the duration of the healer's opportunities for treatment.

## OPENING THE BODY: VIOLATING NATURE AND VIOLATING ART

The adversarial relation between "art" and "nature" that appears in *On the Technē* is not a very conspicuous theme in Greek medical literature, but the concealing tendency of nature that is at its root—perhaps most famously expressed in Heraclitus's "nature has a tendency to conceal itself"[43]—continued to be viewed as posing formidable challenges to "art" (technē) in post-Hippocratic medicine. The various "natures" (see above) admittedly offered many manifest signs, especially through the observable effects of their capacities (dynameis) to act on something or to be acted on by something. But nature both conceals and reveals, it veils and unveils, it closes and discloses. In particular, the invisibility of internal diseases and parts, which prompted the author of *On the Technē* to advocate forcing nature by artificial constraints to disclose its invisible features, was a frequent source of concern and complaint among medical writers, from the Hippocratics to late antiquity. Not all Greek physicians agreed, however, that one should, or even could, forcibly compel nature to reveal the parts, processes, or conditions that it normally conceals.

In the early third century BCE two Greeks, Herophilus of Chalcedon and Erasistratus of Ceos, became notorious for trying to address the problem of nature's concealing habits in a radically new way: by systematic human dissection and by vivisectory experiments on condemned criminals. For present purposes the spectacular new anatomical and physiological knowledge that Herophilus and Erasistratus acquired by means of these new heuristic tools may be left aside.[44] More significant for the relation between "art" and "nature" is the debate triggered by the fact that their dissections and vivisections violated entrenched taboos concerning both cadavers and living human bodies. In various forms the debate continued for centuries, despite the fact that, after the generation of Herophilus and Erasistratus, systematic human dissection and human vivisection apparently were abandoned as abruptly as they were initiated.[45]

The earliest extant evidence of the controversy is preserved by the first-century Roman encyclopedist Aulus Cornelius Celsus, who tried to stake out a middle ground in the debate, even though he drew on Hellenistic sources that were far from impartial.[46] In the proem to the eight medical books in his *Artes*, Celsus reports the following defense of the art of laying bare what nature has hidden:

> Moreover, since both pains and various kinds of diseases arise in the more internal parts [of the body], they believe that no one can administer remedies for these, if he is ignorant of the parts themselves. It therefore is necessary to dissect the bodies of the dead and to examine their viscera and intestines. Herophilus and Erasistratus, they say, did this in the best way by far, when they cut open people who were alive, criminals out of prison, received from kings. And while breath still remained in these criminals, *they* [Herophilus and Erasistratus] *inspected those parts which nature (natura) previously had concealed* [*clausisset*, closed, blocked], also their position, color, shape, size, arrangement, hardness, softness, smoothness, connection, and the projections and depressions of each part, and whether anything is inserted into another thing or whether anything receives a part of another thing into itself.[47]

The detailed enumeration in the last sentence of naturally invisible things—"the position, color, shape, size, arrangement, hardness, softness, smoothness," and so on, of internal parts—that become accessible to the human senses through vivisectory intervention against nature became a key element both in the advocacy and in the rejection of human dissection and vivisection, as will be shown below.

As in the Hippocratic work *On the Technē*, an intrusive intervention against nature is depicted as having been prompted by nature's secretiveness and, in particular, by the fact that nature blocks internal parts and internal diseases from view. Furthermore, in both cases the justification for tampering with nature—here, literally prying nature open by cutting open a healthy human being—is therapeutic efficacy, and hence the benefit of human beings who suffer pain and who might be in danger of losing their lives. Among the arguments attributed to the proponents of such forcible uncovering of what nature has covered, of such dreadful disclosing of that which nature has closed, are the following:

> For, [they say,] when pain occurs internally, neither can a person who has not [previously] come to know in what area each internal organ or

> intestine lies know what hurts the patient, nor can a person cure the [part] which is diseased who does not know what it is. And when someone's viscera are exposed by a wound, one who is ignorant of the color of each part in its healthy state is unable to know what is intact and what is damaged; thus he cannot come to the aid of the damaged parts. External remedies too can be applied more suitably by one acquainted with the locations, shapes, and size of the internal organs. . . . Nor is it cruel, as most people maintain, that remedies for innocent people of all times should be sought in the sacrifice of people guilty of crimes, and of only a few such people at that.[48]

Despite the enormously influential anatomical and physiological revolution entailed by the use of human dissection and vivisection in the early Hellenistic period, detailed attacks on these new methods of prying loose nature's secrets were soon launched, notably by the self-styled medical empiricists (*empeirikoi*), whose third-century BCE leaders included Philinus of Cos, a renegade pupil of Herophilus. The empiricists' objections to human vivisection included considerations of both *natura* and *ars* (a point largely overlooked in modern discussions of this ancient controversy). The objection based on natura is usually epistemological, not moral or aesthetic. The empiricists argued, for example, that any inquiry into nonevident or invisible things is superfluous, because "nature cannot be comprehended" (*quoniam non conprehensibilis natura sit*),[49] as is apparent from the disagreements among those who try to give accounts of nature. Human cognition, being limited, will never be able to understand natura.

The antivivisectionist argument based on ars, by contrast, interweaves epistemological considerations with moral, professional, and pragmatic arguments. All theory, the empiricists argued, including physiological theories about "nature's actions" in the body, and likewise all attempts to achieve knowledge of invisible causes, are "superfluous [for medical practice], and they [the empiricists] think it cruel as well (*etiam crudele*) to cut into the belly and chest of human beings who are still alive, and to impose upon the *ars* which presides over human well-being not only a death for someone but this most atrocious death [*scil.* by vivisection] to boot."[50]

Unlike the discussion in the Hippocratic work *On the Technē*, the violation of ars, not of natura, by an agent of ars is at issue here, as it is in other ancient sources that emphasize the cruelty of human vivisection. The medical "artist" not only acts with savage cruelty to the very persons he is sworn to aid by his "art" but also forces killing on his own life-giving "art" (technē). Furthermore, vivisection does not lead to greater therapeutic efficacy: from a clinical point of view it is "superfluous." Moral and professional objections thus converge with the epistemological assault on vivisection.

The epistemological part of the argument from ars aims in part at establishing the evidential unreliability and therefore inutility of vivisection, in part at showing that some of the things "discovered" by means of human vivisection could in fact be known without using vivisection: "And add to this especially the fact that, of the things investigated with such great violence, some cannot be known at all, while others can be known without recourse to a crime."[51] While the epistemological argument from natura asserts the absolute unknowability of nature,[52] the corresponding argument from ars thus distinguishes between the unknowable and knowable features of nature. It subsequently elaborates that the latter can become known without resorting to the criminal violence of vivisection, for example, through the chance (*casus*) observation of wounded but "still breathing" soldiers, gladiators, or crime victims, especially if such observations are made by expert practitioners of the ars.[53] Here the moral, professional, and epistemological strands in the empiricists' antivivisectionist arguments again are intertwined. Such fortuitous observations of "the location, position, arrangement, shape [of internal organs and structures], and of other similar things" by an observant, skilled physician[54] are, they claimed, preferable to systematic vivisectory investigation. For, they argued, when an expert practitioner gains new knowledge by making such chance observations in the course of treating severely wounded patients, "he learns through compassion what others come to know by means of dreadful cruelty."[55]

Despite such efforts to muster moral and professional arguments against vivisection, the empiricists' criticism of those who open the body in order to gain knowledge of hidden features of nature tended to accord epistemological reservations a central role. One of their major elaborations on the theme of the evidential unreliability of vivisection concerns changes wrought in the body by vivisection itself:

> For when the body has been laid open, the color, smoothness, softness, hardness and all similar things are not such as they are when the body is intact (*integrum*), because, even when bodies have not been violated (*corpora inviolata*), they often change [in appearance] from fear, pain, lack of food, indigestion, fatigue, and a thousand other minor affections. The internal parts, whose softness is greater and for which the light itself is something new, are much more likely to

> change under extremely severe [*scil.* vivisectory] woundings—indeed, under savage massacre [by vivisection].[56]

Precisely those internal things that the vivisectionist aims to observe, they argued, are susceptible to change, even by mere exposure to light in the process of being uncovered, not to mention other changes triggered in internal parts by the trauma associated with incisions in the healthy living body. In short, the observer alters the observed, rendering his own observations invalid.

The changes in the vivisected body are, however, due not only to the effects of light and of vivisectory trauma; a much more radical consequence of vivisection, they argued, is death, which alters the body so fundamentally that vivisection cannot be said to yield any reliable information about a healthy living body:

> It is possible to open up the belly, which is of lesser consequence, while the person keeps breathing. But as soon as the knife actually penetrates to the chest and the transverse septum [midriff] is cut through, . . . the person instantly loses his life. It therefore is only a dead person's vitals and any viscera that can come into the view of the brigand-healer [i.e., the vivisectionist], and these parts are necessarily such as are a dead person's [parts], not as are a living person's. It thus follows that the [vivisecting] healer cruelly slaughters a person, and not that he knows [by vivisection] what our viscera are like when we are alive.[57]

Vivisection of the most significant, truly vital parts, in other words, results in instant death, and death immediately alters the internal parts. Vivisection thus defeats its own purpose: the very act of unveiling by ars that which natura has veiled renders the unveiled object epistemologically and hence clinically useless.

That similar arguments could also be brought to bear against human dissection did not escape the notice of the empiricist opponents of active anatomical and physiological investigation: "Since most things behave differently in the dead, not even dissection of the dead is necessary. And although not cruel, it [dissection of cadavers] nevertheless is shameful [*foeda*, offensive, repugnant to civilized taste or feeling, unclean]."[58] Obviously a few Greeks disagreed with this view, including some who themselves did not practice human dissection. Galen, who, like Aristotle, Marinus, Quintus of Rome, Satyrus, Numisianus, and Pelops, dissected and vivisected only animals, nevertheless overtly acknowledged the value of systematic human dissection. Celsus likewise distanced himself from the empiricists' rejection of human dissection, suggesting that human dissection, unlike human vivisection, does have a valid place in medical education:

> I in fact believe that the *ars* of medicine should be rational, but that hidden causes should be rejected, not from the thinking of the professional practitioner of the "art" (*artifex*), but from the *ars* itself. Moreover, to cut open bodies of people who are alive is both cruel and superfluous, but to cut open the dead is a necessity for those who are learning. For they should know the position and arrangement [of internal parts], and a cadaver presents these to view better than does a living and a wounded person. But as for the remaining things too that can be learned only in the living, one's actual practice itself will reveal them in the course of treatments of the wounded, [though] somewhat more slowly, yet in a considerably more gentle way.[59]

In subsequent phases of the controversy about whether the ars may use dissection and vivisection to pry open things concealed by natura, the nature and purpose of ars itself continued to be a central consideration. John of Alexandria, for example, reports that some antivivisectionists, while accepting the dissection and vivisection of apes and bears, claimed that "the dogmatist [physicians] are distinguished by committing murder, since they perform dissection on human beings who are alive, although medicine is an ars that effects health in human bodies."[60] A similar objection based on ars is sounded by Agnellus of Ravenna (who may have drawn on the same source as John of Alexandria).[61] And Tertullian claimed that Herophilus's efforts to "cut up innumerable persons in order to examine nature (*ut naturam scrutatur*)" turned the doctor into a butcher who hated humans in order to have knowledge of nature, but who failed to acquire such knowledge because of the changes caused by the human subject's "aberrant death in the midst of the artificial processes of dissection" (*morte . . . ipsa inter artificia exsectionis errante*).[62]

That artifices, ranging from very mild to fatally violent interventions, could be—and had been—used by ars to force natura to reveal what it normally conceals was therefore recognized by Greek medical writers from the classical period to late antiquity (John of Alexandria, Agnellus). But it is striking that such exercises of the power of the technē against nature became controversial in Greco-Roman antiquity especially when they were thought to violate the self-understanding and purposes of the technē itself. In later antiquity, the act of using the ars to pry open natura was also seen to be a contravention of humanitas or as a hubristic effort to make all aspects of the body's nature accessible to the mortal agents of ars. In some versions of a fragmentary text known as *Gynaecia* and attributed to the fourth-century North African physician Vindicianus (but that probably drew on Soranus and was reworked in later antiquity), for example, the dissection of cadavers to determine the cause and manner of death is rejected, on the grounds that "*humanitas* prohibits doing this, since all things would be manifest and fully open to those conducting the examination"[63]—a cognitive state of affairs that apparently would be incompatible with the limits that define what it is to be human.

## "NATURE" AS "ART": NATURAL TECHNOLOGY AND TECHNICAL NATURE

In early Hellenistic medicine, as in the works of Galen, nature (physis) itself often was explicitly depicted as working like, acting like, or being like technē, rather than as resisting (or being opposed to) technē. One of the two third-century BCE pioneers of systematic human dissection, Erasistratus, for example, repeatedly characterized nature as *technikē*—that is, as doing things in a way that is characteristic of technē, or as "capable of technē." Here it is not a case of technē imitating physis, but rather of physis displaying characteristics that are best understood by their similarity to technē.[64] Among the views attributed to Erasistratus by the fragmentary ancient evidence are, for instance, the following:

> Nature (physis) is capable of forethought for the living being and of expert craftsmanship (technikē).[65]
>
> Erasistratus marvels at nature as being at once capable of expert craftsmanship (technikē) and of forethought for living beings.[66]
>
> Erasistratus calls nature capable of expert craftsmanship—nature which, right away from the beginning, had both shaped and arranged all parts of a living being well.[67]
>
> Erasistratus sings a hymn to nature, as being capable of expert craftsmanship.[68]

The characterization of the activities of nature in terms of technē has a long history. Plato, Aristotle, and the Stoics all depicted nature as operating in a manner characteristic of technē. But Erasistratus may have been particularly influenced by the Aristotelian use of such analogies. Sarah Broadie, Wolfgang Detel, and others have pointed out that depicting nature as working like technē plays a fundamental role in many of Aristotle's arguments in favor of his teleological principles (as it had, though in quite a different way, in Timaeus's "likely story" of the creation of the natural universe by a supremely skilled divine artisan or craftsman (*dēmiourgós*).[69]

Erasistratus appears to have been familiar directly or indirectly with Aristotle's use of the analogy between nature and technē, which is not to be understood as implying intentional, consciously designing activity. Rather, it implies that results or end states produced by natural processes ("nature" understood here as the inner principle of the behavior and organization of each individual substance and of its parts) are testimony to nature's "skilled," "expert" activity, much as technē becomes fully visible in the end products that realize its goals. Physis and technē thus both display goal-driven, expert activity.

Aristotle's claim that "technē imitates physis"[70] (a view endorsed by his successor Theophrastus,[71] who is reputed to have been Erasistratus's teacher) has received much attention, but in his biological works it is technē that repeatedly provides a model for understanding aspects of nature (physis), notably of the nature of a living thing as end-directed. Even though Aristotle reiterates in his biological works that "the beautiful and that-for-the-sake-of-which is even more in the works of nature than in the works of technē,"[72] and although in the biological sphere too (as in his *Physics*, *Metaphysics*, and elsewhere), he famously differentiates nature from technē with respect to the source of motion,[73] it nevertheless often is technē that provides him with the structural relations, the characteristic activities, the concepts, and the imagery that help clarify the workings of the nature of living things.

"As technē, so physis" thus becomes a recurrent refrain both in Aristotle's biological works and in the fragments of Erasistratus, and after Erasistratus it often recurs in Greek medicine, notably in the works of Galen. Among the analogies Galen shares with Aristotle are, for example, the following: just as technē uses tools, so nature uses natural entities as instruments (*órgana*); as in technē, so in nature a thing is made by something actually existing out of that which is potentially such as the finished product; as in technē, so in nature inferior things come to their completion sooner; as some tools serve multiple purposes in the technai (for example, the hammer and the anvil in the smith's technē), so does *pneuma* in the things formed by nature; and, perhaps most significantly, just as things in technē are always for the sake of something, so are things in nature. Galen and Aristotle agree that "if for-the-sake-of-something exists in technē, it also exists in physis."[74] Galen, in part prompted by his Platonism, expressed strong disagreement with certain features of the Aristotelian and Erasistratean versions of teleology. He objected to any suggestion that there might be limits to teleological explanation; such suggestions are, of course, well attested in the works of Aristotle, Theophrastus, and Erasistratus. Galen likewise objected to their lack of an "intentional" teleology that posits consciously entertained, directed, and executed purposes. But on the figuration of physis as technē they all agreed—and in this respect Galen had much in common not only with Plato but also with his arch enemy Erasistratus and with the early Aristotelians.

Erasistratus seems to have put the figuration of physis as technē to extensive use in his detailed accounts of the workings and functions of parts within the body. A good example is offered by his account of the heart. He depicted the heart as an example of nature's forethought, skill, and purposiveness. But this product of nature's providence also was described in terms remarkably similar to those used to represent a machine created by Hellenistic mechanical technology of his day.

The heart, Erasistratus believed, is an automatic double-action, two-chambered suction-and-force "bellows." This cardiac "bellows pump" is equipped with superbly functional valves that ensure the irreversibility of the flow both of what rushes into the heart's two chambers and what it pumps out.[75] The parallels between Erasistratus's version of the heart—a natural organ created by a "technical nature" (physis technikē)—and a water pump invented by Ctesibius during Erasistratus's lifetime are remarkable.[76]

Like the heart depicted by Erasistratus, Ctesibius's water pump is two-chambered. Furthermore, both the heart and the water pump are depicted as equipped with valves that ensure the irreversibility of the flow into and out of the two chambers.[77] The Erasistratean heart and Ctesibius's pump each has four sets of valves, two controlling the intake, two regulating the outflow from the two chambers. Ctesibius used these valves to achieve the irreversibility of the flow not only of liquid but also of air (as in several of his other machines), and Erasistratus likewise described the heart valves as controlling the unidirectional flow both of air (*pneuma*, "breath") and of a liquid (blood). Pneuma coming from the lungs, he argued, passes through one set of valves into the left chamber of the heart and then is pumped through another set of valves from this left ventricle into the aorta (whence it is "sent" through all the arteries).

Similarly, blood passes from the principal vein (vena cava) through a set of valves into the right chamber of the heart, and from there through another set of valves into the pulmonary vessels. Moreover, Ctesibius's water pump had forking pipes outside the two main chambers, and Erasistratus's vascular system likewise depends on vessels forking off from the aorta and from the vena cava.

The Erasistratean heart thus serves as a double intermediate chamber—that is, on the one hand for pneuma between the lungs and the arterial system, on the other hand for blood (coming from the liver) between the venous system and the pulmonary vessels that carry blood to the lungs.[78] Similarly, Ctesibius's pump centrally depends on the principle of an intermediate valved chamber (*medius catinus*).[79] More significantly, the mechanical principles are similar in the two cases, notably the principle that, when a space expands or dilates, matter will move into it from a contiguous space (and, conversely, a contracting space will expel the matter within it into a contiguous space), provided, of course, that appropriate connections between the contiguous spaces exist. Both in Ctesibius's mechanical technology and in Erasistratus's physiology the principle that matter will instantly move into any space that is being emptied, is based on the recognition that continuous or massed void does not exist naturally.[80]

The parallels between the Erasistratean version of the heart (an "instrument" produced by physis) and Ctesibius's water pump (an instrument created by technē) are numerous and nontrivial. For present purposes we may leave aside the question whether Erasistratus borrowed from Ctesibius or Ctesibius from Erasistratus, or neither from either. More significant is the detailed account by a medical teleologist of a product of physis technikē in terms of the innovative technology of his day: physis as technē, "nature" as "art."

Within ancient medical literature, Galen made the most extensive, and most famous, use of the depiction of nature as craftsman or of nature as operating in ways that are characteristic of technē. But Erasistratus's uses of technē to characterize physis appears to be a significant earlier chapter in the history of this influential analogy.

Numerous further aspects of physis, of technē, and of their relations and interactions in Greek medical literature merit extensive analysis. The examples above merely offer a selective illustration of the diversity of rival medical views concerning the relation between "nature" and "art." Collectively, however, they show that, for all the disagreements among ancient physicians, both "art" and "nature" were central to the self-understanding of Greek "scientific" medicine and to its attempts—never entirely successful—to demarcate the boundaries between itself and popular, so-called prescientific medical traditions. Whether ancient Greek medicine depicted the relation between "art" and "medicine" as mimetic or semiotic, as one of dependence or interdependence, as adversarial or as harmonious, it reveals the remarkable extent to which the concepts of technē and physis became reciprocally defining.

## Notes

1. Hp., *On Regimen* 1.4.1 (VI, p. 474 L. = *Corpus Medicorum Graecorum* [hereafter = *CMG*] I.2.4, p. 126 Joly/Byl); *On the Techne* 4.4 (VI, p. 6 L.; p. 228 Jouanna). See also *On the Nature of a Human Being* 2.2. and 5.2 (VI, pp. 34, 42 L. = *CMG* I.1.3, pp. 166, 176 Jouanna). Cf. the distinction in Hp., *Aff.* 47 (VI, pp. 254–256 L.), between foods that have an evident (*phanerē*) or clear and manifest *dýnamis* and those whose effects on the body are more obscure (*amydrotera*). The Hippocratics, of course, also use *eîdos* in several other senses, including "type" or "class"; see, for example, *On Ancient Medicine* 15.3 (I, p. 606 L.; p. 138 Jouanna), in the sense of "type," "class." On *eîdos* in the Hippocratic Corpus see H. Diller, "Zum Gebrauch von *eîdos* und *idéa* in vorplatonischer Zeit," in *Festgabe für E. Heischkel-Artelt und W. Artelt zum 65. Geburtstag* (Stuttgart, 1971), 23–30; A. E. Taylor, *Varia Socratica* (Oxford, 1911); C. M. Gillespie, "The use of *eîdos* and *idéa* in Hippocrates," *CQ* 6 (1912): 179–203; J. Jouanna, *Hippocrate V.1 = Des vents, De l'art* (Paris, 1988), 132–133 (note 4).

2. *Technē* occurs several times in both the *Iliad* and the *Odyssey* to refer to skill, craft, manual dexterity, and cunning. *Physis* occurs only once in Homer (*Od.* 10.303, significantly, to refer to the *physis* of a *pharmakon*).

3. For example, Hp., *On Ancient Medicine* 3.5 and 13.3 (I, pp. 578, 600 L. = pp. 122, 134 Jouanna); *On the Nature of a Human Being* 5.3 (VI, p. 42 L. = p. 176 Jouanna); *On Sacred Disease* 18.1 (VI, p. 394 L. = p. 32 Jouanna); *On Regimen* 2.38.1, 6 (VI, pp. 530, 534 L. = pp. 160, 162 Joly/Byl). See also H. W. Miller, "*Dynamis* and *Physis* in *On Ancient Medicine*," *TAPhA* 83 (1952), 184–197; H. W. Miller, "The Concept of *dynamis* in *De victu*," *TAPhA* 90 (1959), 147–164; H. W. Miller, "*Dynamis* and the seeds," *TAPhA* 97 (1966), 281–291; G. Plamböck, "*Dynamis* im Corpus Hippocraticum," *Abh. Mainz*, 1964.2, 61–110; H. von Staden, "*Dynamis:* the Hippocratics and Plato," in Konstantinos J. Boudouris, ed., *Philosophy and Medicine*, Studies in Greek Philosophy, no. 29 (Athens/Alimos: International Association for Greek Philosophy, 1999), vol. II, 262–279.

4. I say "constituent part" rather than "element," because it is a striking fact that Hippocratic writers assiduously avoided the problematic Greek word for "element" (*stoicheîon*) when discussing the constitution or composition of the body. Modern

scholars (not to mention ancients such as Galen) repeatedly refer to a Hippocratic theory (or to Hippocratic theories) of "the elements," but Hippocratic authors in fact more cautiously talk of "the things in the body" or "things of which the body is put together," not of "elements" (*stoicheia*). See, for example, Hp., *De natura hominis* 2.3–4 (VI, p. 36 L. = pp. 168–170 Jouanna); *On Regimen* I.2.1 (VI, p. 468 L. = *CMG* I.2.4, p. 122 Joly/Byl).

5. G. Vlastos, *Plato's Universe* (Seattle: University of Washington Press, 1975), 19; Vlastos argues persuasively that this new use of *physis* is visible already in Herodotus, but he makes no more than vague passing reference (p. 18) to the Hippocratics and fails to mention the role of *dynamis* in the deployment of *physis*. See pp. 18–22 for a brief but incisive account of the role of *physis* in the transition from the archaic period to the fourth century BCE. W. A. Heidel, "Peri Physeōs: A Study of the Conception of Nature among the Presocratics," *Proceedings of the American Academy of Arts and Sciences* 45, 1910, 77–133, and D. Holwerda, *Commentatio de vocis quae est physis vi atque usu praesertim in Graecitate Aristotele anteriore* (Groningen, 1955), offer valuable observations on uses of *physis* in pre-Aristotelian Greek texts. See also F. Heinimann, *Nomos und Physis* (Basel, 1945); M. Michler, "Die praktische Bedeutung des normativen Physis-Begriffes in den hippokratischen Schriften *De fracturis–De articulis*," *Hermes* 90 (1962): 385–401; A. Pellicer, *Histoire sémantique de physis* (Paris, Thèse, 1967); Daniela Manetti, "Valore semantico e risonanze culturali della parola *physis* (*De genitura, De natura pueri, De Morbis IV*)," *PP* 28 (1973), 426–444; H. Patzer, *Physis* (Habil.-Schrift Marburg, 1939; repr. Stuttgart 1993); L. Ayache, "Hippocrate laissait-il la nature agir?," in J. A. López Férez, ed., *Tratados Hipocráticos* (Madrid: UNED, 1992), 19–35; Maria Teresa Gallego Pérez, "*Physis* dans la Collection hippocratique," in R. Wittern and P. Pellegrin, eds., *Hippokratische Medizin und antike Philosophie: Medizin der Antike*, vol. I (Hildesheim: Olms, Weidmann, 1996), 419–436; G. E. R. Lloyd, *Methods and Problems in Greek Science* (Cambridge: Cambridge University Press, 1991), 417–434 ("The Invention of Nature"); G. E. R. Lloyd, *The Revolutions of Wisdom* (Berkeley: University of California Press, 1987), 187–190, 322–324; Simon Byl, "Liste des fréquences de *physis* et classement des oeuvres hippocratiques," in Antoine Thivel and Arnaud Zucker, eds., *Le normal et le pathologique dans la Collection hippocratique*, 2 vols. (Nice: Faculté des Lettres, Arts et Sciences Humaines de Nice, 2002), vol. 1, 45–54; Valeria Ando, "La *physis* tra normale e patologico," in Thivel and Zucker, *Le normal et le pathologique dans la Collection hippocratique*, vol. I, 97–122; Maurizio Giambalvo, "Normale versus anormale? Lo statuto del patologico nella *Collezione hippocratica*," in Thivel and Zucker, *Le normal et le pathologique dans la Collection hippocratique*, vol. I, 55–96; Amneris Roselli, "Dalla *dikaia physis* dei trattati chirurgici alla *dikaiosyne tes physeos* di Galeno," in Thivel and Zucker, *Le normal et le pathologique dans la Collection hippocratique*, vol. II, 731–752; Ivan Garofalo, "La nature d'Hippocrate chez les Alexandrins," in Thivel and Zucker, *Le normal et le pathologique dans la Collection hippocratique*, vol. II, 753–768.

6. For example, Hp., *On Ancient Medicine* 13.3 (I, p. 600.4–5 L. = p. 134.15–16 Jouanna); *On Regimen in Acute Diseases* 37.1 [10 L.] (VI, pp. 298.8–300.7 L. = pp. 50.24–51.11 Joly). A similar notion, without the use of *idios*, is expressed in a number of other passages—for example, Hp., *On Regimen* 2.39, 2.62 (VI, pp. 534–536, 576–8–9 L. = *CMG* I 2,4, pp. 162, 184.18 Joly/Byl); *Morb.* 4.39 (VII p. 558.20

L. = p. 92.20 Joly). But cf. also Hp., *On Ancient Medicine* 19.7 (I p. 620.4–5 L. = p. 145.14–16 Jouanna).

7. Modern scholars have sharply disagreed about the date of this treatise, some placing it as early as 440 BCE, others as late as 350 BCE. The later fifth century seems most plausible. See Jacques Jouanna, *Hippocrate,* II.1: *De l'ancienne médecine* (Paris: Les Belles Lettres, 1990), 84–85, for a brief summary of the debate about its date.

8. For example, Hp., *On Ancient Medicine* 3.4–6, 12.1–2, 14.3, 16.1 (I, pp. 576–578, 596–598, 600–602, 606–608 L. = pp. 121–123, 132–133, 135–136, 139 Jouanna).

9. Jacques Jouanna, *Hippocrate,* V.1: *Des Vents, De l'Art* (Paris: Les Belles Lettres, 1988), 190–191, offers convincing arguments in favor of the conclusion that "Le traité de l'*Art* se situe vraisemblablement dans le dernier quart du Ve siècle"; see also Jacques Jouanna, *Hippocrate* (Paris: Les Belles Lettres, 1992), 532. But Heidel, "*Perì phýseōs*" (note 5 above), p. 116, note 146, places it in the fourth century.

10. *On the Technē* 11.1–7 (VI, pp. 18–22 L. = pp. 237–239 Jouanna).

11. Hp., *On Regimen* I.2.1 (VI, p. 468 L. = pp. 122–125 Joly/Byl); see also II.39.1 (VI, p. 534 L. = p. 162 Joly/Byl).

12. Hp., *On Ancient Medicine* 20.3–6 (I, pp. 622–624 L. = pp. 146–147 Jouanna).

13. By "variety" I mean diversity or the quality of being varied; by "variability" I mean the quality of being susceptible to inconstancy or to change in characteristics (or in behavior or response) from one time to another. See H. von Staden, "The Rule and the Exception: Celsus on a Scientific Conundrum," in Carl Deroux, ed., *Maladie et maladies dans les textes latins antiques et médiévaux,* Actes du Ve Colloque international "Textes médicaux latins," (Brussels: Collection Latomus, vol. 242, 1998), especially 123–126; H. von Staden, "Hos epi to poly: 'Hippocrates' between generalization and individualization," in Thivel and Zucker, *Le normal et le pathologique dans la Collection hippocratique,* vol. I, 23–43.

14. Hp., *On Joints* 8 (IV, p. 94 L. = II, p. 120 Kühlewein).

15. Hp., *On Joints* 71 (IV, p. 292 L. = II, pp. 226–227 Kühlewein).

16. See, for example, Aristotle, *An. Post.* 1.30.87b19–25, 2.12.96a8–12; *Metaph.* 6.2.1027a20–21; *On Generation of Animals* 1.19.727b29–30; Jonathan Barnes, *Aristotle's Posterior Analytics* (Oxford: Oxford University Press, 1975), 184, 229; G. E. R. Lloyd, *Revolutions of Wisdom,* 321–323; Lloyd, *Methods and Problems in Greek Science,* 423–424. It is striking that Aristotle (like the Hippocratics) characterizes "for-the-most-part" scientific propositions as belonging to the sublunar sphere, including the biological world, whereas universal propositions tend to be applicable to the celestial sphere. This distinction continues to be reflected in the modern tendency to see biology as a messier, "softer" science and physics as "hard" science.

17. See Von Staden, "*Hos epi to poly:* 'Hippocrates' between generalization and individualization," in Thivel and Zucker, eds., *Le normal et le pathologique dans la Collection hippocratique,* vol. I, 23–43.

18. For example, Hp., *On Regimen in Acute Diseases* 8.1 (II, p. 240 L. = p. 39 Joly); *Law* 1 (IV, p. 638 L. = *CMG* I.1, p. 7 Heiberg); *On the Technē* 1.1–3 (VI, p. 2 L. = pp. 224–225 Jouanna).

19. *On the Technē* I.1 (VI, p. 2 L. = p. 224 Jouanna).

20. See *On the Technē*, passim; *Oath* 8 (IV, p. 632 L. = *CMG* I.1, p. 5 Heiberg); *Law* 1 (IV, p. 638 L. = *CMG* I.1, p. 7 Heiberg); *On Sacred Disease* 1.5 (VI, p. 356 L. = p. 5 Jouanna); *Prognostic* 1 (II, p. 112 L. = I, p. 79 Kühlewein); *Prorrhetic* II. 2 (IX, p. 10 L.); *On the Physician* 1 (IX, p. 204 L.); *On Decorum* 1, 3, 4, 18 (IX, pp. 226, 228, 230–232, 244 L. = *CMG* I.1, pp. 25, 26, 29 Heiberg); *Precepts* 4 (IX, p. 256 L.; p. 31 Heiberg).

21. E.g., *On Regimen in Acute Diseases* 8.1–2 (II, pp. 240–242 L. = p. 39 Joly); *On Ancient Medicine* 1.1, 2.2 (I, pp. 570, 572 L. = pp. 118, 119–120 Jouanna); *On the Technē* 2.1–4.4, 5.3, 5.5–6.4, 8.1 (VI, pp. 2–10, 12 L. = pp. 225–230, 232 Jouanna); *Morb.* I.1 (VI, pp. 140–142 L. = pp. 98–100 Potter); *Precepts* 9 (IX, p. 265 L. = *CMG* I.1, p. 33 Heiberg).

22. *On the Technē* 3.1–3 (VI, pp. 4–6 L. = pp. 226–227 Jouanna). See Renate Wittern, "Die Unterlassing ärztlicher Hilfeleistung in der griechischen Medizin der klassischen Zeit," *Münchener medizinische Wochenschrift* 121, no. 21 (1979): 731–734; Renate Wittern, *Grenzen der Heilkunst: Eine historische Betrachtung* (Stuttgart: Robert Bosch Stiftung, 1982); H. von Staden, "Incurability and Hopelessness: The Hippocratic Corpus," in P. Potter, G. Maloney, and J. Desautels, eds., *La maladie et les maladies dans la Collection hippocratique: Actes du VIe colloque international hippocratique*, (Quebec: Editions du Sphinx, 1990), 75–112.

23. *Prognostic* 1 (II, p. 110 L. = I, p. 78 Kühlewein).

24. *On the Technē* 13.1; cf. 8.1–6, 11.7 (VI, pp. 26, 12–14, 22 L. = pp. 241, 232–234, 239 Jouanna).

25. *On the Technē* 8.1–4 (VI, pp. 12–14 L. = pp. 232–233 Jouanna).

26. *Prorrhetic* II.8, II.19 (IX, pp. 26, 46 L.).

27. *Epidemics* I.11 (II, pp. 634–636 L. = I, pp. 189–190 Kühlewein). The larger context in which this principle is embedded here is worth keeping in mind: (1) the practitioner of the technē must state past, present, and future; (2) with reference to diseases the practitioner should know to help or not to harm; (3) the technē works through three things: the disease, the diseased person, and the healer; (4) the healer (*iētrós*) is a servant of the technē; (5) the diseased person opposes the disease along with the healer.

28. *On Regimen in Acute Diseases* 9.1 (II, p. 244 L. = pp. 39–40 Joly).

29. For example, *Prognostic* 1 (II, 112 L. = I, p. 78 Kühlewein).

30. For instance, Hp., *On Regimen* I.11.1–I.12.2 (VI, pp. 486–488 L. = *CMG* I.2.4, pp. 134–137 Joly/Byl).

31. Hp., *On the Technē* 12.1 (VI, pp. 22–24 L. = p. 240 Jouanna). See also chapter 11.1–7 (VI, pp. 18–22 L. = pp. 237–239 Jouanna).

32. Hp. *On the Technē* 12.2 (VI, p. 24 L. = p. 240 Jouanna).

33. Hp., *On the Technē* 12.3 (VI, p. 24 L. = p. 240 Jouanna).

34. Hp., *On the Technē* 12.4–5 (VI, pp. 24–26 L. = pp. 240–241 Jouanna).

35. In this use, it at times serves a Hippocratic rhetoric of certainty and dogmatism, as Geoffrey Lloyd has suggested; see G. E. R. Lloyd, *The Revolutions of Wisdom* (Berkeley: University of California Press, 1987), 114–123.

36. For example, Herodotus 1.116; Antiphon 6.25; Thucydides 1.99; Polybius 15.28.2; Herodas 5.5.

37. T. Gomperz, *Die Apologie der Heilkunst* (Leipzig, 1910), 57, 140.

38. Hp., *On the Technē* 12.1, 12.3, 12.5 (VI, p. 24.2, 24.8, 24.17 L. = pp. 240.5, 240.12, 241.4 Jouanna).

39. For example, Antiphon, 1.6, 1.10, 2.4, 2.8, 5.32, 5.40, 5.42, 6.25; see also Isocrates 17.53; Isaeus 8.12; Demosthenes 30.37; Aristotle, *Rhetoric* 1.15.1376b31–1377a7. See Gerhard Thür, *Beweisführung vor den Schwurgerichtshöfen Athens: Die Proklesis zur Basanos, Sitzungsberichte der Österreichischen Akademie der Wissenschaften*, philosophisch-historische Klasse, 1977, vol. 317, passim (e.g., pp. 15–24 on the torture of nonslaves, and pp. 25–27, 159–193, 287–312 on the torture of slaves); Michael Gagarin, "The Torture of Slaves in Athenian Law," *Classical Philology* 91 (1996): 1–18. On the Platonic Socrates' allusion to similarities between the examination of witnesses under torture and the Pythagoreans' empirical investigations in acoustics (*Republic* 7.531b2–8), see G. E. R. Lloyd, *Magic, Reason and Experience* (Cambridge: Cambridge University Press, 1979), 145–146.

40. See Lloyd, *Magic, Reason and Experience*, 59, 63, 79–80, 84–86, 99–100, 244, 250–252, 259, 262–263; Lloyd, *Revolutions* (note 5 above), 79–82; Lloyd, *Demystifying Mentalities* (Cambridge: Cambridge University Press, 1990), 8–9, 58–66, 76–78, 96–97, 141–142.

41. See Thür, *Beweisführung vor den Schwurgerichtshöfen Athens*; Gagarin, "The Torture of Slaves in Athenian Law" (note 39 above).

42. Hp., *On the Technē* 12.6 (VI, p. 26 L; p. 241 Jouanna).

43. Heraclitus, fr. B123 (Diels/Kranz).

44. See H. von Staden, *Herophilus: The Art of Medicine in Early Alexandria* (Cambridge: Cambridge University Press, 1989), chapters 6–7; Ivan Garofalo, *Erasistrati Fragmenta* (Pisa: Giardini, 1988).

45. Animal dissection and vivisection, however, which started no later than the classical period in Greece, continued to be performed with varying degrees of frequency and systematicity at least until the time of Galen, whose *On Anatomical Procedures*

offers vivid accounts and detailed justifications of numerous dissections and vivisections of a variety of animals.

46. Celsus's source concerning this controversy appears to have been a work or works by an adherent (or adherents) of the self-named empiricist school of medicine, which adamantly opposed all dissection and vivisection, also of animals. On the question of Celsus's sources see Friedrich Marx, *A Cornelii Celsi quae supersunt* (*Corpus Medicorum Latinorum* [hereafter = *CML*] I, Leipzig/Berlin, 1915), LXXVI–XCIV; Philippe Mudry, *La Préface du De Medicina de Celse* (Rome: Bibliotheca Helvetica Romana XIX, 1982), 83–205 (passim). See also Werner Deuse, "Celsus im Prooemium von 'De Medicina': Römische Aneigung griechischer Wissenschaft," *ANRW* II 37.1, 819–841; Fabio Stok, "La scuola medica Empirica a Roma: Problemi storici e prospettive di ricerca," *ANRW* II 37.1, 600–645; Fabio Stok, "Celso e gli Empirici," in Guy Sabbah and Philippe Mudry, eds., *La médecine de Celse: Aspects historiques, scientifiques et littéraires* (Saint-Étienne: C. N. R. S., 1994), 63–75; H. von Staden, "*Media quodammodo diuersas inter sententias*: Celsus, the 'Rationalists,' and Erasistratus," in *La médecine de Celse*, 77–101.

47. Celsus, *Medicina*, pr. 23–24 (*CML*, p. 21 Marx).

48. Celsus, *Medicina*, pr. 25–26 (*CML* I, p. 21 Marx). On the last sentence and related passages see Lloyd, *Methods and Problems in Greek Science* (note 5 above), 356–359.

49. Celsus, pr. 27 (*CML* I, p. 22 Marx). It should not be overlooked that Celsus subsequently distanced himself from this empiricist view, arguing that while the study of the nature of things (*natura rerum*) cannot in and of itself turn one into a medical practitioner, it renders one more apt and perfect in the medical *ars* (pr. 47, p. 25 Marx). He also attributed Erasistratus's failure to recognize that "nothing is due to one cause alone" to the Hellenistic physician's insufficient study of "the nature of things" (*naturae rerum*) (pr. 59, p. 27 Marx). For Celsus's historiographic uses of *contemplatio/cognitio rerum naturae* see pr. 6 and 9 (p. 18 Marx). See also H. von Staden, "Celsus as Historian?," in Philip J. van der Eijk, ed., *Ancient Histories of Medicine: Essays in Medical Doxography and Historiography in Classical Antiquity*, Studies in Ancient Medicine, vol. 20 (Leiden: Brill, 1999), 251–294; H. von Staden, "*Media quodammodo diuersas inter sententias*" (note 46 above).

50. Celsus, pr. 40 (*CML* I, p. 23 Marx).

51. Celsus, pr. 40 (*CML* I, p. 23 Marx).

52. See Celsus, pr. 27 ("non conprehensibilis natura sit"); any nonevident cause, such as causes hidden by nature, likewise is said to be *inconprehensibilis* (pr. 31). See also pr. 40 ("alia non possint *omnino* cognosci").

53. Celsus, pr. 43 (*CML* I, p. 24 Marx).

54. Celsus, pr. 43 (*CML* I, p. 24 Marx).

55. Celsus, pr. 43 (*CML* I, p. 24 Marx).

56. Celsus, pr. 41 (*CML* I, pp. 23–24 Marx).

57. Celsus, pr. 42–42 (*CML* I, p. 24 Marx).

58. Celsus, pr. 44 (*CML* I, p. 24 Marx).

59. Celsus, pr. 74 (*CML*, p. 29 Marx).

60. Iohannes Alexandrinus, *Commentaria in librum De sectis Galeni*, 5ra35–72 (pp. 57–58 Pritchet).

61. Agnellus Ravennas (?), *In Galeni De sectis commentaria* 23 (p. 92 Westerink et al.).

62. Tertullian, *De anima* 10.4 (p. 13 Waszink).

63. Vindicianus, *Gynaecia*, praef., cod. L, in Karl Sudhoff, *Archiv für Geschichte der Medizin* 8 (1915): 417–418.

64. See Friedrich Solmsen, "Nature as Craftsman in Greek Thought," *Journal of the History of Ideas* 24 (1963): 473–496; rpt. in Solmsen, *Kleine Schriften*, vol. 1 (Hildesheim, 1968), 332–355.

65. Galen, *On Natural Faculties* 2.2 (II, p. 78 K = III, p. 157 Helmreich).

66. Galen, *On Bloodletting against Erasistratus* 4 (XI, p. 158 K).

67. Galen, *On Natural Faculties* 2.3 (II, p. 81 K; III, pp. 159–160 Helmreich).

68. Galen, *On the Usefulness of Parts* 4.15 (III, p. 315 K; I, p. 231 Helmreich).

69. Plato, *Timaeus* 24a5–7, 28a6, 29a3–7, 31a4, 40c1, 41a7, 41d4, 42e8, 46e4, 47e4, 59a5, 68e2, 69c3–4, 75b5, 76d6, 80e4.

70. Aristotle, *Physics* 2.2.194a21–22; cf. 2.8.199a15–18.

71. Theophrastus, *On Stones* 60; *On the Causes of Plants* 2.18.2; *Metaphysics* 74a4–5.

72. Aristotle, *On Parts of Animals* 1.1.639b19–21 (cf. 14–18). See also *Physics* 2.8.199a8–20, b26–30.

73. Aristotle, *On Generation of Animals* 2.1.735a2–3; cf. 1.18.724a31–35.

74. See *Physics* 2.8.199b29–33; *On Generation of Animals* 1.23.730b19–22, 2.1.734a28–32, b19–22, 28–31, 3.11.762a15–18, 4.2.767a10–22, 4.6.775a20–22, 5.8.789b7–12; *On Generation and Corruption* 2.9.335b26–33; *On Parts of Animals* 1.1.639b14–16, 641b10–15; *Physics* 2.1.193a31–b5, 2.8.199a33–b4; *Meteorologica* 4.3.381a4–12.

75. Cf. Galen, *Whether by Nature There Is Blood in the Arteries* 8.4–5 (IV.733–734 K; 178–180 Furley/Wilkie); Galen, *On Anatomical Procedures* 7.11, 7.16 (II.624, 646 K); Galen, *On the Usefulness of Parts* 6.12, 7.8 (III.465, 537–540 K; I.339 Helmreich); Galen, *On Affected Places* 5.3 (VIII.316 K); Galen, *On the Usefulness of Respiration* 2.1, 2.10, 5.1 (IV.474–475, 482, 520 K; 82–84, 94, 120 Furley/Wilkie).

76. Vitruvius, *On Architecture* 10.7.1. For another example of the interaction between Hellenistic medicine and mechanical technology see H. von Staden, "Andréas de Caryste et Philon de Byzance: Médecine et mécanique à Alexandrie," in *Sciences exactes et sciences appliquées à Alexandrie (IIIe siècle av. J.-C.–Ier siècle ap. J.-C.)*, Actes du Colloque international de Saint-Étienne, publiés par Gilbert Argoud et Jean-Yves Guillaumin (Centre Jean-Palerne, Mémoires XVI, Université de Saint-Étienne, 1998), 147–172.

77. Vitruvius, *On Architecture* 10.7.1. For a similar device see also Heron, *Pneumatica* 1.28 (I, pp. 130–136 Schmidt, with figure 29).

78. See, for example, Galen, *On the Opinions of Hippocrates and Plato* 6.6.4–11 (V. 548–550 K; *CMG* V 4,1,2, p. 396 De Lacy).

79. Vitruvius, 10.7.1.

80. Erasistratus seems, however, to have acknowledged the distinction between continuous void and disseminate or dispersed, interstitial void, and to have made allowance for the latter in the case of the movement of liquids through the body (perhaps under the influence of his older contemporary Strato of Lampsacus, who succeeded Theophrastus as leader of the Peripatos in 288 or 286 BCE). For Erasistratus's views see, for example, Galen, *Nat. fac.* 2.1 and 2.6 (II. 75–76, 95–99 K.; *Scr. min.* III. 155–156, 170–173 Helmreich); Anonymus Londinensis XXVI.48c, XXVII.6–7 and 25–39. On the relation between Erasistratus, Strato, and the Alexandrian mechanicians see Luciana Repici, *La natura e l'anima: Saggi su Stratone di Lampsaco* (Turin, 1988), 85–90, and a more controversial earlier analysis by Hermann Diels, "Über das physikalische System des Strato," *SB Berlin* 1893, phil.-hist. Kl., 101–127, especially 105–117 (rpt. in Diels, *Kleine Schriften zur Geschichte der antiken Philosophie*, ed. W. Burkert, Hildesheim, 1969, 239–265). Cf. J. T. Vallance, *The lost theory of Asclepiades of Bithynia* (Oxford: Oxford University Press, 1990), 9, 62–79, 84–85, 123–130.

# 3

## The Three Pleasures of Mimēsis According to Aristotle's *Poetics*

Francis Wolff

*English version revised by Linda Blake*

In the seventeenth and eighteenth centuries, modern thinkers thought that the essence and ultimate finality of arts consisted in "imitating nature." In so doing, they were often appealing to the authority of ancient philosophers, and in particular to Aristotle. Yet it is well known that they were misinterpreting the ancients. When the latter spoke of *technē* (pl: *technai*), they did not mean what we now call the arts (in the sense of "fine arts"); they were referring to technique or "know-how." Therefore, the ancient theory according to which "art imitates nature" is by no means an aesthetic theory; it is at most an explanation of processes of production for artifacts in general.

However, I believe the ancients, and specifically Aristotle, *did* indeed create a concept to subsume all and only those activities we call "artistic," identifying their essential unity. He called them "imitating arts" (*technai mimētikai*). The coexistence in Aristotle's work of this concept with the idea that "art (in general) imitates nature" has been a matter and the source of great confusion in recent centuries. To develop these ideas, I will focus on the following interrelated points: first that Aristotle introduced a hybrid concept of *technē mimētikē* in his *Poetics*, second that even if every art imitates nature, it is not *this* relationship that defines "imitating arts," and third that the constitution of this complex concept allows us to explain the different pleasures people feel through works of art, and in particular to distinguish the pleasure that springs from their imitative function (the pleasure of representation), from that derived from their artistic form (aesthetic pleasure).

### Art, Imitation, and "Imitative Arts"

Aristotle invented the concept of "mimetic arts" (*technē mimētikē*),[1] which is very akin to our concept of "fine arts," by combining two heterogeneous concepts: the concept of "art" (*technē*) and the concept of *mimēsis* (which I will translate by "imitation" or "representation").

The Greek word *technē* refers to a skill in some field, and to the knowledge on which it is based (the "know-how"). In *Nicomachean Ethics*, Aristotle claims this *technē* to be the virtue of *making*, the *poein*. Making, contrary to doing (*prattein*) and to contemplating (*theorein*), does not have its "end" in itself, but in a product, the work. For example—these are the two favorite examples given by Aristotle—medicine "makes" (or "produces") health, just as building "makes" (or "produces") a house. The virtue by which something is well made is based on true knowledge of it: this is *technē*.[2] It can therefore be compared to nature (*physis*). Both *technē* and *physis* are "principles of movement," that is, the ultimate cause of any change; like natural processes, technical operations may be explained by their end result. But while a natural being develops of its own accord and carries in itself its own completion, the principle and end of an artificial operation lie not in the operation but outside it, respectively in the artist (the principle) and in the product (the end result). While nature is the principle of a finalized movement that lies in the moved thing,[3] *technē* is the principle of a finalized movement that lies outside the moved thing.[4]

Let's take a closer look at the meaning of *mimēsis*. Aristotle does not give any definition of it. This word commonly refers to the act of miming someone (or something), and more generally to every kind of imitation or representation. In Aristotle's view, *mimēsis* is a natural human activity in the sense that it is both spontaneous before being technical ("artistic"), and universal, even if only the most talented individuals accomplish it perfectly.[5] Just as most arts are not mimetic, so most mimetic arts are not technical.

By crossing this concept of *mimēsis* with the concept of *technē*, Aristotle invented a new way to categorize experience: mimetic arts lie at the intersection of regulated activities (as opposed to informal, spontaneous, or improvised) and of imitating activities (as opposed, especially, to "productive" ones). Coming under this new concept of *technē mimētikē* (representative art) we find "literature" (Aristotle's anonymous invention[6]), painting, sculpture, music and dance—that is, what we call the "fine arts" today, a very astonishing collection for the ancients, and especially for the Platonic minded.[7] But differences do exist between our modern concept of "fine arts" and the Aristotelian concept of representative arts. To speak of fine arts is primarily to refer to *works* (novels, statues, paintings, symphonies, ballets, and so on). But by *technē mimētikē*, Aristotle primarily means *acts*; he does not refer to things made, but to modes of making. In fact, art is some kind of methodical making, and *mimēsis* also designates making, the making believe, the imitating act (either of someone, or of some action or emotion); it is not a product (a painting, a statue, a tragedy)—these are called the *mimēmata*. This is why, if we view painting—or sometimes poetry—as the preeminent fine art (for its result is a self-sufficient and self-subsistent product), the ancients and in particular Aristotle, considered drama to be the preeminent *technē mimētikē* (for its result is not a product but an action that depends on agents and that does not subsist beyond the time of its representation).

The homogeneity of the concept is thus the following: *technē mimētikē* is methodical know-how (that is its "genus") used to produce representations (that is its "specific difference"), as opposed to nonproductive arts or to nonrepresentative productions). And yet even though they are both considered to be "arts" from the standpoint of "production" and not of "contemplation," *technē* and *mimēsis* are not on the same plane of experience. *Technē* is a particular determination of productive activity (artistically achieved as opposed to being naturally achieved). *Mimēsis*, on the contrary, defines a particular relationship between the product and nature (it mimes nature). The concept of *technē mimētikē* is consequently characterized by two relationships, the relationship a person has with it (*technē*), and the relationship it has with the world (*mimētikē*). What the work owes to art, it owes in connection with its producer; what is "mimetical" in the work it owes to the work's connection with the reality represented,[8] producer and reality being both prior and exterior to the work itself. And these two relationships—that is, the technical and "mimetical" relationships—move in opposite directions with respect to nature. To say that a production belongs to art, is to say twice over that it is *not* dependent on nature: it is made by humans, (as opposed to natural beings, which do not depend on humans), and it has been achieved by humans according to rules of art (as opposed to "natural" activities, that is, those that are not learned, not transmitted, not regulated). But to say that a production belongs to *mimēsis* is to say that it *is* dependent on nature, because it represents a natural reality that exists independently from humankind.

## The Two "Imitative" Relationships between Art and Nature

### Imitative Arts and the Imitation of Art

We must now distinguish two kinds of relationships between art and nature through mimēsis. Imitating nature is the purpose of only *some* arts, the *technai mimētikai*, those that create works representing something that exists naturally. But all arts qua arts mime nature since, as the famous Aristotelian expression puts it, "art imitates nature."[9] This does not necessarily mean that art creates an image of nature, but that it proceeds like nature. For Aristotle, this statement is not only intended to clarify how art proceeds, but also how nature proceeds. What he means is that we can understand nature if we consider art as its explicative model, since art takes nature as its real model in its *modus operandi*. It does this in two ways: first, like natural production, every artistic production "informs" a material, that is, it organizes several parts into a whole, which would indicate that the material cause is subordinated to the formal cause; second, in natural production as in artistic production, the successive operations are subordinated to their end, that is, in both cases the chronological order of productive causality is opposed to the rational order of real causality, meaning the efficient cause is subordinated to the final cause.

We can see the extent to which modern thinkers misunderstood the Aristotelian expression in having made it the slogan, justification, and essence of fine arts. In reality, the expression "art imitates nature" is not prescriptive, but descriptive. It does not indicate what art should *intend*, nor what any particular work of art should *represent*, it simply characterizes an operating mode. It does not define the static relationship between works and their model from the point of view of the spectator, but compares two dynamic processes of production from the point of view of the maker. Moreover, it does not concern fine arts ("imitative arts"), but art or skill in general. For it is not qua imitative but qua arts that imitative arts imitate nature.

We must therefore distinguish two ways of "making like nature": it is both making *the way* nature makes, and this is what any art does, whether it is imitative or not, and it is also making (or mimicking) *what* nature makes—and this is what any mimēsis does, whether it is artistic or not. "Mimetic art" is mimetic twice over: being an art, it imitates natural *production*, and being mimetic, it imitates the natural *product*. We may infer from this what might be called the *double* four-causes theory of mimetic art: the first theory considers art as a process of making, and is developed in Aristotle's *Physics*; the second theory considers art as a process of imitation (*mimēsthai*) and is developed in Aristotle's *Poetics*.[10]

## The Double Four-Causes Theory of Mimetic Arts

In *Physics* (II, 3) Aristotle presents the famous "four causes" (or "types of explanation") theory, and illustrates it with artistic production. Being a product of making, the statue is explained in a fourfold manner by the matter out of which it has been made, by the form given to this matter, and by the acts of the maker who aimed at producing an achieved work. In the same manner, the first four chapters of the *Poetics* can be read as a four-causes theory of *mimēsis*. In this case, any mimetic production (the statue or even a tragedy) is not considered as a product of *technē* but as an effect of *mimēsis*. Indeed, in the very beginning of the *Poetics* (ch. 1, 1447a13–17), Aristotle states that imitations (*mimēseis*) "represent with different media, or represent different things, or represent in different ways." He thus announces the subject matter of the first three chapters of the treatise, in which he deals with the material (ch. 1), the formal (ch. 2), and the efficient causes (ch. 3) of *mimēsis*.

**Matter** In general, the *matter* of something is the answer to the question "what is it made of?," which means "what has it been made from?" and "what constitutes its parts?"[11] But a *technē mimētikē* work has two different kinds of matter, depending on whether it is considered the product of *technē* (as theorized in the *Physics*) or the product of *mimēsis* (as theorized in the *Poetics*). The matter of a work of art is for example bronze, wood, and even sound, which is shaped by the artist. The matter of mimetic works varies in relation to the types of representation and designates the medium "in which" (1447a17) the representation is made: each imitation has its own medium. This question, dealt with by Aristotle in chapter 1 of the *Poetics*, leads him to an initial classification of the fine arts in relation to the medium: sculpture, the medium of which is shapes (*schēmata*, 1447a19), painting, the medium of which is shapes and colors, and so forth for literature (art of language), instrumental music (art of rhythm and melody), dancing (art of rhythm and shapes), and singing (art of melody and language).

**Form** In general, the *form* of something is the answer to the question "what is it?" Once again, we can distinguish the form of a mimetic work *qua* work of art and *qua* mimetic work. The first is the "configuration of the whole as such" that the working artist has in mind. The other differentiates the *mimēseis* according to *what* is represented (1447a17). The form of the representation is thus a reality coming from the artist's mind. *What* the *mimēsis represents*—this will be its form. So generally, every *mimēsis* is the representation of at least a possible if not an actual reality (see ch. 25, 1460b7–11). More specifically, each mimetic art has its own form of representation, and this leads Aristotle to make a second classification of them in chapter 2 of the *Poetics*.

Summarizing this chapter with chapter 5 of book VIII of the *Politics*, we can infer the following: the intent of every imitation is to mime acting persons; music represents their temperaments, literature represents actions,[12] and the visual arts represent persons in action.

There are consequently two ways of identifying the "form" of a mimetic work of art: what is it that the artist has done, that is, what is it as a work of art? (Answer: A statue of Zeus.) And what is it that the work represents, that is, what is it as a mimetic work? (Answer: Zeus hurling his lightning bolts.)

**Efficient Cause** Let's now examine the third cause, the "efficient" cause, which again is not the same depending on whether we consider the work *qua* art or *qua* representation. The efficient cause of *technē* is the artist (Polycleitos or Sophocles) or the gestures he or she makes according to the rules of art. The efficient cause of *mimēsis* is not who or what makes the work, but what in it *makes it represent* something. And of course this depends on the *way* (*Poet.* 1, 1447a17) of representing. This is dealt with in chapter 3 of the *Poetics*. For example (it is the only one given by Aristotle), there are two main ways of representing actions as the poet does: the narrative way (1448a21–23)—as in the epic form, where the motor of representation is somehow external to the moved thing (*Poet.* 3, 1448a19–24)—and the dramatic way, where the motor of representation is somehow internal, since acting individuals (the "actors") are representing by their acts other acting people (the "agents").

**Final Cause** What is the fourth cause, the *telos*? Any art *qua* art, whether it is imitative or not, aims at producing and achieving a work. This is true for medicine, which produces health, for building, which produces a house, for painting, which produces a drawing and for poetry, the work of which is "the plot." But any *mimēsis* qua *mimēsis*, whether it is artistic or not, has a different end: it is simply *pleasure*. This is the main concept of chapter 4 of the *Poetics*, which establishes the "natural causes of the birth of poetical art as such." Considered from the point of view of their final cause, there are no differences among the mimetic arts, whatever differences they may have in terms of media, objects, and ways of representation. There is only one reason why representation (*mimēsis*) exists: it is simply because people naturally enjoy both making and contemplating representations.

We must now analyze this pleasure, which is the single final cause of all the imitative arts. Again, we will discover that this pleasure is twofold: the first kind of pleasure corresponds to imitating (the processes of) nature as every art does; the second corresponds to imitating nature as only mimetic arts do.

## Pleasure: The Finality (End) of Every Imitation

For Aristotle, pleasure is simply a natural state of being signifying the perfect accomplishment of a natural function in a living being.[13] This raises two kinds of questions about a representative work of art. Does it give pleasure to those who perform it (the artists)? Does it give pleasure to those for whom it is performed (the spectators)?

### The Artist's Standpoint

From the artist's standpoint, we must again distinguish what depends on art from what depends on mimetic activity. According to the definition of pleasure that implies an "achieved process," there cannot be any true pleasure in making, but only in having made—except when "making" and "having made" are coincident:[14] the musician (not the composer, but the interpreter) finds pleasure in playing music because it is an activity whose result cannot be separated from its making.[15] The true pleasure of a painter or a poet does not come from making a work (painting, tragedy) or from making it well, but from having made it, and from having made it well, just as for a doctor, pleasure does not come from curing the patient but from seeing him or her healthy. (But if there is no pleasure in art *qua* art for the artist, there must be pleasure in the artist's work as a work for the audience. We will examine the spectator's pleasure a bit later.)

But if we consider *technē mimētikē* qua mimetic, this completely changes things for the artist himself or herself. *Mimēsis*, whether it is artistic or not, is a natural activity, the mere practice of which is pleasant.[16] This is the pleasure the agent of any mimetic activity feels, whoever he or she is, whether an inexperienced child or a skilled artist.

Moreover, not only does imitating a natural activity give pleasure to anyone who does it, and humans indulge in it specifically for this reason, but the mimetic arts are also natural beings having a natural history. Arts are born naturally and grow spontaneously from informal mimetic activities.[17] The *Poetics* (chapter 4) describes the three stages of the natural history of poetic art: the embryonic stage (1448b5), corresponding to a universal innate tendency to imitate and to a natural instinct for melody and rhythm (1448b20); followed by the birth of art (1448b22), corresponding to the individual innate gift of some people for improvisation;

and finally, the maturity of the arts, marked by their differentiation into genres (1448b24–26). Mimetic arts are thus doubly natural: as mimetic they at first coincide with an anthropological disposition for imitation, but since "nature never acts in vain," they naturally grow to a technical stage, because as technai (arts), they satisfy another natural human tendency: in fact, not any special need (as for the other arts), but merely a disinterested delight, or as Aristotle puts it in his *Metaphysics*,[18] some pleasure. Pleasure then is afforded to humans not only by mimēsis, by the act of representing, but also by its result, the mimēma. The work of mimetic art does not afford the artist with any natural pleasure of art, but with the natural pleasure of imitating. Let's now consider the problem from the standpoint of the spectator for whom the work is destined. The spectator's pleasure depends on both the "artistic" (technical) aspect of the work, and on its mimetic aspect. A work pleases because it is well made and because it represents well.

**The Spectator's Standpoint**

The different pleasures of contemplating a work of mimetic art are explained by Aristotle in chapter 4 of the *Poetics*:

> Men have . . . a tendency to enjoy representations. Practical experience provides proof of this: we enjoy looking at the most refined images of things we can't bear to see in reality, for example the shapes of animals absolutely vile or of corpses, because learning is not only a pleasure for philosophers but also for other men . . . . The reason why men enjoy seeing an image is, that in contemplating it they find themselves learning or inferring, and saying perhaps, "Ah this is that." For if you happen not to have seen the original, the pleasure will be due not to the imitation as such, but to execution (*apergasia*), or the colouring, or some such other cause. (*Poet.* 4, 1448b6–19)

This text points to at least two principal types of pleasure evoked by a work of art. The first is associated with the mimetic function and depends on the way the spectator connects the representation to the thing represented. The second type is explained in the last sentence quoted above and depends on the image itself. But I believe that this final dense sentence distinguishes two kinds of pleasure of the second type of pleasure: the pleasure afforded by colors, and that afforded by the quality of execution (*apergasia*). There are thus three kinds of pleasure: the pleasure that is purely sensory (the colors), the pleasure felt by contemplating the perfection of execution—both based on image alone—and the intellectual pleasure resulting from comprehending the *relationship* between the representation and the thing represented. I will first attempt to define these three pleasures, then to show, setting the first one aside because it is linked to the natural activity of our senses, that the other two depend on *technē mimētikē*, the first stemming from the proper effect of art (*technē*), the latter from the effect of *mimēsis*.

**The Sensory Pleasure Provided by a Work of Art** I will not spend much time on this type of pleasure because it depends neither on *technē* nor on *mimēsis*. Aristotle gives the example of the colors in a painting. According to his view, color only embellishes drawing, which is the representative element of painting. This is also the case with poetry. In the spectacle of a tragedy, rhythm, melody, and singing are not used in their mimetic function but only as sensory ornaments of the plot, which, being expressed through language, the very medium of poetry, is the only representative element of tragedy. Rhythm, melody, and singing afford only a sensory, additional pleasure, being both "extratechnical" and extramimetic. Aristotle calls them "seasonings" or "embellishments" (*hēdusmata*).[19] The "seasonings" of tragic language are rhythm, melody, and singing, that is, representation media that are foreign to literature—the proper medium of which is language. They belong to other mimetic activities (music and dance). In the tragic spectacle, these media are not used in their mimetic function but only as sensory ornaments. They play the same role as color does in relation to drawing in painting: the drawing represents and color embellishes. In theater, language represents and music embellishes. The "seasonings" add sensory pleasure that is deeply felt,[20] but they are called *atechnon*, because strictly speaking they are neither "art" nor "representation."[21]

This pleasure is derived solely from the activity of the senses. According to the definition of pleasure, the senses afford pleasure when they are used naturally, when there are no limits to their activity, and when they are exercised merely for the sake of themselves. This pleasure is at most the sign of a natural activity.[22]

**The Pleasure of Artistic Representation** Let us now consider the pleasure of representation. How does it function? This pleasure can be understood by comparing the text of the *Poetics* with a portion of the *Rhetoric* (I, 11, 1371b4–11). According to Aristotle, it is possible to feel very great pleasure through the contemplation of unpleasant images if as a result we can make a deduction ("this is that") and in so doing, learn something.

The problem is to understand what "this is that" means. It is clear that *this* means the representation (whatever it may be, picture, statue, tragedy, and so on), and that *that* is its model. But exactly what do they stand for? It is not, for example, "the image of Coriscus" and "Coriscus"; what could we learn in merely recognizing Coriscus? But it could mean something like: "these colored patterns on the wall are Coriscus." This interpretation sounds better. Here we have a real task of deduction from the matter of the representation to its form. The shift from the thing representing to the thing represented is not immediate: I must be able to perceive the shapes as a whole, and to recognize in a two-dimensional drawing the three-dimensional patterns of Coriscus. Pleasure follows from this: "yes, it's him, it's *really* him." (We may compare this pleasure with the pleasure we feel when we look at a family picture album.) And learning also follows from this. We learn to see Coriscus himself in his represented figure or profile; we see him, so to speak, as presented to our mind and recognition alone, to pure contemplation; his form is detached from anything else, his body hidden. Here lies, in my opinion, the learning and pleasure of *mimēsis*. In visual *mimēma*, we learn to see things, to merely *see* them, by identifying them. The process by which we recognize beings through their portraits is the perceptive transposition of the scientific question: "what is this?" Thus, each thing, each person, can be subsumed by its various representations; we learn to identify the thing or person via its representation in different media. We therefore learn perceptively what they are; this is exercising our senses for the sake of pure pleasure.

This interpretation of "this is that" is certainly relevant but still insufficient. If it can be said in the case of painting that "this" designates shapes and colors, it cannot be said that "that" designates Coriscus. The true form of *mimēsis* is not nature in general, nor humans in particular, but more specifically, acting humans. The true meaning of "this is that" in this case must be something like: these drawn and colored patterns on the wall are "Coriscus fishing" or "Socrates conversing" or "a mother mourning for her son." Not only do we learn to recognize specific people, we also learn to recognize the specific way they express their affections. It is through such representations that we learn to know and understand the human heart.

And this also happens in tragedy, which, through the plot, represents action. We can understand how the general pleasure of comprehending, which is evoked by any spectacle of imitations, becomes a specific pleasure when evoked by tragic imitation. A tragedy is mimetic of an action in which noble individuals appear, an action that is one. Its strictly imitative motor is thus the story (the *mythos*; see *Poet.* 6, 1450a3), which is to tragedy what drawing is to painting, its "soul" (6, 1450a39–b3): it is the "soul," strictly speaking, that mimes the action. Therefore, the strictly imitative pleasure is that which comes from the narrative and not from the personalities of the heroes. Tragic pleasure is linked to the work of tragic narrative, which is to "represent" (*mimētikēn*) frightening and pitiful events (13, 1452b31–33). The "frightening" and the "pitiful" are objective attributes of the succession of events the story represents.[23] The spectator who feels fright and pity thus experiences the appropriate affections when confronted with the situations presented, a sign that the spectator's emotive activity is functioning properly, apart from the fact that there is nothing really painful actually occurring, since tragedy is only representation. *We thus feel the pleasure of emotive activity without the pain.* The conclusion drawn from this in the next chapter is: "it's not just any pleasure that poetry must produce, but the pleasure that is proper to poetry. But since the pleasure the poet should afford is that which comes from pity and fright through imitation, it is evident that this quality must be impressed upon the events" (14, 1453b11–14).

But we will not go into a detailed analysis of tragic pleasure here. Suffice it to say that drama, like painting, both being considered as *mimēma*, brings cognitive pleasure that, however complex, is still linked to identification: we learn to know and recognize what beings, and especially what humans are, with their different temperaments, drawn for us by means of an exposé of their features. Like all visual arts, painting answers the question "what is it?" Considered as a *mimēma*, drama also brings cognitive pleasure linked to the succession of acts and events that constitute the plot. But drama does not answer the question "what is it?" (nor the "is there?" question), it only answers the "what?" and "why?" questions,[24] "what?" meaning "what's going to happen next?" and "why?" meaning "why did it happen?" For the spectator, understanding the plot is always based on two constantly repeated questions, one of which is prospective: "what will happen next?" (which can never been answered with certainty); the other of which is retrospective: "why did it happen?" (which can always be answered with certainty but with a constantly changing response). Drama relies on our experience of the world in order to expand our knowledge of it.

The third type of pleasure for the spectator of a work is also cognitive, but it is not evoked by *mimēsis*, but by art itself.

**The Pleasure of Artifice** There is another pleasure afforded by mimetic works of art that is neither sensory nor linked to mimetic function: it is the effect of art itself. We feel pleasure in admiring the artist's labor, the perfection of his or her work: *apergasia*. This unusual word[25] would appear to designate the quality of an image well done, implicating know-how and skill that distinguish the artist from the amateur.

This third pleasure can best be understood by comparing it with the pleasure Aristotle describes in *Parts of Animals*. In this text, Aristotle demonstrates that one can feel pleasure in contemplating ugly animals the moment one recognizes the work of nature in them, just as one admires a sculptor's or a painter's work. This pleasure is not based on connecting the image to its model (its formal cause), but to the agent's art (its efficient cause). The image is well done qua image. We admire the artist, whether a human being or nature. This admiration has the capacity to convert displeasure into pleasure. We can enjoy contemplating the image of an unpleasant being, admiring how the artist has managed to give it an organic unity, and we can also enjoy contemplating its natural model, provided that at the same time we admire how nature has succeeded in giving a living being its organic unity. The pleasure we feel in freely contemplating any natural being (dealt with in Aristotelian biological treatises) must therefore be compared with the pleasure we feel in contemplating a technically successful artifact (dealt with in the *Poetics*: the *apergasia*); this is *not* the pleasure of recognition (which comes from *mimēsis*). This is why we find a theory relating to this third type of pleasure throughout the *Poetics*, the pleasure afforded by the arrangement of parts into a whole, resulting from the artist's talent and the work of *technē*.

Chapter 7 of the *Poetics* in particular describes the formal qualities of "the arrangement of events" that a tragedy must have in order to afford this artistic pleasure, which is set apart from the mimetic element of the plot, affording the intellectual pleasure of recognition and understanding. A tragedy must have consistent parts constituting a single whole, as opposed to chance, in order to imitate finality. The idea of totality is complemented with the idea of unity. The plot must develop a unified action that forms a whole. And the pleasure, which results from apprehending the unity of the whole, is greater when it results from a higher number of parts. This is the *maximal* principle (the maximum number of elements in the biggest unity), provided the totality does not exceed a specific limit, defined by the second rule: not too big, nor too small, the standard being, in space, the spectator's range of vision (1450b39–51a1) and, in time, the spectator's memory.

We can finally comprehend what this pleasure is: it results from what we call beauty. Beauty is the effect of *technē*, and more specifically, it comes from the work of composition achieved by the artist. Aristotle bases the rules of art (*technē*) on the general principles, which are justified and altogether illustrated in natural beings, and, in particular, in living beings. For living beings too are the organized unity of different elements answering to a single finality, life—that is, the preservation and perpetuation of being as such. The rules of beauty are not, in consequence, rules of mimetic work as such (what we would call the work of Art), but rules of the artistic work as such (what we would call the technical or artificial). A work of art is not beautiful because it imitates nature (for imitated nature can itself be ugly), but because it has been made with art—that is, it has been made by imitating nature, in the sense that any art imitates natural processes by proceeding rationally, by subordinating the material cause to the formal cause, and by organizing the maximum number of elements into a whole. In this, nature acts like art.

## Conclusions

It is generally thought that the concept of the fine arts, which groups together arts having no other purpose than that of pleasure and disinterested contemplation, is an eighteenth-century invention. In reality it goes back to antiquity, to Aristotle's concept of imitative arts. But if the extension and the finality of imitative arts are the same as those of the fine arts, Aristotle's concept is viewed from the standpoint of the artist and not from the standpoint of the spectator.

It is generally thought that for the ancients, the natural was opposed to the artificial, and that between the two there was a simple and asymmetrical relationship of imitation. We have seen that this is not the case. "*Every art* imitates nature" does not mean that it reproduces it, it means that the two modes of making are isomorphic. Without nature, it would be impossible to proceed artificially, without art it would be impossible to understand how nature naturally proceeds.

It is generally thought that the "imitation of nature" is the definition par excellence of artificial activity. For Aristotle it is nothing of the sort, first of all because imitating nature is a natural activity, at least for humans, secondly because such activity only becomes "artistic" at the end of a historical process, itself conceived as a natural process of development.

The relationships between the artificial and the natural become more and more complex in the concept of "imitative art," which intertwines art in general as imitator of nature, and imitation as a natural activity of humans; *mimēsis*, while being a natural activity, does represents nature, but art, while imitating the natural process, does not represent nature. The imitative arts are thus imitative twice over. As arts they imitate the natural production process, and as mimetic they imitate the natural product. From this twofold perspective results a twofold theory of the imitative work of art: as a product of art, it contains, paradoxically, all the characteristics of natural production (the same type of material cause subordinated to the formal cause, the same type of efficient cause subordinated to the final cause), and inversely as an effect of mimetic activity, it contains matter, forms, efficiencies, and singular ends that are specific to imitation and that are not those of nature, nor those of art, but bear witness to the autonomy of the sphere of representation. This twofold nature of imitative arts is also translated into the twofold nature of the pleasure imitative works give to the artist and to the spectator: the pleasure that depends on their artistic dimension and the pleasure that depends on their mimetic dimension, disregarding a third pleasure, which is purely sensory and which points to the fact that our senses are functioning properly. What is amazing and once again reveals the distance between the ancient and modern positions, is that the aesthetic pleasure of representation is first of all a pleasure of knowledge and recognition. Conversely, the pleasure linked to the contemplation of the beauty of a work of art is the same pleasure one experiences in viewing any well-made artificial object, and it is again the same pleasure one experiences in viewing any natural object (considered only for its own sake). Nature and art afford their contemplators the same pleasure.

## Notes

1. There is such an expression in Plato, but it is depreciatory: *technē mimētikē* is a type of "productive" *technē* (*poiētikē*) designed to produce images, not realities (Plato, *Soph.* 265a–b).

2. It is a "state involving true reason (*logos*) concerned with production" (Aristotle, *Nic. Eth.* VI, 4, 1140a21).

3. See Aristotle, *Phys.* II, 1, 192b20.

4. See Aristotle, *Metaph.* XII, 7, 1070a7.

5. See *Poet.* 4, 1448b424.

6. See *Poet.* 1, 1447a28–b13.

7. In Plato there is no *technē mimētikē* in this specific sense for at least three reasons: (1) he generally refuses to give to poetry (and to rhetoric for similar reasons), the title of *technē*, that is, methodical know-how; (2) there is no *single* concept designating linguistic *mimēsis* in general: some forms of *mimēsis* are highly valued (the "Socratic dialogue" for example), others are completely disregarded; (3) Plato never puts music (nearly always highly valued) on the same level as the other imitative arts (which are almost always disregarded).

8. See S. Halliwell, *Aristotle's Poetics* (Chapel Hill: University of North Carolina Press, 1986), 51.

9. This expression can be found in *Physics* II, 2, 194 a 21 (see also *Meteor.* 381b6, *Protrept.* B 13, 14, 23), but it is in another passage of *Physics* (II, 8, 199a8–20; see also *Part. An.* I, 1, 640a16 et *Pol.* VII, 17, 1337a1–3) that Aristotle develops the idea.

10. Based on a remark made by V. Goldschmidt, *Temps physique et temps tragique chez Aristote* (Paris: Vrin, 1982), 209–210.

11. See *Phys.* II, 3, 194b23.

12. A noble action in tragedy, a vulgar action in comedy (*Poet.* 6, 1449b24).

13. See Aristotle, *Nic. Eth.* X, 1174b23–33.

14. See Aristotle, *Metaph.* IX, 8, 1050a23 sq.

15. See Aristotle *Eth. Nic.* X, 4, 1175a13–15. This text refers to the playing musician and not to the composer in the modern sense of the word. It is probably the same for the actor as opposed to the dramatic poet: "two-stage" arts (creation/interpretation) give pleasure to the artist (the interpreter), whereas "single-stage" arts do not give pleasure to the artist as such.

16. See *Poet.* 4, 1448b5–7.

17. On the natural history of the arts, see in *Pol.* VII, 10, 1329b25sq., the periodic reinvention of arts propelled by need.

18. See *Metaph.* I, 981b15, and 981b18, 22.

19. By "seasoned language I mean language that contains rhythm, melody and singing" (6, 49b28–31). Elsewhere, singing is said to be the most important of tragedy's "seasonings."

20. In tragedy there is another "seasoning," the "spectacle" (*hopsis*), having to do with what we would call today artistic direction or stage management, and concerning

which Aristotle explicitly states that the control of its effects is not, strictly speaking, "art." Refer to his comments concerning music, which he places on the same plane as "spectacle." Tragedy possesses all that which an epic possesses, plus "music" and what would be considered "spectacle," whence springs the deepest-felt pleasures.

21. Indeed, the proper effects of tragedy (for example pity) can be produced by extratragic means and in particular by means of the "spectacle." But producing these effects by this means cannot be considered "art," it is a matter of artistic direction. It is the same as producing fear, or rather the monstrous.

22. See Aristotle *Eth. Nic.* X, 4, 1174b14 sq. on vision, and also X, 3, 1174a14.

23. Similarly in 9, 1452a1–3: "the purpose of representation (mimēsis) is not only an action and its achievement, but also frightening and pitiful events."

24. On these four types of questions ("is there?" and "what is?," "what?" and "why?") and their relationships, see Aristotle, *Anal. Post.* II, 1–2.

25. See Plato, *Prot.* 312d.

# 4

## Art and Nature in Ancient Mechanics

Mark J. Schiefsky

### 4.1 Introduction

In this chapter I discuss the art-nature relationship in antiquity with special reference to one important ancient art or technē: mechanics. Although different ancient authors express different views about the goals, methods, and scope of mechanics, they tend to agree in conceiving it as a technē, an art or science, involving a combination of various kinds of theoretical and practical knowledge. In a range of sources from the third century BC to the third century AD, mechanics is typically described as a technē that includes a wide range of fields, from the theory and practice of building machines for lifting heavy objects to the construction of artillery engines and automata.[1]

This chapter addresses two primary questions. First, how did ancient writers on mechanics conceive of the relationship of their technē to nature or physis? Second, what did these views on the art-nature relationship imply about the ability of mechanics to provide knowledge of nature? According to one line of modern interpretation, advocated by F. Krafft in a series of influential publications, mechanics in antiquity was universally viewed as bringing about effects that were "contrary to nature" (*para physin*); as a result, it is claimed, the study of mechanics could yield no knowledge of nature. Krafft often describes the goal of ancient mechanics as the "tricking of nature" ("Überlistung der Natur"), and claims that it was only in the early modern period—in the work of Galileo in particular—that the view arose that mechanics follows nature and its laws rather than working against it.[2] What exactly Krafft means by the tricking of nature is not immediately clear, but the opposition he sets up between tricking nature and following natural laws, as well as his emphasis on the association between mechanics and wonder-working or trickery that can be found in a number of ancient authors, suggest that he understands the tricking of nature to amount to the production of a break or suspension in the order of nature, where this is understood as the domain of a regular, predictable sequence of cause and effect. On this view the effects brought about by mechanics would be "beyond nature" (*para physin*) in the sense of being *super*natural, and mechanics itself would amount to a kind of magic.[3] In other contexts Krafft's view seems to be the more restrained one that mechanics acts "contrary to nature" by virtue of the fact that it is concerned with forced rather than natural motion in the Aristotelian sense. Since the subject matter of mechanics is limited to forced motion, mechanics is completely distinct from physics, the subject of which is natural motion; once again the conclusion is that the study of mechanics cannot yield any knowledge of nature.[4] To support this conclusion, Krafft appeals to the general claim that because the Greeks viewed nature as an organic unity, any artificial intervention in natural processes could only distort the behavior of nature rather than helping to reveal it.[5] The possibility that mechanics might provide knowledge of nature through its use of mathematics is ruled out by Aristotle's view of the relationship of mathematics to physics, which allegedly made it impossible for mathematics to explain any features of the natural world.[6]

I have attempted to set out Krafft's interpretation in some detail because it illustrates a number of assumptions that have strongly colored scholarship on ancient mechanics and indeed on the art-nature relationship in general, assumptions that are deeply problematic if not demonstrably false. To take one example, the notion that the Greek (or, less vaguely, the Aristotelian) concept of nature in itself implied that artificial intervention could yield no knowledge of nature has recently come under well-justified and persuasive criticism.[7] There is ample evidence from all periods of Greek science for the idea of a close analogy between art and nature, and for the use of analogies between art and nature as a heuristic device for learning about the latter. As far as mechanics itself is concerned, the notion that ancient mechanics was universally viewed as acting "contrary to nature" (*para physin*) is largely based on an interpretation of the earliest extant text devoted to the subject, the *Mechanical Problems* (*Mēchanika Problēmata*) commonly ascribed to Aristotle in antiquity and early modern times, but now generally thought not to be by him.[8] This text opens with a striking passage that describes art in general, and mechanics in particular, as bringing about effects that are *para physin*:

> Among things that occur according to nature [*kata physin*], we wonder at those whose cause is unknown; among things that occur *para physin*, we wonder at those that come about by means of art [*technē*] for the benefit of humankind. For in many cases nature acts in a way opposed

> to what is useful for us. For nature always acts in the same way and simply, while what is useful changes in many ways. Whenever, then, it is necessary to do something *para physin*, because of the difficulty we are at a loss [*aporia*] and have need of art [*technē*]. For this reason, we also call that part of art that assists in such situations [*aporiai*] a device [*mēchanē*]. For as the poet Antiphon said, so it is: "By means of art we gain mastery [*kratoumen*] over things in which we are conquered by nature." Instances of this are those cases in which the lesser master [*kratein*] the greater, and things possessing a small inclination move great weights, and practically all those problems that we call mechanical. These are not entirely identical with physical problems nor entirely separate from them, but they have a share in both mathematical and physical speculations: for the "how" in them is made clear through mathematics, while the "about what" is made clear through physics.[9]

A central goal of the present paper is to clarify what exactly the author of the *Mechanical Problems* means in claiming that art produces effects that are *para physin*. As already noted, this Greek phrase can be understood in a number of distinct senses. In section 4.3 I will argue that the author's meaning is best captured by the idea that art brings about effects that go "beyond nature" in the sense that they would not be possible without the intervention of art. These effects, though they go beyond nature and are in that sense *un*natural, need not be viewed as acting *against* nature—whether "nature" is understood in the global sense as a universal order of cause and effect, or as the specific principle of change present in a particular substance or object. In section 4.4 I will argue that despite the apparent conflict between art and nature described at the beginning of the *Mechanical Problems*, the author's notion of the art-nature relationship is in fact well illustrated by Aristotle's idea that art imitates nature and brings to completion what nature cannot. Moving beyond the *Mechanical Problems*, the notion that ancient mechanics was universally viewed as operating "contrary to nature" is further undermined by a consideration of certain branches of mechanics such as the building of automata, which were explicitly described as imitating nature. In section 4.5 I will argue that there are two primary respects in which ancient mechanics could provide knowledge of nature. First, machines could serve as models for understanding processes going on in the natural world. Second, mechanics could yield knowledge of nature through the use of mathematics: mechanics considers certain mathematical properties of physical bodies and uses those properties to explain their behavior.

To place these observations on mechanics in a broader context, I will begin in section 4.2 with a brief discussion of the art-nature relationship in two important traditions of early Greek thought, Hippocratic medicine and Aristotelian philosophy. Both of these lay great weight on the idea that art imitates nature and can therefore be used as a means of gaining knowledge of it. Finally, in section 4.6 I discuss a number of attempts from the early modern period to resolve the tensions present in the opening passage of the *Mechanical Problems*, attempts that show some striking similarities to the interpretation I propose in section 4.3.

## 4.2 Background

Before turning to mechanics in particular, it is important to note that the idea of a conflict between art and nature is quite uncommon in antiquity. Ancient authors were much more likely to stress the parallels between art and nature than to portray them as opposed to one another. Two important traditions in early Greek thought provide substantial evidence for this claim: Hippocratic medicine and Aristotelian philosophy. In this section I therefore present brief and necessarily oversimplified accounts of some of the ways the art-nature relationship was conceived of in these traditions.

The close connection between art and nature in Hippocratic medicine has two principal aspects. First, a large number of Hippocratic writers conceive of the art or technē of medicine as a systematically organized set of procedures based on a body of theoretical knowledge of human nature (physis). To be sure, different authors had different conceptions of what this knowledge of human nature amounted to. But many Hippocratics agreed that medicine's claim to be a systematic art or technē depended on its having a basis in *some* general theory of human physis. One consequence of the view that medical practice should be based on a theory of human physis was the idea that the aim of treatment is to restore a patient to his or her natural condition. Therapy could thus be understood as a matter of "following nature" by assisting the patient's return to a natural or normal state; indeed in some passages nature is viewed as possessing the ability to bring itself back to health, implying that the doctor's task is only to assist it in this process. Second, there is ample evidence in the Hippocratic literature for the procedure of drawing analogies with artistic or technological processes as a means of learning about human physis. Again and again the Hippocratics appeal to analogies drawn from the realm of technē to elucidate obscure phenomena such as the growth of the embryo or the behavior of the humors in the body.[10] In one remarkable passage, the author of the treatise *On Regimen* explicitly recommends drawing analogies with technē as a method of studying the obscure realm of human physis:

> Human beings do not understand how to observe the invisible through the visible. For though they make use of arts [*technai*] that are similar to human nature [*physis*], they are unaware of it. The mind of the gods taught them to imitate [*mimeisthai*] their own [activities], and though they know what they are doing they are ignorant of what they are imitating. . . . But I will show that the well-known arts are like the affections of human beings, both visible and invisible.[11]

The author goes on to suggest analogies between a wide variety of technological procedures and invisible processes taking place inside the human body. Though some of these analogies are highly obscure, it is clear that many of them appeal to the idea that artistic processes are either closely similar or identical to those that occur in the human body; it is because of this that an understanding of technē can be used to gain knowledge of human physis.[12]

Turning to Aristotle, it is well known that he distinguishes between natural things and artifacts on the ground that the former possess an internal principle of change while the latter do not (*Phys.* 192b8–32). But it would be a serious mistake to conclude from this that Aristotle viewed artifacts as inherently inferior to the products of nature, or held that artificial interference with the behavior of a natural object can reveal nothing about its nature.[13] Aristotle often emphasizes the similarity of natural and artificial processes. In *Physics* B8 he famously appeals to the analogy between art and nature to support the claim that nature is inherently teleological or end directed:

> Further, where there is an end, all the preceding steps are for the sake of that. Now surely as in action, so in nature; and as in nature, so it is in each action, if nothing interferes. Now action is for the sake of an end; therefore the nature of things also is so. Thus if a house, for example, had been a thing made by nature, it would have been made in the same way as it is now by art; and if things made by nature were made also by art, they would come to be in the same way as by nature. The one, then, is for the sake of the other; and generally art in some cases completes [*epiteleitai*] what nature cannot bring to a finish, and in others imitates [*mimeitai*] nature. If, therefore, artificial products

> are for the sake of an end, so clearly also are natural products. The relation of the later to the earlier items is the same in both.[14]

Two points about this crucial passage deserve special emphasis. First, the context makes clear that Aristotle's notion of imitation goes far beyond mere mimicry. His main purpose is to argue that nature, like art, operates in a goal-directed manner, and his point is that both natural and artificial processes take place in an ordered sequence in which each stage comes about for the sake of the next. What Aristotle has in mind is thus a close analogy between natural and artificial processes.[15] Second, the passage gives a precise meaning to the notion that art goes beyond nature. In producing an artifact such as a house or a ship, art brings about results that nature itself cannot. But because of the close analogy between art and nature, this is really just a matter of bringing to completion what nature leaves unfinished. Art, by acting in a natural way—the way nature *would* act if it could generate the products of art—is able to produce results that do not come about by nature alone.[16]

Aristotle often contrasts art and nature as external and internal principles of change, respectively, as in the following passage from the *Nicomachean Ethics*:

> All art is concerned with coming into being, that is, with contriving and considering how something may come into being which is capable of either being or not being, and whose origin is in the maker and not in the thing made; for art is concerned neither with things that are, or come into being, by necessity, nor with things that do so in accordance with nature [*kata physin*] (since these have their origin in themselves).[17]

To the extent that the art-nature distinction is simply a distinction between external and internal sources of change, it allows for a close parallel between natural and artificial processes. This parallel is emphasized in Aristotle's extended account of artificial and natural generation in *Metaphysics* Z7–9. In both cases a form is realized in matter; in artificial generation the form is originally present in the soul of the artisan, while in natural generation it originates in the parent. In a sense, then, the form is imposed from outside in natural generation as well. A passage from the *De generatione animalium* makes a similar point: "Art is the principle and form of the thing that comes to be; but it is located elsewhere than in that thing, whereas the movement of nature is located in the thing itself that comes to be, and is derived from another natural organism which possessed the form in actuality."[18] Although Aristotle contrasts art and nature here, this remark in fact comes at the conclusion of an extended analogy between them. Just as hot and cold do not suffice of themselves to temper iron in the way necessary to make a sword but must act in accordance with the form specified by art, so in the formation of the embryo the action of hot and cold will only generate the right parts if directed by the form supplied by the parent (734b28–735a2). The upshot is to stress the similarity of natural and artificial generation, even if the source of change in the former is internal rather than external.

In some Aristotelian passages the boundary between artificial and natural processes is quite unclear. Important evidence of this is provided by *Meteorology* IV, a text whose authorship is disputed but that (like the *Mechanical Problems*) clearly belongs in the tradition of Aristotelian philosophy and science. The book discusses various transformations between natural substances, especially those brought about by hot and cold. Chapters 2 and 3 focus on the process of concoction or *pepsis* (defined at 379b18–19 as "perfection brought about by a thing's own natural heat") and its various species, which are said to be ripening, boiling, and roasting. In the author's usage, *pepsis* is thus a generic term that subsumes artificial as well as natural processes. The author builds on ordinary usage, in which *pepsis* refers to a wide variety of natural changes such as the ripening of fruit and digestion as well as artificial processes such as cooking. Two passages in the discussion of *pepsis* make explicit reference to the art-nature relationship. First, in concluding his discussion of boiling the author remarks that "Such, then, is what is called concoction [*pepsis*] by boiling: and it makes no difference whether it takes place in artificial or natural instruments [*organois*], for the cause will be the same in all cases."[19] Second, in the discussion of roasting (defined at 381a23–24 as "concoction by extrinsic dry heat") he states:

> Now roasting and boiling are artificial processes, but, as we have said, the same general kind of thing is found in nature too. The affections produced are similar though they lack a name; for art imitates [*mimeitai*] nature. For instance, the concoction [*pepsis*] of food in the body is similar to boiling, for it takes place in a hot and moist medium and the agent is the heat of the body.[20]

While the first of these passages suggests that there is no difference between artificial and natural processes, the second seems to indicate that they are closely similar rather than identical: boiling and roasting have natural counterparts for which there is no name but that are similar to them in kind or *eidos*. But whatever the author's exact conception of the relationship between artificial and natural processes, these passages clearly imply that the study of the former can yield an understanding of the latter, whether by analogy or direct inference.[21] The view that artificial processes are either closely similar or identical to natural ones underlies Aristotle's extensive appeal to analogies between technē and physis, which are especially common in the biological works. Because the processes of art are similar or identical to those of nature, art can serve as a model for the understanding of nature.[22]

To be sure, the art-nature relationship was not always described as one of imitation. Near the end of the Hippocratic treatise *On the Art*, the author states that in the case of diseases that affect the internal organs, medicine makes use of visible symptoms to diagnose the patient's condition. But when nature does not reveal such symptoms of its own accord, the art has discovered means of compelling it to do so:

> Whenever nature itself will not yield these sources of information of its own accord, medicine has found means of compulsion [*anagkai*], whereby nature is constrained [*biastheisa*], without being harmed, to give them up; when it has been released it makes clear, to those who know about the art, what ought to be done.[23]

By administering certain foods and drinks or prescribing certain exercises, the doctor can force the patient's physis to give evidence of its condition. The language of this passage, which suggests the image of nature undergoing torture, could hardly be more striking. Yet while it portrays the art-nature relationship as one of antagonism rather than imitation, the passage does not suggest that the secrets nature is compelled to reveal are in any way *unnatural*. The point is simply, as Heraclitus put it, that "nature loves to hide."[24] Nature does not reveal its secrets without being subject to compulsion, but that does not make them any less a manifestation of nature. Here, then, we have a clear case where artificial modification and even compulsion must be used in order to gain knowledge of nature; there is no suggestion that such compulsion in any way distorts the information that it provides.

In conclusion, the notion that art imitates nature plays an important role in both Hippocratic medicine and Aristotelian philosophy. This notion in turn rests on the close similarity between natural and artificial processes. As Aristotle put it, art can go beyond nature, but in doing so it acts in the way that nature would act, if it could br[illegible] such results. Both Aristotle and the Hippocratics make extensive use of technical analogies as a means of learning about physis. There is, then, no a priori reason to suppose that mechanics could yield no knowledge of nature simply because it was viewed as a technē. Indeed the evidence discussed in this section suggests exactly the opposite conclusion.

## 4.3 Art and Nature in the *Mechanical Problems*

With these considerations in mind we may turn to the interpretation of the art-nature relationship in the *Mechanical Problems*. The author begins (above, p. 68) by distinguishing between two kinds of phenomena that excite wonder: those that come about "according to nature" (*kata physin*) and whose cause is unknown, and those that come about *para physin* by means of art (*technē*), for the benefit of human beings. Nature often acts in a way that is opposed to what is beneficial for human beings; this is because it always acts in the same way, while human needs vary widely. Nature's constant, unvarying activity leads to difficulties; confronted with these, human beings are often at a loss (*aporia*) and must make use of art (*technē*). Hence the part of *technē* that provides assistance in such situations is called a "device" (*mēchanē*). A quotation from the poet Antiphon sums up the author's view of the art-nature relationship: art enables human beings to triumph over matters in which they are conquered by nature.

Before going further it is essential to distinguish a number of senses that might be conveyed by the phrase *para physin* in Greek. At least four are relevant. First, *para physin* may refer to something that is "beyond nature" in the quite strong sense of beyond the order of nature, understood as a domain of the orderly sequence of cause and effect: in other words, the supernatural. On such a view mechanics would amount to a kind of magic. Second, *para physin* might refer to something that is contrary to a *specific* nature, understood as an internal principle of change in the Aristotelian sense. As is well known, Aristotle uses the phrase *para physin* to refer to motions that are contrary to the natural motion of an element: fire moves upward *kata physin* and downward *para physin*.[25] In some Aristotelian contexts the distinction between *para physin* and *kata physin* motion amounts to a distinction between motion caused by external force or compulsion (*bia*) and motion caused by an internal principle of change.[26] It is in any case clear that *para physin* in this sense involves no break or rupture in the order of nature; indeed Aristotle himself points to the existence of natural motions that are *para physin*, for example, lightning that moves downward despite its fiery nature.[27] Third, *para physin* in Aristotle often refers to what is unusual, while *kata physin* refers to what occurs usually or for the most part. This usage is particularly common in Aristotle's biological works, where *para physin* is often used of monstrous births.[28] Again, however, it is clear that Aristotle does not understand such phenomena as lying outside the domain of nature. A passage of the *De generatione animalium* makes the point especially clearly:

> A monstrosity [*teras*] belongs to the class of things contrary to nature [*para physin*], although it is not contrary to nature in its entirety but only to nature as it holds for the most part. As for the nature which is always and by necessity, nothing occurs contrary to that: unnatural occurrences are found only among those things that happen as they do for the most part, but which may happen otherwise. For even in these cases, something comes about that is contrary to this particular order, but never in a merely random fashion; thus it seems less of a monstrosity [*teras*], because even that which is contrary to nature [*para physin*] is, in a way, in accordance with nature [*kata physin*] (i.e., whenever the formal nature does not master the material nature).[29]

Though *para physin* by virtue of being unusual, monstrous births are also *kata physin*, both in the sense that they come about in nature and in the sense that they can be explained in terms of the behavior of specific natures: for they come about when the formal nature fails to master the material, and are thus in accordance with the nature of the elements.[30] Fourth, *para physin* might be taken to refer to what goes "beyond nature"—not in the strong sense that it lies outside nature understood as the domain of a regular sequence of cause and effect, but in the sense of something that nature cannot or does not do without the intervention of art. In this section I aim to show that it is this last sense that best captures the author's conception of mechanical phenomena and their relationship to nature. Art in general, and mechanics in particular, go beyond what unaided nature can achieve—but they do not lie outside the domain of nature, and they do not even necessarily involve forced motion in the Aristotelian sense of motion contrary to a thing's specific nature.

With these distinctions in mind we may now turn to the interpretation of the *Mechanical Problems* itself. At the opening of the text art is described as bringing about effects that are *para physin*, and art and nature are opposed to one another. But it is immediately evident that this is in fact an opposition between nature and human needs: art acts to benefit human beings, while nature often behaves in ways that are "opposite" or "contrary" (*hypenantion*) to what is useful for humankind. And this in turn is explained in terms of another opposition: nature always acts in the same way and "simply" (*haplōs*), while what is useful varies in many ways (*pollachōs*). Thus the opposition between art and nature comes down to a contrast between the varied complexity of art and the simplicity and constancy of nature. Art is more complex than nature and brings about effects that are beneficial for human beings; in so doing it acts *para physin*, but this does not imply that it works *against* nature. The contrast between the simplicity of nature and the complexity of art rather suggests that art works by making creative use of nature's constant, regular behavior to bring about beneficial effects. To be sure, the behavior of nature leads to difficulty, which can only be overcome by means of art: "Whenever, then, it is necessary to do something *para physin*, because of the difficulty we are at a loss (*aporia*) and have need of art (*technē*)." But there is no need to understand this difficulty as the result of nature's active resistance to the efforts of art; the point is simply that it is challenging to *adapt* or *direct* nature's constant course toward the diversity of human needs. Rather than describing a struggle between art and nature, the opening sentences of the *Mechanical Problems* in fact suggest the ability of art to go *beyond* nature by producing more complex effects that are also beneficial for human beings.

The notion that art aims at an end that is beneficial for human beings is deeply rooted in the tradition of Greek thought about technē. The same is true of the idea that technē enables human beings to escape from a situation of need or helplessness (*aporia*), and that it does so by means of a "device" or *mēchanē*. In Greek literature of the fifth and fourth centuries BC the term *mēchanē* commonly refers to a means, device, or stratagem, whether physical or intellectual, that enables a person to escape from a situation of difficulty. Conversely, the term *amēchania* refers to helplessness in such a situation. Two familiar passages from fifth-century tragedy provide good illustrations of this complex of ideas: (1) In the *Prometheus Bound* attributed to Aeschylus, the character Prometheus describes how human beings originally lived a helpless existence, lacking all the benefits of technology, and how he enabled them to rise above this primitive state by discovering the arts (*technai*): astronomy, counting, writing, the domestication of animals, seafaring, medicine, prophecy, and mining (447ff.). Yet despite these discoveries he lacks the ability to free himself from imprisonment at the hands of Zeus:

"Wretched I am: such are the means [*mēchanēmata*] I discovered for mankind, yet I have no skill [*sophisma*] to rid myself of the present calamity" (469–471). Prometheus's *mēchanēmata* are "means" that assist human beings in a condition of need.[31] (2) The first stasimon of Sophocles's *Antigone* (332ff.) celebrates human intelligence as the faculty that enabled human beings to discover the arts and overcome the dangers of a hostile natural world. Man the "skillful" (*periphradēs*) has devised means of capturing all manner of animals; by means of "skillful contrivances" (*mēchanai*) he is able to overcome (*kratein*) wild beasts (342ff.). With the aid of the arts (*technai*) man is "all-resourceful" (*pantoporos*) and goes to meet nothing in the future "without resource" (*aporos*); medicine even has the ability to deal with "unmanageable" (*amēchanōn*) diseases, and its power is limited only by the certainty of death (360–364). It is the inventive ingenuity associated with *mēchanē* that enables human beings to discover the arts: "Clever (*sophon*) beyond hope is the contrivance of his art (*to mēchanoen technas*), and he advances sometimes to evil, at other times to good."[32] In these and many other passages, *mēchanē* and related terms convey the ideas of inventiveness, flexibility, and adaptability to different circumstances—qualities that fit very well with the contrast between the uniformity of nature and the diversity of human needs stressed at the opening of the *Mechanical Problems*. A *mēchanē* is a means or instrument that enables human beings to make creative use of circumstances at their disposal to attain an end that would otherwise be impossible to achieve.[33]

In some contexts *mēchanē* carries strong connotations of trickery or deception.[34] Clearly the kind of inventive ingenuity associated with the term was sometimes viewed as suspicious or devious. Yet already in Herodotus and the Hippocratic writers *mēchanē* is used of a physical device or machine, with no connotations of trickery or deceit; in such contexts, however, it is often implicit that the device in question makes it possible to attain a goal that could not otherwise be achieved.[35] To conclude from the fact that *mēchanē sometimes* has connotations of trickery or deception that the author of the *Mechanical Problems* viewed mechanics as an art of tricking nature would be quite unjustified. There is no reason to import the idea of "trick" or "ruse" into the present passage: a *mēchanē* is simply a device or means that enables human beings to direct the constant, regular course of nature toward their more complex needs.[36]

The author continues (847a19–21) with a quotation from the poet Antiphon: by means of art human beings triumph (*kratein*) over things in which they are conquered (*nikōmetha*) by nature. The language of triumph (*kratein*) and victory (*nikōmetha*) suggests the image of art struggling with and conquering a hostile nature. But Antiphon does not actually state that art conquers nature, and in fact the idea expressed by his remark is rather more precise: technē enables human beings to overcome natural disadvantages and thus to attain control and mastery over the natural world. The background to Antiphon's remark is a view of the origins of culture that was widely shared in the fifth century BC, according to which human beings are described as struggling to overcome the dangers of a hostile environment. In one scenario that can be traced back to fifth-century sources, human beings are said to be inferior to animals in physical characteristics such as speed and agility, but to make up for this by their intellect and the development of technology.[37] A similar picture is suggested by the first stasimon of Sophocles's *Antigone*: through its amazing intelligence, humankind has developed the ability to overcome the dangers of the natural world; it "masters" (*kratein*) wild beasts by means of "devices" (*mēchanai*). The idea that art enables human beings to overcome natural limitations and to achieve control and mastery over the natural world does not in itself imply anything about *how* it does so. In particular, it does not imply that art causes a break or rupture in the natural order of cause and effect (*para physin* in sense 1 as defined above), or even that it acts against the natural tendencies of particular objects or substances (*para physin* in sense 2). Art "conquers" nature by enabling human beings to modify nature's behavior in beneficial ways, and thus to overcome their natural limitations. Once again the idea is that art goes *beyond* nature by producing beneficial effects that nature would not if left to itself. It is notable that in contrast to some fifth-century sources, the author of the *Mechanical Problems* does not portray nature as a hostile force that threatens the very existence of humanity. For him the opposition between nature and human needs arises solely from the contrast between the constancy and uniformity of the former and the complexity and diversity of the latter.[38]

Up to this point the author's remarks have been concerned with art in general rather than mechanics in particular. Next, at 847a22ff., he attempts to define the character of mechanics more precisely. Mechanical problems are those in which "the lesser master [*kratei*] the greater, and things possessing a small inclination [*ropē*] move great weights." This builds on the idea of overcoming natural disadvantages expressed in the Antiphon quote. Though the strength of human beings is limited by nature, these limits can be exceeded by means of mechanics, which makes it possible to move great weights with only a small force.

Additionally, this way of describing mechanical effects implies that they are *para physin* in the third sense distinguished above, that is, unusual or paradoxical. It is normal or natural for a greater force to "master" a lesser, and for large weights to be moved by large forces; mechanics brings about a reversal of this situation. What distinguishes mechanical effects is not so much that they involve forced motion in the Aristotelian sense (*para physin* in sense 2); the emphasis, rather, is on the fact that they bring about a reversal of the normal relationship between forces.[39]

After a brief remark on the relationship between mechanics, mathematics, and physics (to which I will return below) the author turns to the discussion of circular motion, which he claims to be the fundamental explanation of practically all mechanical phenomena. The remainder of the introductory section (847a28–848a37) establishes an association between wonder and the strange or *atopon*, and an opposition between both these ideas and the notions of nature (*physis*) and cause (*aitia*). The author begins by mentioning the lever as a paradigm example of a mechanical device: it enables a large weight to be moved by a small force, even though its own weight must be moved in addition. Again, what is emphasized is not that the lever causes a weight to move in a way that is contrary to its natural inclination (*para physin* in sense 2); the point is rather that it reverses the normal (i.e., natural) relationship between forces (*para physin* in sense 3). Hence the operation of the lever appears strange or *atopon* (847a28–b15). But in fact it can be explained by reference to the circle:

> Now the primary cause [*tēs aitias tēn archēn*] of all such phenomena is the circle; and this is reasonable, for it is in no way strange [*atopon*] that something wondrous [*thaumaston*] should result from something more wondrous [*thaumasiōteron*], and the most wondrous thing [*thaumasiōtaton*] is the combination of opposites with one another. The circle is made up of such opposites, for to begin with it is composed of both the moving and the stationary, whose natures [*physis*] are opposed to one another. So that when one reflects on this, there is less reason to wonder [*thaumazein*] at the oppositions that occur in connection with it.[40]

The lever is both "strange" (*atopon*) and "wondrous" (*thaumaston*). In part, its wondrous character is explained by something even more wondrous: the circle, in which opposites are combined with one another. Yet the circle itself is said to be much *less* wondrous once its nature as a combination of opposites is understood. Since the circle is the ultimate explanation of all mechanical phenomena, the upshot is that a machine like the lever is no longer strange or wondrous once its operation has been explained in terms of circular motion. The wondrous properties of the circle include (1) the presence of both concave and convex in the circumference (847b23–848a3) and (2) the fact that it moves simultaneously in opposite directions, both forward and backward (848a3–10; the idea seems to be that a point on a circle moves in the opposite direction from the point diametrically opposed to it when the circle rotates; cf. 848a19ff.). But the most important reason why the circle is a source of wonders in fact has nothing to do with this mysterious union of opposites: as it rotates, a point farther from the center moves more quickly than one closer to the center (848a14–19). At 848a11–14 the author expands on his earlier remark that the circle is the primary cause of all mechanical phenomena: "The things that occur with the balance are referred to the circle, and those that occur with the lever to the balance; while practically everything else concerned with mechanical motions is referred to the lever." This suggests a highly systematic procedure of explaining all mechanical phenomena in terms of circular motion. Since the operation of the lever is explained by reference to the balance, and that of the balance by reference to the circle, it is only necessary to show how a particular mechanical phenomenon can be analyzed as the operation of a lever in order to explain it in terms of the circle. In this sense the circle is the fundamental explanation of all mechanical phenomena, making their character as "wonders" only apparent.

The final part of the introductory section (848a19–37) describes a mechanical device whose operation depends on the fact that the circle combines opposite movements. The exact nature and function of this machine are unclear. However, it is apparent that it involves a number of circles placed along a line and in contact with one another, so that when the first one rotates it causes the second to turn in the opposite direction, and so forth. The section concludes with the following important remark: "Craftsmen, seizing on this natural property [*physis*] of the circle, construct an instrument by concealing the principle [*archē*], so that only the wondrous character [*to thaumaston*] of the machine [*mēchanēma*] is apparent, while its cause [*aition*] is unclear."[41] Here again the ideas of nature, principle, and cause are associated with one another and opposed to wonder. The operation of this machine is wondrous only if the causal principle (*archē*) that explains it is concealed. Moreover the machine is constructed not by working against the nature (*physis*) of the circle but by making creative use of it. If there is trickery in mechanics, it is not trickery of nature but trickery of the ignorant observer who is unaware of the true explanation of mechanical phenomena.[42]

Thus, while the author of the *Mechanical Problems* emphasizes the wondrous character of mechanical phenomena, it should be clear that this is not because he thinks that they fall outside the domain of nature understood as the realm of the regular progression of cause and effect (i.e., *para physin* in sense 1). Such phenomena are wondrous because they are unusual or paradoxical (*para physin* in sense 3), and their wondrous character disappears once they have been explained. The author's attitude toward wonders is similar to that of other early Greek thinkers committed to the view that all phenomena can be explained in terms of natural causes. Within this framework, wonders constituted a class of unusual, problematic phenomena that could in principle be explained, and that provided a stimulus to scientific inquiry.[43] The idea is of course Aristotelian. In the *Metaphysics* Aristotle identifies the wonder felt in response to phenomena such as eclipses as the starting point of philosophy and science; he goes on to mention automata as a paradigm example of a device that is wondrous only to those who do not understand the cause (*aitia*) that explains its operation. Once this cause is grasped, such devices are no more wondrous than the incommensurability of the diagonal to a person trained in geometry; indeed nothing is more wondrous to a geometer than the notion that the diagonal is commensurable with the side.[44] The notion that the wonder of mechanical phenomena disappears once they have been explained tends to break down the barrier between the artificial and the natural set up in the opening sentence of the *Mechanical Problems*. For it turns out that the effects produced by art, just like natural phenomena, are wondrous only insofar as their causes are unknown.[45]

After this introduction, the author discusses thirty-five "problems" or puzzling mechanical phenomena. Many of these are drawn from the experience of everyday life or technological procedures; others are more theoretical in nature. By far the longest discussion is that of problem 1, which is devoted to explaining the alleged fact that larger balances are more accurate than smaller ones. The reason for this is the property of the circle mentioned already at 848a14–19: of two points lying on the same radius, the point farther from the center travels faster as the circle rotates. In problem 1, however, the author attempts to explain why this is so. His discussion raises a number of difficulties that I cannot discuss here. What is important for present purposes is that he analyzes circular motion into a combination of "natural" (*kata physin*) motion along the tangent and motion *para physin* toward the center. The basic idea underlying his explanation is simple: because the motion of a point on a longer radius is less deflected (*ekkrouesthai*) toward the center than the motion of a point on a shorter radius, it travels more quickly—that is, it covers a greater distance in the same time (849a6ff.). What is the sense of *para physin* in this passage? Nothing speaks for sense 1: there is no break in the natural order of cause and effect here, only the constraint of a body's natural motion in a way that is subject to precise geometrical analysis. The closest match is evidently with sense 2: *para physin* referring to forced motion in the Aristotelian sense. Yet in some respects the author's analysis seems remarkably un-Aristotelian. It has often been supposed that because the subject under discussion in problem 1 is the motion of the balance, the author conceives of motion along the tangent as *kata physin* because it is directed downward, that is, in the direction of the natural tendency of a heavy body. But in fact it becomes clear in the sequel that this analysis is a very general one that is meant to apply to *all* cases of circular movement—including the rotation of a circle that moves in a plane parallel to the ground, for example, a potter's wheel (cf. ch. 8, 851b19–21; 852a1ff.). The author's idea seems to be that an object moving in a circle will naturally fly off at a tangent, *however* the circle may be oriented with respect to the ground. At most, then, we may say that *para physin* is used here of motion that is caused by external force or compulsion. The upshot is that mechanical effects that are explicable in terms of circular motion involve a certain amount of external force or compulsion.

In the remaining thirty-four problems the author largely follows the program stated in the introduction: mechanical phenomena are analyzed in terms of the lever, the balance, or by direct application of the circular-motion principle. It is notable that the author does not use the phrase *para physin* again after problem 1, not even in connection with the lever. Moreover in a number of cases the mechanical effect in question can be understood as assisting rather than working against a body's natural motion. Problem 8 discusses why circular and round bodies are easiest to move (851b15ff.). In the case of a circle standing upright on a horizontal surface (i.e., a wheel), one reason is that the circle is moved in a direction in which it already inclines (*repei*) by virtue of its weight (851b27–33). In the case of a circle that turns in a plane parallel to the ground (such as a potter's wheel), the author makes a clear reference back to the analysis of circular motion in chapter 1: movement along the tangent is natural (*kata physin*), and the mover pushes a circle along the tangent (852a7–13). In both cases, then, to set a circle in rotation is to cause it to move in a way in which it is already naturally disposed to move; machines that take advantage of the ease with which round and circular bodies can be moved thus assist their natural movement rather than working against it.[46] Problem 19 asks why an ax placed on a block of wood does not split the wood if a heavy weight is placed on it, but does if one raises it and strikes the wood. The reason is that "a heavy object takes on the movement of weight more when it is moving than when it is at rest" (853b19–20). That is, the force generated by the blow augments the natural downward tendency of the weight of the ax. In such cases, the mechanical effect in question is *para physin* only in the sense that it goes beyond what unaided nature can achieve (sense 4) and involves some external force or compulsion (sense 2).

In sum, the author's conception of the art-nature relationship is best understood as the claim that art goes beyond nature in bringing about effects that unaided nature cannot (*para physin* in sense 4). This is supported by the work's opening, where the apparent opposition between art and nature in fact amounts to a contrast between the complexity of the former and the simplicity and uniformity of the latter. Nothing in the opening suggests that art produces a break or rupture in the order of nature (*para physin* in sense 1), or even that it works against the natural tendencies of bodies (*para physin* in sense 2). Rather, the author describes art as going beyond what unaided nature can achieve by making creative use of the behavior of nature itself. This calls for the kind of inventive ingenuity associated with the term *mēchanē*, understood as a "device" that makes it possible to achieve results that could not otherwise be attained. Insofar as art is able to "conquer" nature, it is by the creative combination of natural regularities to produce effects that are beneficial for human beings. What distinguishes mechanical effects in particular is their unusual or paradoxical character, as manifest in a reversal of the normal or natural relationship between forces (*para physin* in sense 3). The remainder of the introduction and the author's detailed discussion of thirty-five mechanical problems confirms this analysis. Mechanical phenomena are wondrous because of their unusual or paradoxical character, but this disappears once they have been explained, and it implies no break in the order of nature. Circular motion is the fundamental explanation of all mechanical phenomena; it is analyzed in precise geometrical terms as a combination of motion *kata physin* along the tangent and *para physin* toward the center. This analysis suggests that the author conceives of mechanical phenomena as involving some external force or compulsion (*para physin* in sense 2). But several of the mechanical phenomena the author discusses involve the use of force to augment rather than to counter an object's natural tendencies. The claim that mechanics involves some external force or compulsion is simply a way of indicating that mechanics, as a technē, modifies the natural behavior of the objects with which it deals. And this is best understood as the claim that mechanics goes "beyond nature" in the sense that I have explained.

As a confirmation of this interpretation, I note that Latin translators and commentators of the early modern period tend to render the phrase *para physin* in the *Mechanical Problems* as *praeter naturam*.[47] The basic meaning of the Latin word *praeter* is "beyond"; though it can also mean "against" or "contrary to," it is much more likely to have the former meaning, especially since the phrase *contra naturam* was available to express the latter idea (and Latin *contra* can only mean "against"). From at least the thirteenth century on, medieval and early modern writers recognized a category of what may be called "preternatural phenomena"—phenomena that, though they went "beyond nature" (*praeter naturam*) in being unusual or exceptional, did not lie outside the order of nature. The early moderns' tendency to render *para physin* as *praeter naturam* in the *Mechanical Problems* suggests that they placed mechanical phenomena in this category; in doing so, they faithfully rendered the author's idea that mechanics goes beyond nature without breaking the natural order.[48]

This interpretation of the author's view of the art-nature relationship has the further advantage of ascribing to him a view that captures the actual relationship of technology to nature much better than the notion of struggle against a hostile nature. Though machines do bring about effects that go beyond what unaided nature can achieve, it is implausible to suppose that this is because they consistently strive to thwart nature's activity. While *some* machines may cause bodies to move in ways that are contrary to their natural tendencies, it is by no means the case that all do. And even if a machine does produce such motions, there is no reason to think that it does so by forcing all its components to act against their natural tendencies. It is much more plausible to suppose that in building a machine a craftsman takes advantage of the natural properties of its components, combining them in such a way that they produce an effect that nature cannot. The essence of the machine, and indeed of technē in general, lies in the creative combination of natural regularities. This seems to me an insight that not only captures the view expressed by the author of the *Mechanical Problems*, but one that is also substantially correct.[49]

## 4.4 Mechanics and the Imitation of Nature

Now that we have clarified the conception of the art-nature relationship in the *Mechanical Problems*, we can return to the question broached at the end of section 4.2: does mechanics occupy a distinctive position in its relationship to nature vis-à-vis the other technai in antiquity? As far as the *Mechanical Problems* is concerned, this question must be answered in the negative. In fact, Aristotle's remark at *Phys.* 199a15–17 that art either imitates nature or completes what nature leaves unfinished provides an excellent illustration of the art-nature relationship as the author of the *Mechanical Problems* conceives of it. As we have seen in section 4.2, the notion of imitation presupposed by Aristotle's remark goes far beyond mere mimicry. Art's imitation of nature must be understood in terms of a close analogy between artificial and natural processes: art imitates nature by acting in an analogous manner. And Aristotle's remark also gives a precise sense to the notion that art goes beyond nature: art brings about results that nature cannot and thus remedies nature's deficiencies, but it does so by acting in the way that nature would act if it could bring about such results. The role of art is to build on and extend nature's activities in a way that is beneficial for human beings. Art goes beyond nature and thus may be said to "conquer" it, but only by imitating nature itself. This is precisely the idea expressed at the opening of the *Mechanical Problems*: art makes creative use of natural regularities to go beyond what nature itself can achieve. The author seems to identify the use of external compulsion or force as one respect in which mechanics goes beyond nature. But as noted in section 4.2, for Aristotle the art-nature distinction often amounts to no more than a distinction between an external and internal principle of change; the actual process of change is identical, whether the cause is internal or external. Given this, the notion that mechanics involves external force or compulsion by no means stands in the way of a close analogy between mechanical and natural processes. Mechanics brings about effects that unaided nature cannot, but it does so by acting in a way that is analogous to that of nature itself.[50]

In this connection it is relevant to note that the idea that mechanics imitates nature is associated with two branches of ancient mechanics in particular: the building of automata and of armillary spheres to represent the motions of the heavenly bodies. The notion that these branches of mechanics imitate natural motions appears in both Pappus and Proclus.[51] The idea of mechanics as imitating nature is also prominent in the following passage from the late Latin writer Cassiodorus (from a letter addressed to Boethius):

> All fields of study—the combined efforts of the wise—seek to know the power of nature, insofar as they are able; a machine is the only thing that seeks on the contrary to imitate [*imitari*] and, if it may be said, in some respects even strives to surpass [*superare*] it. For this art is distinguished by having made Daedalus fly, by having caused an iron Cupid to hang in the temple of Diana without any fastening; even today it makes mute things sing, lifeless things live, immobile things move. The mechanician is, if it may be said, almost on a par with nature; by throwing open what is hidden, altering what is manifest, sporting with what is wondrous, he simulates so well that what is not doubted to be artificial [*compositum*] is judged to be the real thing.[52]

Here the notion of conquering or surpassing nature (*superare*) appears alongside that of imitation (*imitari*). The passage suggests that the mechanician can surpass nature by imitating it, in the sense that he creates an imitation that is more lifelike than nature itself.[53]

In Cassiodorus it is clear that what the mechanician creates is but an imitation of nature: the observer who is not in on the trick is fooled, though the craftsman himself is not. But other ancient sources raise at least the theoretical possibility of building a machine that not only mimics or simulates a living thing, but that actually replicates the very processes going on inside living things. Evidence for this is provided by Gregory of Nyssa's *On the Soul and Resurrection*, a work that is cast as a dialogue between Gregory himself and his sister Macrina. Plagued by grief at the death of a close friend, Gregory comes to Macrina in search of consolation; the two then engage in a lengthy discussion that raises fundamental questions about the nature of the soul. Throughout Macrina speaks from the perspective of a committed Christian whose views are thoroughly informed by the heritage of Greek philosophy. She gradually brings Gregory from a state in which he is tempted by Epicurean materialism to a full acceptance of the Christian doctrine of resurrection of the flesh. At one point early in their conversation, Gregory asks whether the existence of a certain kind of machine does not make the appeal to an immaterial soul superfluous:

> What . . . if someone were to say that, just as there is something material common to the perceptible nature of the elements, but great differences between each form of matter according to its distinctive character . . . there is a faculty [*dynamis*] compounded from these according to proportion, a faculty that brings about intellectual impressions and movements by virtue of its natural peculiarity and power [*dynamis*]? Indeed, we often see such things brought about by the makers of machines [*mēchanopoioi*], at whose hands matter arranged in an artistic fashion [*technikōs*] imitates nature [*mimeitai tēn physin*]: the matter displays resemblance not only in its shape, but it also moves, and feigns [*hypokrinetai*] a kind of voice . . . and we discern no intellectual faculty [*noētē dynamis*] bringing about these individual things.[54]

The success of the builders of automata in imitating life prompts Gregory to ask whether the kinds of behavior normally attributed to the intellectual faculty of the soul cannot be explained by virtue of some "motive faculty [*kinētikē dynamis*] in the nature of the elements within us."[55] Macrina responds that the construction of a machine requires technē, which in turn implies the existence of a creative intelligence and a soul:

> To know in this way how to take in hand and arrange lifeless matter, so that the art [*technē*] incorporated into machines practically becomes a soul to the matter, and thus the matter acts out [*kathypokrinetai*] motion and sound, and shapes, and similar things: this is a proof that there is something in the human being by means of which it is naturally able to conceive of these things through its faculty [*dynamis*] of contemplation and discovery, and first to construct machines [*mēchanēmata*] in thought, then bring them into existence through art [*technē*] and display its thought in matter.[56]

Macrina continues with a remarkable passage that describes the elementary principles of pneumatics that the maker of any such machine needs to know. But, she claims, if the capacity for creating such wondrous effects were present in the nature of the elements themselves, machines would arise spontaneously.[57] Besides showing that the notion that mechanics imitates nature was sometimes understood as implying a very close analogy between them, Gregory's objection and Macrina's response provide a beautiful illustration of the idea that the essence of a machine lies in the creative arrangement of its components. The operation of a machine is fully natural in the sense that it can be explained by the natural behavior of the substances of which it is made, once they have been placed into a certain arrangement. The contribution of technē lies in the creation of such an arrangement. Even if the processes that go on in the living thing could be replicated in a machine, there is still a place for an immaterial soul: for machines do not make themselves.

Finally on the subject of imitation, a passage from book 10 of Vitruvius suggests that human beings learned to construct machines from the study of nature itself:

> Now all machinery [*machinatio*] has its origin in nature, and is founded on the teaching and instruction of the revolution [*uersatio*] of the world. Let us but observe and contemplate the natural system formed by the sun, the moon and the five planets; unless these revolved by skillful contrivance [*machinata uersarentur*], we would not have had the alternation of day and night, nor the ripening of fruits. Thus, when our ancestors had seen that this was so, they took examples from nature, and by imitating them [*imitantes*] and being led on by divine facts, developed applications that are serviceable in life. Some things, with a view to greater convenience, they worked out by means of machines and their revolutions [*uersationibus*], others by means of engines; and so, whatever they found useful in practice, they took care to improve by means of study, arts, and doctrines established step by step.[58]

The first remarkable feature of this passage is its emphasis on the natural character of mechanics: far from any notion that machines somehow break the order of nature, mechanical motions are produced by nature itself. Second, the passage identifies the imitation of nature as the driving factor in the development of technology. By imitating nature's own movements, human beings learn to build machines that go beyond nature in ways that improve the human condition. The passage in fact provides an excellent illustration of Aristotle's dictum that art imitates nature and completes what nature leaves unfinished. Moreover, as in Aristotle, the imitation in Vitruvius is more than mere mimicry. Like the author of the *Mechanical Problems*, Vitruvius claims that the operation of machines can be understood in terms of circular motion.[59] The present passage implies not only that human beings found examples of machines in nature, but also that they discovered the basic principle that explains their operation from the study of nature. By learning the principles underlying the motions found in nature, human beings gain the understanding needed to build machines that go beyond anything that nature produces on its own.[60]

## 4.5 Mechanics and the Study of Nature

I now return to the question of how mechanics could yield knowledge of nature. The evidence considered above suggests two ways it could do so.

First, mechanics like any other technē could provide models for the understanding of natural objects and processes. Aristotle, for example, often draws parallels between mechanics and nature. In the biological works nature is frequently described as a craftsman that displays the inventive ingenuity associated with the terms *mēchanē* and *mēchanasthai*.[61] At *De motu animalium* 701b1ff. Aristotle likens the motion of animals to that of automatic puppets (*automata*), a type of mechanical device. At *De motu animalium* 698b21ff. he appeals to the fact that it is impossible to move a boat unless one rests against an external fixed point to show that the motion of animals and of the cosmos as a whole is only possible in relation to an external fixed point or object. While each of these parallels must be examined on its own terms, taken as a group they suggest that Aristotle did not hesitate to draw analogies between mechanics and nature, or between forced and natural motion, and that such analogies played an important role in the study of nature.

Second, mechanics could provide knowledge of nature through the use of mathematics. Important evidence for this idea is provided by the opening of the *Mechanical Problems*, immediately after the author describes such problems as those in which "the lesser master the greater": "These are not entirely identical with physical problems nor entirely separate from them, but they have a share in both mathematical and physical speculations: for the 'how' [*to hōs*] in them is made clear through mathematics, while the 'about what' [*to peri ho*] is made clear through physics."[62] This is the author's most explicit attempt to clarify the relationship between mechanics, physics, and mathematics. The contrast he has in mind between the "how" (*to hōs*) and the "about what" (*to peri ho*) is not immediately clear. But in fact it seems to express a conception of mechanics close to that of Aristotle himself. In several passages Aristotle mentions mechanics as a science like optics, harmonics, and astronomy, each of which uses mathematics to study physical bodies. In *Posterior Analytics* A9 Aristotle states that geometrical proofs can be applied to mechanics and optics, just as the proofs of arithmetic can be applied to harmonics (76a23–25). In *Posterior Analytics* A13 mechanics (*ta mēchanika*) is said to be subordinate to (literally "under," *hypo*) stereometry (a branch of geometry), just as optics is to geometry, harmonics to arithmetic, and "study of the phenomena" (*ta phainomena*) to astronomy (78b36–39). Aristotle's conception of subordinate sciences raises many problems, but two features of it are clear and relevant in the present context. First, it is mathematics that provides the explanation in the subordinate sciences. For example, a proposition in optics can be demonstrated by means of geometry, and a proposition in harmonics by means of arithmetic (*A. Po.* 75b14ff.). The subordinate science is concerned with the fact (*to hoti*) while mathematics supplies the explanation (*to dioti*). Second, the subordinate sciences deal with the mathematical properties of physical bodies. In *Metaphysics* M3, 1078a14ff. Aristotle states that neither harmonics nor optics conceives of (*theōrei*) its subject matter qua pitch or qua visual ray; rather, they conceive of their subject matter qua numbers or lines, and the same holds true for mechanics (*mēchanikē*). In this chapter Aristotle is attempting to clarify the relationship of mathematical objects to physical bodies. His point is that since mathematical properties belong to physical bodies, mathematicians in fact study the properties of such bodies. But they do not study them qua properties of physical bodies; there are many things that can be said about such bodies simply as (qua) lines or planes, and it is such things that mathematicians study (see 1078a5ff.). Harmonic theorists or students of optics incur no falsity in abstracting away the physical properties of pitch or the visual ray and considering only their mathematical attributes for the purposes of analysis. *Physics* B2 adopts a somewhat different perspective on the question of the subject matter of the subordinate sciences. It begins by raising the general questions of how mathematicians differ from physicists, and whether astronomy is part of physics (193b22–30). As in *Metaphysics* M3 mathematicians are said to study the attributes of physical bodies, though not qua attributes of such bodies (193b31–35). At 194a7–12 Aristotle contrasts geometry with optics, harmonics, and astronomy, which he calls the "more physical" (*physikōtera*) branches of the mathematical sciences. Geometry studies (*skopei*) a physical line, but not qua physical; optics studies a mathematical line, but qua physical, not qua mathematical. Prima facie Aristotle's position here is opposed to that stated in *Metaphysics* M3, for here he insists that the "more physical" branches of mathematics consider their objects qua physical bodies. But there is no real conflict if we take his point to be that the difference between mathematics and the subordinate sciences lies in the need for the latter to consider mathematical attributes as they occur in particular arrangements of physical bodies. The aim of the subordinate sciences is to explain the behavior of particular kinds of physical objects; Aristotle's idea seems to be that they make use of mathematics to construct models that explain this behavior. Geometers are concerned with anything that can be demonstrated about lines and planes simply by virtue of their nature as lines and planes. But astronomers are concerned only with those mathematical objects that accurately model the heavenly bodies and their motions; similarly students of optics are concerned only with those arrangements of mathematical lines that correspond to the physical rays that cause optical phenomena. The apparent tension with the *Metaphysics* passage is really just a matter of a difference of emphasis. In *Metaphysics* M3 Aristotle's emphasis is on the mathematical aspect of the subordinate sciences—that is, on the point that they treat certain physical bodies as mathematical lines and numbers and incur no falsity in doing so. *Physics* B2 does not deny this; rather it makes the point that astronomers or students of optics do not study just *any* mathematical lines, but the ones involved in the explanation of certain physical phenomena. Though mechanics is not mentioned in *Physics* B2, there is no reason to suppose that Aristotle did not consider it one of the "more physical" mathematical sciences.

In stating that mathematics supplies knowledge of the "how" (*to hōs*), the author of the *Mechanical Problems* indicates that it provides the explanatory element in such problems, just as Aristotle claims that explanation in the subordinate sciences is provided by mathematics. And in claiming that the "about what" (*to peri ho*) is given by physics, the author echoes Aristotle's idea that the subordinate sciences study physical bodies from a particular point of view. Mechanics uses mathematics to explain the motion of physical bodies, but it is distinct from mathematics in that it is concerned only with the movements undergone by physical bodies in certain arrangements and under certain conditions. This picture of the place of mechanics in relation to physics and mathematics is supported by the use of geometry in the *Mechanical Problems* itself. Consider once again the discussion of problem 1. The author begins by asking why (*dia tina aitian*) larger balances are more accurate than smaller ones. The origin or principle (*archē*) of this is said to be the fact that a point on a larger circle travels faster than one on a smaller circle, when both are moved by the same force (848b1–9); this, in turn, is because a point on the circumference of a circle is moved by two motions (848b9–10). After explaining and justifying this idea at some length (in part by making use of geometrical representations of bodies in motion), the author states the basic physical idea that underlies his analysis: if two bodies are moving under the influence of the same force, the one that is more deflected (*ekkrouesthai*) travels more slowly (849a6–19). A point on a smaller circle is more deflected than one on a greater circle—that is, it has a greater component of its motion directed toward the center (*para physin*) than along the tangent (*kata physin*). To establish this fact the author appeals to an elaborate geometrical construction in which lines in the geometrical figure correspond to different components of the body's motion; he uses this to prove that the line corresponding to the *para physin* motion is larger in the case of the smaller circle (849a21–849b19). Finally the author states that "the reason [*aitia*] why the point further from the center is moved more quickly by the same force is now clear" (849b19–22), and proceeds to apply this result to the original question of why larger balances are more accurate (849b22–850a2). Thus, it is the use of geometry to represent mechanical movements that supplies the explanatory element in mechanics, for geometry explains *why* the point on the larger radius moves more quickly. More generally, the procedure followed throughout the *Mechanical Problems* may be described as the use of geometrical models to explain physical phenomena. The key step in the author's analysis of all the problems he discusses is the application of a simple model: the lever, the balance, or the circle. It is the application of the model that supplies the explanation. For example, consider the beginning of problem 4: "Why do the rowers in the middle of the ship contribute most to its movement? Is it because the oar acts like a lever [*mochlos*]? For the thole-pin is the fulcrum (for it is fixed), and the sea is the weight, which the oar drives back; the sailor is the mover of the lever."[63] Treating the oar as a lever explains why the rowers amidships exert the greatest force, but the explanation of how the lever operates depends ultimately on the geometrical analysis set out in problem 1. Thus the author may be viewed as using geometry to explain the behavior of a certain set of physical phenomena in a way that closely matches Aristotle's description of the subordinate sciences.

I note in conclusion that none of Aristotle's fleeting references to mechanics (the three passages discussed above) suggests that mechanics is concerned with forced rather than natural motion. All of Aristotle's subordinate sciences deal with the mathematical aspects of physical bodies; all occupy a middle ground between physics and various branches of mathematics. Thus, though distinct from physics, mechanics can provide knowledge of physical objects, and in this way can contribute to the knowledge of nature. It is true that there is a contrast to be drawn with the early modern period, when mechanics comes to be regarded as a part of physics itself rather than a closely related discipline. But the difference here results more from Aristotle's notion of mathematical abstraction than from any idea that mechanics is inherently "contrary to nature."

## 4.6 Early Modern Reflections

In the early modern period the *Mechanical Problems* became the focus of an extensive commentary tradition to which a number of humanists, mathematicians, and engineers made significant contributions.[64] The work's striking and problematic opening, with its important implications for the understanding of the art-nature relationship, naturally attracted a good deal of attention, and the remarks of a number of commentators provide close parallels to the analysis given above. In this section I present four illustrations of this in chronological order, each drawn from a late sixteenth-century source.[65]

1. Alessandro Piccolomini's highly influential paraphrase of the *Mechanical Problems* first appeared in Latin in 1547; a second edition appeared in 1565, and the work was translated into Italian in 1582.[66] Piccolomini reads the opening of the *Mechanical Problems* as claiming that art operates differently from nature, since it seeks to further human ends. But art also imitates nature, since it acts in just the way nature would act if it could achieve the effects brought about by art:

> For art, although it imitates nature and assists it, nevertheless brings about many things in a way different from it, with a view to our needs. But it must not be considered any less an imitator [*imitatrix*] of nature for that reason, since it brings about whatever it does in the way that nature itself would, if it did so. For while nature always acts in the same way in whatever it does, if unimpeded, art, because what is useful, necessary, and helpful for us is varied, also proceeds in a varied manner: in whatever path it takes, it also follows nature. If then it should come about that we are not content with the simplicity of nature (which indeed because of its simplicity is in no way diverse) on account of our varied needs, and thus undertake something against it, then it struggles and contends against us and imposes difficulty on our efforts. Hence in order that we might overcome this difficulty (or rather hesitation) we have need of art, by which we bring about our work, either destroying completely or at least demolishing to some degree the efforts of nature.[67]

Even if unimpeded nature always acts in the same way and art acts in different ways in different situations, art must be understood as following nature in all cases. Such opposition as there is between art and nature (it is more a question of "hesitation" on nature's part than active resistance) results solely from the fact that nature always acts in the same way, while the needs of human beings vary widely.

2. Giuseppe Moletti, Galileo's predecessor as professor of mathematics at Padua, lectured on the *Mechanical Problems* in 1581; his notes are preserved in manuscript form at the Bibliotheca Ambrosiana in Milan (MS S.100 *sup.*, folios 1–199 *passim*).[68] The lectures are organized around particular topics, and several pages are dedicated to the question "whether the art of mechanics is found in the works of nature" (*An in operibus naturae ars mechanica reperiatur*) (fol. 22). The problem with which Moletti deals is a familiar one by now: how to reconcile the notion that art imitates nature with the ability of art to bring about effects that nature cannot, and with the ability of some arts (especially mechanics) to conquer nature. Moletti's strategy is to argue that mechanics conquers nature by applying principles that it has learned from nature itself. Thus the circular-motion principle so important in the *Mechanical Problems* was discovered by observing the circular motion of the heavenly bodies; the lever was discovered by observing the joints of animals. In general the art of mechanics is found everywhere in nature, meaning that mechanics operates on fully natural principles.

3. Guidobaldo del Monte's 1588 paraphrasis of Archimedes's *On the Equilibrium of Planes* (*In duos Archimedis Aequeponderantium libros paraphrasis*) begins with a general discussion of mechanics that draws heavily on the opening of the *Mechanical Problems*.[69] Like Moletti, Guidobaldo's concern is to explain how mechanics can conquer nature while also imitating it. Nature appears to be conquered by art, though in fact it contributes greatly to its own defeat; mechanics is "as much an imitator of nature as a mighty opponent" (*tam naturae emula, quam oppugnatrix valida*).[70] The branch of mechanics in which art seems most clearly to conquer nature is that which is concerned with weights and equilibrium, and Guidobaldo turns immediately to this most challenging case. Referring to *Physics* B8 as well as the opening of the *Mechanical Problems*, Guidobaldo claims that the operation of art can be conceived of in three ways: art either imitates nature, brings to completion what nature cannot, or produces results that are beyond or contrary to nature (*praeter naturam*). But in each of these cases, art must be understood as imitating nature. First, it is evident that there are many arts whose goal is the

imitation of nature, such as sculpture. Second, when art brings to completion what nature cannot, as in medicine, it functions as a tool (*instrumentum*) of nature, and thus brings about a natural effect; in such cases art acts in exactly the same way nature would, if it brought about such effects (*eodem modo operatur, ac si natura rem ipsam absque artis ope perficere posset*). Third, in cases where art conquers nature and brings about a result that is *praeter naturam*, it does so by producing a certain disposition or arrangement (*dispositio*) of bodies, the same arrangement that nature would produce if it intended to bring about those effects. Once bodies are placed in this arrangement their motions follow naturally, since it is natural for the arrangement itself to move in a certain way—even though individual bodies may move in ways contrary to their own individual natures. Thus in the case of the lever, it is natural for a large weight to be raised by a smaller one if the center of gravity lies between the fulcrum and the smaller weight; the larger weight moves upward naturally (*naturaliter*), even though such motion is contrary to its own individual nature.

4. Finally, a passage in Henri de Monantheuil's 1599 commentary on the *Mechanical Problems* describes the goal of art as "turning" (*convertere*) or "accommodating" (*accommodare*) the constant behavior of nature to human needs.[71] The difficulty of achieving this stimulates human beings to invent the arts. Following closely the passage of Vitruvius discussed above, Monantheuil argues that mechanics was discovered by the observation and imitation of nature (*per imitationem rerum a natura procreatarum*).[72] As in Moletti, the notion that mechanics is discovered by the imitation of nature is combined with the idea that mechanical processes occur in nature itself. Monantheuil describes mechanics as an art that goes *praeter naturam*, "beyond nature," in a way that contributes to human needs.

As has often been noted, Galileo at the beginning of *Le Mecaniche* (1593) insists that mechanics does not operate by tricking nature; nature cannot be deceived, and mechanical effects come about in a fully natural way.[73] Scholars have sometimes pointed to this claim as marking a radical break with previous views of mechanics, and in particular with the *Mechanical Problems*.[74] But the evidence presented in this section shows that the notion that mechanics operates "according to nature" was quite common in the early modern period. Prompted by the striking opening of the *Mechanical Problems*, early modern authors attempted to reconcile the text's apparent opposition of art and nature with their view that all processes must take place in a natural way. What is more, the discussion in sections 4.2–4.4 shows that it would be a grave mistake to suppose that in doing so, these writers were introducing a radically new idea not present in the ancient texts on which they drew.[75] Rather than a hallmark of the rejection of Aristotle that supposedly characterized the crucial phase in the rise of modern science, the notion that mechanics operates according to nature is in fact a deeply Aristotelian idea.

## Notes

I would like to thank the following individuals for their advice and assistance of various kinds during the preparation of this chapter: Jenny Attiyeh, Malcolm Hyman, Bill Newman, and Gisela Striker. Much of this chapter was written in the fall of 2003 during a sabbatical year at the Max Planck Institute for the History of Science in Berlin; I would like to express special thanks to Jürgen Renn, director of Department I of the Institute, for his generous invitation and support.

1. For representative discussions of the various branches of ancient mechanics see the opening chapters of Pappus, *Collectio* 8 (pp. 1022ff. Hultsch) and Proclus, *In prim. Eucl. Elem.*, pp. 41.3ff. Friedlein. Similar divisions go back to the early Hellenistic period, as shown by the topics covered in Philo of Byzantium's lost *Mechanical Syntaxis*: the theory of levers, harbor construction, siegecraft, the construction of artillery, pneumatics, and the construction of automata. See Gille 1980, 104–105.

2. See Krafft 1967, 27: "Die 'Mechanik' galt . . . in der Antike und im Mittelalter als eine Kunst und hatte als Kunst dort einzugreifen, wo die Natur das vom Menschen Erwünschte nicht von sich aus vollbringen konnte. Das Künstliche ist damit etwas Nicht-Natürliches, ja etwas Naturwidriges. Im Falle der 'Mechanik', der sich Mittel und Werkzeuge zur 'Überlistung' der Natur bedienenden Kunst der Antike, ist diese Auffassung dann endgültig erst von Galileo Galilei und seinen Nachfolgern überwunden worden, die aufzeigten, warum man auch mit mechanischen Hilfsmitteln nur etwas erreichen kann, wenn man der Natur folgt, sich nach ihren 'Gesetzen' richtet. Auch bei der Anwendung einer Maschine geschehe nichts Widernatürliches."

3. For the connection between mechanics and magic see Krafft 1973, 7: "Listen und Tricks, Zauberkunst und Magie, versteckte Mittel sind die Werkzeuge der 'Mechaniker'; sie lassen der Natur nicht ihren Lauf, sie überlisten sie und bewirken damit etwas gegen sie."

4. Cf. Krafft 1970, 137: "'Physik' (*physikē epistēmē*) behandelt, wie schon das Wort selbst zeigt, die 'Natur' (*physis*) der Dinge, ihr 'Wesen' und ihre 'natürlichen', ihrer 'Natur' gemäßen (Eigen)-Bewegungen, also solche Körper, die, wie Aristoteles es definiert, das Prinzip ihrer naturgemäßen Bewegung in sich selbst haben, während die 'Mechanik', wie wir es besonders aus der Einleitung der *Mechanischen Probleme* und dem Fehlen einer Behandlung des 'natürlichen' Falles (Freien Falles) ersehen konnten, 'naturwidrige', d. h. künstliche Bewegungen behandelt, Bewegungen, zu denen die 'Natur' eines Körpers nicht selbst in der Lage ist, ja die der 'Natur' eines Körpers entgegen wirken und für deren Zustandekommen sich der Mensch auf

kunstvolle Weise 'mechanischer' Hilfsmittel (*mēchanai*) bedienen muß. Die 'Mechanik' (*mēchanikē technē*) ist eine 'Kunst', die nach antiker Auffassung deshalb auch zur wissenschaftlichen Erkenntnis der 'Natur', des 'natürlichen' Geschehens nichts Wesentliches beizutragen vermag." For similar remarks see Krafft 1973, 11; 1967, 12–13. Krafft (1970, 138) also claims the existence of an "unüberbrückbare Schranke" between physics and mechanics in antiquity.

5. Krafft 1970, 157: "Da dann seit Platon und Aristoteles die 'Natur' als eine organische Einheit, als ein Ganzes, der Kosmos als ein Lebewesen aufgefaßt wird, kann nämlich die 'Natur' auch eines einzelnen Dinges nicht isoliert betrachtet werden. Greift der Mensch künstlich (mit oder ohne zusätzliches 'Mittel': *mēchanē*) in den organischen Zusammenhang und den 'natürlichen' Ablauf ein, so kann er nach dieser Auffassung nicht mehr 'Natur' erkennen (Naturwissenschaft treiben), sondern nur noch menschlichen 'Kunst'. Das Experiment trägt als ein solcher 'künstlicher' Eingriff deshalb ebensowenig zur Erkenntnis der 'Natur', zur 'Naturwissenschaft', bei wie eine 'Kunst' und wird als 'naturwissenschaftliches' Erkenntnismittel abgelehnt oder gar nicht erst in Betracht gezogen." Cf. Krafft 1967, 28.

6. Cf. Krafft 1970, 159: "Die Mathematik gehört nach Aristoteles einem anderen Seinsbereiche als die 'natürlichen' Dinge an. Sie kann deshalb nichts zur Erkenntnis der 'Natur', der Ursachen 'natürlicher' Vorgänge, beitragen und im sublunaren Bereich nicht einmal der Beschreibung dieser Vorgänge dienen." For similar remarks see Krafft 1967, 30–31.

7. See especially Newman 1997 and 2001.

8. I will not go into the question of authorship here; what is important for my purposes is simply that the *Mechanical Problems* clearly lies in the tradition of Aristotelian science.

9. *Mech.* 847a11–28. Translations are my own unless otherwise indicated. For the Greek text of the *Mechanical Problems* (*Mech.*) see Apelt 1888.

10. See especially the treatises *On Generation*, *On the Nature of the Child*, and *On Diseases IV*, with the commentary of Lonie 1981.

11. Hp. *Vict.* 11–12, 6.486–488 Littré.

12. On this passage see Schneider 1989, 208–212; he rightly insists that the analogies imply the existence of structural similarities between technē and physis.

13. See Newman 1997 for another critique of these traditional assumptions.

14. Arist. *Phys.* 199a8–20. Translation Barnes.

15. Cf. *Phys.* 199b28–29: if the art of shipbuilding were present in the wood, a ship would come to be by nature in the same way that it does by art.

16. For a similar interpretation of Aristotle's remark that art imitates nature or brings to completion what it cannot see Schadewaldt 1960a, 900–901: "Dieser Satz meint natürlich nicht, daß die Technik lediglich die Natur nachäffe (obgleich die Technik,

wenn ich recht unterrichtet bin, in vielen Zweigen auch heute wirklich zunächst einmal die Natur befragt). Er meint, wenn man ihn richtig faßt: daß die Technik *in Analogie zur Natur* handelt, daß sie in ihrem eigenen Tun die zunächst einmal grundwissenschaftlich erforschten Prinzipien der Natur befolgt. In immer neuen Kombinationen und Gestaltungen auf Grund der zunächst einmal erforschten Prinzipien der Natur gelangt die Technik auf ihre partielle Weise zu neuen Kombinationen, die auf den Menschen abgestellt sind. In diesem Sinne sagt auch Aristoteles, daß 'die Technik weiter und zu Ende führe, was die Natur von sich aus bisher noch nicht bewerkstelligt' hat. Dieses Weiter- und Zu-Ende-Führen ist das, worin die Technik in ihrer Weise sich mit Recht der Natur 'überlegen' fühlen darf. Im ganzen aber beruht eben diese ihre 'technische' Überlegenheit darin, daß sich die Technik zunächst einmal der Natur einschmiegt, ihr gehorcht und auf sie hinhört, so wie das treffend Bacon gesagt hat: *natura parendo vincitur*" (original emphasis). See also Schneider 1989, 214–215.

17. Arist. *EN* 6, 1140a10–16. Translation Barnes.

18. Arist. *GA* 735a2–4.

19. *Meteor.* IV, 381a9–12.

20. *Meteor.* IV, 381b3–9.

21. On the influence of these passages in early modern alchemy, see Newman 2001.

22. For Aristotle's use of technological analogies see Bourgey 1975; Lloyd 1966.

23. Hp. *De arte* 12, 240.10–13 Jouanna, 6.24 Littré.

24. Fr. 123 Diels-Kranz (DK).

25. Cf. *Phys.* 230b10ff.

26. Cf. *Cael.* 300a23: *to de biai kai para physin tauton.*

27. *Meteor.* 342a12–16. For the general point that *para physin* motion does not violate the order of nature see Schneider 1989, 260–261; Micheli 1995, 30.

28. Cf. *GA* 772a36–37, where monstrous births are said to come about "contrary to what happens for the most part and usually" (*para to hōs epi to polu kai to eiōthos*).

29. Arist. *GA* 770b9–17.

30. Cf. *Phys.* 197b32–37, where Aristotle states that the difference between chance (*tychē*) and spontaneity (*to automaton*) is especially clear in the case of things that come about by nature: "The difference between spontaneity and what results by chance is greatest in things that come to be by nature [*physei*]; for when anything comes to be contrary to nature [*para physin*], we do not say that it came to be by chance, but by spontaneity. Yet strictly this too is different from the spontaneous proper; for the cause of the latter is external, that of the former internal" (trans. Barnes). Here again we have an example of phenomena that, though they come about by nature, are also *para physin*. Both chance and spontaneity are concerned

with the production of endlike results; the difference between them is that the former is restricted to results that could be brought about by human choice (cf. 197a36ff.). Thus what Aristotle is referring to are endlike results that arise in nature but are produced in an unusual or exceptional fashion; Ross (1936, 524) is probably correct to argue that the reference is to spontaneous generation, in which normal offspring are produced in a way that deviates from the usual process of generation from seed.

31. See also *Prom.* 476–477: "Hear the rest and you shall wonder all the more at the arts [*technai*] and resources [*poroi*] I have devised"; 506: "All the arts [*technai*] have come to human beings from Prometheus." *Prom.* 109–110 describes fire, Prometheus's key discovery, as a "great resource [*poros*]" and "teacher of all the arts."

32. Soph. *Ant.* 365–367.

33. For other examples of *mēchanē* as a "means" or "Hilfsmittel" see Aesch. *Sept.* 208f.; Hdt. 2.181; Eur. *Phoen.* 890, *Hel.* 1034; Hp. *VM* 19; Aristoph. *Thesm.* 765; Plat. *Apol.* 39a. At *Gorg.* 512b Plato cites the engineer or *mēchanopoios* along with the general and the sea captain as people whose function is to rescue individuals from situations of difficulty. For the connection between *mēchanē* and *poros* see Eur. *Med.* 260, *Hel.* 813; Plat. *Crat.* 409d, 425d. For the adaptability of *mēchanē* see the use of the term with the verb *plekein* "to weave" (for example, Eur. *Andr.* 66, 995; *Or.* 1423) and the adjective *poikilos* "varied" (for example, Soph. *OC* 762).

34. See LSJ s.v. II, noting that the sense of trickery or deception is characteristic of the plural *mēchanai* ("shifts, devices, wiles").

35. See Hdt. 2.125 for *mēchanē* as a crane or other lifting device used in the building of the pyramids; Hp. *Fract.* 30, *Art.* 72, for *mēchanē* as a device (for example, a lever) used to treat fractures and dislocations. Hp. *Fract.* 30 states that mechanical methods should only be used when human strength is insufficient to deal with a fracture. For *mēchanē* of a construction see Aeschylus' reference to Xerxes' bridge across the Hellespont as *laoporoi mēchanai* (*Pers.* 105–106; *Pers.* 722 remarks that Xerxes bridged the Hellespont by means of *mēchanai*). In Thucydides *mēchanē* is regularly used of a siege engine (2.76, 3.51, 4.13, 8.100). On these passages see Schneider 1989, 220: "Der terminus technicus wahrt den älteren Wortsinn insofern, als es sich um Instrumente handelt, ohne die der Mensch einen bestimmten Zweck nicht erreichen kann." Elsewhere *mēchanē* refers to a theatrical device used to make gods appear in the air; see Plat. *Crat.* 425d, where it is remarked that tragic poets resort to a *mēchanē* of this type when they are at a loss (*aporōsin*) about how to resolve the plot of the drama. The verb *mēchanaomai*, though it sometimes means to plot or intrigue, is often used quite neutrally to mean prepare or contrive in a skillful manner (for example, Hdt. 1.123, where it is paired with the verb *epitechnasthai*, "to devise with skill") or by the use of mechanical devices (for example, Hp. *Fract.* 30, 31; *Art.* 74).

36. On the meaning of *mēchanē* and related terms see especially Schadewaldt 1960b, 914, accepting the derivation of *mēchanē* from the Doric word *mēchos* ("means, expe-

dient, remedy," not "trick") and taking the basic sense of the term to be "'Abhilfe', 'kluger Ausweg', 'klug ergriffenes Mittel', mit dem man irgendeiner Sache beikommt." Krafft's remarks (1967, 15f.; 1970, 140f.; 1973, 5—the latter two discussions add nothing substantial to the first) are now superseded by those of Schneider (1989, 217–222) and, most recently, Micheli (1995, 9–20). Micheli (1995, 28) also quite rightly emphasizes that *mēchanē* at the opening of the *Mechanical Problems* has no connotations of deception or trickery.

37. See Anaxagoras fr. 21b DK and the story of Prometheus and Epimetheus as told in Plato's *Protagoras* (320cff.).

38. See Schneider 1989, 236; 256 n. 177 (rightly insisting, against Vernant 1957, that the *Mechanical Problems* does not describe a "combat de la *technē* contre le *physis*").

39. See Piccolomini 1565, 7–8: "Haec itaque, cum ui quadam contingant, praeter naturam dicuntur fieri, atque horum causam, hoc est ipsam uiolentiam, cum uel ignoramus, *uel minorem arbitramur quam effectus appareant*, miramur ocius" (my emphasis).

40. *Mech.* 847b15–23.

41. *Mech.* 848a34–37.

42. See Micheli 1995, 34. For an example of similar trickery of the ignorant observer see 849b34–850a2: sellers of purple attempt to distort the readings of balances by changing the suspension point or the weight of one of the arms. For a further example of the association of the strange or *atopon* and wonder (*to thaumaston*) cf. 855a28ff. (a description of the paradoxes involved in the "wheel of Aristotle"), where the author goes on to give the cause (*aitia*) at 855b32ff. I do not mean to deny that trickery and the production of wonder were important goals in certain branches of ancient mechanics. Thus Pappus states that the practitioners of a number of branches of ancient mechanics were known as "wonder workers" (1024.24–1026.2 Hultsch): "The ancients also call the wonder workers [*thaumasiourgoi*] mechanicians. Some practice their art [*philotechnousin*] by means of air, as does Hero in his *Pneumatics*; others seem to imitate the motions of living things through ropes and cables, as does Hero in his *Automata* and *Balances*; still others make use of bodies carried on water, as Archimedes does in his *Floating Bodies*, or water clocks, as Hero does in his *Water Clocks*, which is evidently connected with the theory of the sundial [*gnōmonikē*]." In Proclus's account (*In prim. Eucl. Elem.*, 41.3ff. Friedlein) the production of wonders (*thaumatopoiikē*) is a major component of the science of mechanics; this includes the manufacture of (1) devices based on air, as described by Hero and Ctesibius (i.e., pneumatics), (2) devices based on inclinations (*ropai*), with equilibrium producing rest and lack of equilibrium motion, and (3) devices that imitate (*apomimeisthai*) living things through ropes and cables (i.e., the building of automata). As both Pappus and Proclus indicate, the production of wonders was particularly associated with the building of automata and pneumatic devices. It is clear from Hero's remarks in the introductions to his works on automata and pneumatics that the production of stunning or remarkable effects was an explicit goal of

these branches of the mechanical technē; see *Aut.* 1, *Pneum.* 1, and, for a later example, the passage from Cassiodorus discussed in section 4.4. What I wish to emphasize is simply that there was nothing mysterious or inexplicable about the operation of any of these mechanical devices for those with the requisite knowledge, and that the written literature on mechanics was dedicated to transmitting such knowledge. See the remarks of Aristotle discussed in the next paragraph, as well as ps.-Alex., *Quaest.*, Pr. 32–34 Ideler: "For the craftsman [*technitēs*] who has constructed a mechanical device [*ergon ti mēchanikon*] knows all the causes [*aitiai*] of its activities, while the layperson [*idiōtēs*] is completely ignorant of these causes."

43. See Lloyd 1979, 51–52, referring to the notion that all phenomena have a natural cause: "'Marvels' [*thaumata*] and 'monsters' [*terata*] then pick out phenomena that are unusual but in principle intelligible, even if not yet understood."

44. Arist. *Metaph.* 983a12–20.

45. See Micheli 1995, 141–143, arguing (I believe incorrectly) that the wonder of mechanical phenomena is not due solely to ignorance of their cause.

46. See problem 11, which asks why heavy weights are more easily carried on rollers than on carts. The reason, in part, is that "a weight resting on rollers moves at two points of them, the ground supporting from below and the weight pressing from above; for the circle is revolving at both these points, and is impelled in the direction it travels" (852a34–37).

47. Noted by Micheli (1995, 26 n. 21). For representative examples see Piccolomini 1565, Monantheuil 1599, and del Monte 1588; see also section 4.6.

48. For the category of preternatural phenomena see Daston and Park 1998, 121–122, noting its appearance in Aquinas; the adjective *praeternaturalis* is found in Albertus Magnus (*Metaph.* II.xi; cf. OED s.v. "preternatural"). By the late sixteenth century the term *preternatural* had entered English, where it referred to something "that is out of the ordinary course of nature; beyond, surpassing, or differing from what is natural; non-natural" (OED s.v. a). Though originally distinct in meaning from supernatural, it was sometimes synonymous with it; see OED s.v. b. For the distinction between the preternatural and the supernatural see the definition of the former in Webster's *Revised Unabridged Dictionary* of 1913: "Beyond or different from what is natural, or according to the regular course of things, but not clearly supernatural or miraculous; strange; inexplicable; extraordinary; uncommon; irregular; abnormal; as, a preternatural appearance; a preternatural stillness; a preternatural presentation (in childbirth) or labor."

49. See the general remarks of Schadewaldt (1960a, 897) on the way technology can be said to surpass nature: "Die Technik greift die Stoffe und Kräfte der selbstgewachsenen Natur auf, formt sie durch Trennen und Neu-Verbindungen um und gestaltet sie in charakteristischer Weise in diesen neuen Kombinationen auf die Bedürfnisse des Menschen hin. Mit anderen Worten: Die Technik *kultiviert* die Rohstoffe, die die Natur ihr liefert, und mit diesem Kultivieren, Veredeln auf den Menschen hin mag sie wohl auch die Natur in bestimmter Weise 'übertreffen'."

50. The importance of *Physics* 199a15–17 for the interpretation of the *Mechanical Problems* is noted and emphasized by Schadewaldt (1960b, 916–917), Schneider (1989, 263), and Micheli (1995, 30–31). The ancient commentators held that in remarking that art brings to completion what nature cannot Aristotle had in mind the specific example of medicine, understood as an art that perfects an internal principle of change (see Simplicius, *In Aristotelis physicorum libros commentaria, Commentaria in Aristotelem Graeca* [CAG] 9, 378.6ff.; Philoponus, *In Aristotelis physicorum libros commentaria*, CAG 16, 310.16ff.; 316.21ff.). On this view the remark would not be applicable to mechanics, insofar as mechanics involves the modification and rearrangement of material bodies rather than the perfection of their internal principle of change. But first, what Aristotle says is that art is able to bring to completion (*epitelein*) what nature cannot, not that art is able to bring nature *itself* to a greater state of perfection. Nature is viewed as deficient, and art as making up for these deficiencies; see Arist. *Pol.* 1336b40–1337a3 ("The poets who divide ages by sevens are in the main right: but we should observe the divisions actually made by nature; for the deficiencies [*to prosleipon*] of nature are what art [*technē*] and education seek to remedy [*anaplēroun*]") and *Protrepticus*, fr. 13 (quoted below). Moreover, I see no reason to suppose that Aristotle is thinking of medicine at *Physics* 199a15ff. There is no mention of medicine in the immediate context; the example Aristotle gives is housebuilding, which involves the arrangement and fitting together of natural substances rather than a perfection of their internal principle of change. In fact housebuilding or shipbuilding (cf. 199b28–29) provides an excellent example of art going beyond nature in a way that is suited to human needs. Medicine, by contrast, would be a poor example to illustrate the ability of art to bring nature to a state of greater perfection, for medicine (understood as treatment of the sick) simply restores a patient's nature to its original state of health rather than making it more perfect. Finally we may consider fragment 13 of Aristotle's *Protrepticus*, which provides a sort of commentary on the thought of *Physics* 199a15ff.: "Natural things come to be for the sake of something and always exist for the sake of something better than artificial things: for nature does not imitate [*mimeitai*] art, but art imitates nature, and it exists to render aid and make up for [*anaplēroun*] what nature leaves undone [*ta paraleipomena tēs physeōs*]. For nature is evidently able to bring to completion [*epitelein*] some things by itself without any aid, while other things it is barely or not at all able to bring to completion, as is the case with generation: for some seeds will sprout in whatever ground they are sown without any cultivation, while others have need of the art of farming. Similarly, some animals attain their full nature by themselves, while human beings have need of many arts for their survival—both when they are first born and again for their later nourishment." Here art is pictured as helping nature (understood as the internal principle in the seed, or in a human being) to attain a greater state of completion than it could if unaided. But the main point lies in art's ability to modify the behavior of the natural world in beneficial ways—ways that enable human beings to make up for their natural deficiencies. Again, the background is provided by accounts of cultural history in which human beings are pictured as inferior to animals in qualities such as strength and speed, but as making up for this by means of intelligence and technē. The locus classicus for this idea is the story of Prometheus and Epimetheus as told in Plato's *Protagoras* (320cff.): thanks to Epimetheus's forgetfulness, human beings come into the world "naked, unshod,

unbedded, and unarmed" (321c); Prometheus gives them the technai to compensate for these disadvantages.

51. For the building of automata as imitation, see Pappus's reference to "those who seem to imitate [*mimeisthai*] the motions of living things through ropes and cables, as does Hero in his *Automata* and *Balances*" (1024.26–28 Hultsch) and Proclus's mention of the manufacture of devices that operate "through ropes and cables, which imitate the motions and impulses of living things" (*In prim. Euc. Elem.*, 41.13–15 Friedlein). Pappus refers to armillary spheres as an "image" (*eikōn*) of the heavens; cf. also Cic. *Tusc.* 1.63 and Cassiodorus, *Variae* 1.45: *Parva de illa (sc. arte mechanica) referimus, cui caelum imitari fas est.*

In Pappus the notion of mechanics as imitation is combined with the idea that it brings about effects that are *para physin*. See 1022.8–12 Hultsch (from the introduction to Pappus's account of mechanics in general): "[Mechanics] examines bodies at rest and in motion, and their locomotion in general; it not only assigns causes of natural [*kata physin*] motion, but also by forcing bodies from their natural places contrary to their natures [*para physin*] it devises means of setting them into opposite movements." Here, however, the phrase *para physin* is clearly used in the Aristotelian sense of motion that is produced by compulsion and opposed to a body's natural inclination (sense 2 as distinguished above). Notably, Pappus describes mechanics as concerned with both the natural and the forced motion of physical bodies. And elsewhere in Pappus's account the phrase *para physin* is used in connection with only one branch of mechanics, the art of those who "raise large weights contrary to nature [*para physin*] by means of machines, moving them with a smaller force [*dynamis*]" (1024.14–16 Hultsch). For Pappus the notion that mechanics acts *para physin* no more implies a break in the order of nature than it does for the author of the *Mechanical Problems*; both forced and natural motions fall under the scope of mechanics, and both can be explained in precise geometrical terms. There is no real tension between the notion that mechanics imitates nature and the notion that it brings about effects that are *para physin* in the sense that Pappus describes.

52. Cassiodorus, *Variae* 1.45 Mommsen: "Universae disciplinae, cunctus prudentium labor naturae potentiam, ut tantum possint, nosse perquirunt: mechanisma solum est quod illam ex contrariis appetit imitari et, si fas est dicere, in quibusdam etiam nititur velle superare. Haec enim fecisse dinoscitur Daedalum volare; haec enim ferreum Cupidinem in Dianae templo sine aliqua illigatione pendere; haec hodie facit muta cantare, insensata vivere, immobilia moveri. Mechanicus, si fas est dicere, paene socius est naturae, occulta reserans, manifesta convertens, miraculis ludens, ita pulchre simulans, ut quod compositum non ambigitur, veritas aestimetur." I am grateful to Bill Newman for the reference and to Malcolm Hyman for discussing the translation with me.

53. The idea of art that seems more real than the real thing has a long tradition, going back to references in early Greek poetry to the mythical craftsman Daedalus and his stunningly faithful depictions of living things. See Morris 1992, 217ff.; I am grateful to Bill Newman for this reference.

54. *De an. et resurr.* 33C–36A Migne.

55. *De an. et resurr.* 36A10–12 Migne.

56. *De an. et resurr.* 36B Migne.

57. *De an. et resurr.* 36C–40A Migne.

58. Vitruv. 10.1.4, pp. 5–6 Callebat/Fleury: "Omnis autem est machinatio rerum natura procreata ac praeceptrice et magistra mundi uersatione instituta. Namque animaduertamus primum et aspiciamus continentem solis, lunae, quinque etiam stellarum naturam; quae ni machinata uersarentur, non habuissemus interdum lucem nec fructum maturitates. Cum ergo maiores haec ita esse animaduertissent, e rerum natura sumpserunt exempla et ea imitantes inducti rebus diuinis commodas uitae perfecerunt explicationes. Itaque comparauerunt, ut essent expeditiora, alia machinis et earum uersationibus, nonnulla organis, et ita quae animaduerterunt ad usum utilia esse, studiis, artibus, institutis, gradatim augenda doctrinis curauerunt."

59. See Vitruv. 10.1.1, p. 4 Callebat/Fleury: "Machina est continens e materia coniunctio maximas ad onerum motus habens uirtutes. Ea mouetur ex arte circulorum rotundationibus, quam Graeci *kyklikēn kinēsin* appellant."

60. In this connection it should be noted that Aristotle himself mentions the circular-motion principle that is so important in the *Mechanical Problems* (i.e., that of two circles rotating around the same center, the larger moves more quickly) in connection with celestial motions; see *Cael.* 289b34ff. Cf. Krafft 1970, 67f.; Schneider 1989, 261.

61. *GA* 717a30, 745a31; *PA* 652a31, 652b21, 653b34, and so on.

62. *Mech.* 847a24–28.

63. *Mech.* 850b10–13.

64. For an overview of this literature and those who contributed to it see Rose and Drake 1971.

65. See Micheli 1995, 144ff.

66. On Piccolomini and his influence see Rose and Drake 1971, 81–85.

67. Piccolomini 1565, 8: "Ars enim quamuis naturam imitetur, ac illam adiuuet, multa tamen ut usui nobis esse possint, aliter operatur ac illa facit. Neque minus ob id, naturae imitatrix existimari debet, quippe quae eo modo efficit quicquid facit, quo illamet efficeret, si illud faceret. Nam, tametsi natura eundem semper in quo uis opere suo, si non impeditur, seruet modum ars uero, propterea quod nostra utilitas, usus, atque commoditas multiplex est, multiplici etiam & illa itinere progrediatur: in quo uis tamen itinere, naturam & ipsa sequitur. Natura igitur, ex sua simplicitate nequaquam multiformis, si quando euenit, ut eius simplicitate, nos propter multiplices nostros usus non contenti, aduersus ipsam quippiam moliamur, tunc illa contra nitens ac contendens, nostris conatibus difficultatem imponit. quam difficultatem, seu potius haesitationem quandam ut superemus, arte quadam indigemus, qua tandem naturae conatus, uel penitus confringentes, uel aliqua saltem ex parte, demolientes, nostrum opus perficiamus."

68. See Rose and Drake 1971, 93; Nobis 1969. My characterization of Moletti's views is based on my own partial transcription of the relevant folio pages as well as the account of Nobis (1969, 331–332).

69. In the following I paraphrase the argument found on pp. 2–5 of del Monte 1588; an online version of this text is available at http://archimedes.mpiwg-berlin .mpg.de.

70. The same idea is alluded to near the beginning of the preface to Guidobaldo's influential *Mechanicorum Liber* of 1577, where mechanics is described as operating "against nature, though emulating its laws" (*adversus naturam vel eiusdem emulata leges*).

71. Monantheuil 1599, 8. Rose and Drake (1971, 100) describe Monantheuil's work as "the most complete and erudite of the sixteenth-century commentaries on the *Mechanica*, though not the most original in outlook." An online version of the text is available at http://archimedes.mpiwg-berlin.mpg.de.

72. Monantheuil 1599, 9: "Homo enim inquit, animaduertens Solis, Lunae, & reliquorum planetarum continentes motus, & machinationes naturales, sine quibus non habuisset in terra lucem, & fructuum maturitates hinc exempla sumpsit, & ea imitans, inductus rebus diuinis, commodas vitae perfecit explicationes. Itaque comparauit, vt essent expeditiora alia machinis, & earum versationibus: alia organis, quaeque obseruauit ad vsum vtilia esse studijs, artibus, institutis, doctrinis gradatim augenda curauit: hinc tandem extat ars quaedam generalis quae difficultati faciendorum praeter naturam ad vtiltiatem hominum succurrit."

73. Galilei 1960, 147: "These deceptions (sc. the deceptions of mechanicians who attempt to achieve impossible effects) appear to me to have their principal cause in the belief which these craftsmen have, and continue to hold, in being able to raise great weights with a small force, as if with their machines they could cheat nature, whose instinct—nay, whose most firm constitution—is that no resistance may be overcome by a force more powerful than it. How false such a belief is, I hope to make most evident with true and rigorous demonstrations that we shall have as we go along." It should be noted that Galileo attributes this view to practitioners, not to those who proposed the kinds of theories about the nature of mechanics with which I am concerned in the present section.

74. See Krafft 1970, 160–168.

75. For one example of this see Hooykaas 1963, who exaggerates the originality of Monantheuil's position. Krafft (1970 n. 362a) recognizes the contribution of Moletti, but evidently thinks it exceptional.

## REFERENCES

Apelt, O. 1888. *Aristotelis quae feruntur De plantis; De mirabilibus auscultationibus; Mechanica; De lineis insecabilibus; Ventorum situs et nomina; De Melisso, Xenophane, Gorgia*. Leipzig: Teubner.

Barnes, J., ed. 1984. *The Complete Works of Aristotle: the Revised Oxford Translation.* Princeton, NJ: Princeton University Press.

Bourgey, L. 1975. Observation and experiment in analogical explanation. In M. Schofield, J. Barnes, and R. Sorabji, eds., *Articles on Aristotle,* 175–182. vol. I. London: Duckworth.

Callebat, L., and P. Fleury, eds. 1986. *Vitruve: De l'architecture, Livre X.* Paris: Les Belles Lettres.

Daston, L., and K. Park. 1998. *Wonders and the Order of Nature: 1150–1750.* New York: Zone Books.

Galilei, G. 1960. *On Motion and On Mechanics.* Madison: University of Wisconsin Press.

Gille, B. 1980. *Les mécaniciens Grecs: La naissance de la technologie.* Paris: Éditions du Seuil.

Hooykaas, R. 1963. *Das Verhältnis von Physik und Mechanik in historischer Hinsicht.* Beiträge zur Geschichte der Wissenschaft und der Technik 7. Wiesbaden: Franz Steiner.

Krafft, F. 1967. "Die Anfänge einer theoretischen Mechanik und die Wandlung ihrer Stellung zur Wissenschaft von der Natur." In W. Baron, ed., *Beiträge zur Methodik der Wissenschaftsgeschichte.* Beiträge zur Geschichte der Wissenschaft und der Technik 9, 12–33. Wiesbaden: Franz Steiner.

Krafft, F. 1970. *Dynamische und statische Betrachtungsweise in der antiken Mechanik.* Boethius: Texte und Abhandlungen zur Geschichte der exakten Wissenschaften 10. Wiesbaden: Franz Steiner.

Krafft, F. 1973. Kunst und Natur: Die Heronische Frage und die Technik in der klassischen Antike. *Antike und Abendland* 19: 1–19. Berlin: De Gruyter.

Lloyd, G. E. R. 1966. *Polarity and Analogy: Two Types of Argumentation in Early Greek Thought.* Cambridge: Cambridge University Press.

Lloyd, G. E. R. 1979. *Magic, Reason, and Experience: Studies in the Origin and Development of Greek Science.* Cambridge: Cambridge University Press.

Lonie, I. M. 1981. *The Hippocratic Treatises, "On Generation," "On the Nature of the Child," "Diseases IV": A Commentary.* Ars Medica II. 7. Berlin: De Gruyter.

Micheli, G. 1995. *Le origini del concetto di macchina.* Biblioteca di Physis 4. Florence: Olschki.

Monantheuil, H. d. 1599. *Aristotelis Mechanica Graeca, emendata, Latina facta, et commentariis illustrata.* Paris: Jeremia Perier.

Monte, G. del 1588. *In duos Archimedis Aequeponderantium libros paraphrasis.* Pesaro.

Morris, S. 1992. *Daidalos and the Origins of Greek Art*. Princeton, NJ: Princeton University Press.

Newman, W. 1997. Art, nature and experiment among some Aristotelian alchemists. In E. Sylla and M. McVaugh, eds., *Texts and Contexts in Ancient and Medieval Science: Studies on the Occasion of John E. Murdoch's Seventieth Birthday*, 305–317. Brill's Studies in Intellectual History 78. Leiden: Brill.

Newman, W. 2001. Corpuscular alchemy and the tradition of Aristotle's *Meteorology*, with special reference to Daniel Sennert. *International Studies in the Philosophy of Science* 15 (2): 145–153.

Nobis, H. M. 1969. Über zwei Handschriften zur frühneuzeitlichen Mechanik in italienischen Bibliotheken. *Sudhoffs Archiv* 53: 326–332.

Piccolomini, A. 1565. *In Mechanicas Quaestiones Aristotelis, paraphrasis paulo quidem plenior*. Venice: Traianus Curtius.

Rose, P. L., and S. Drake. 1971. The Pseudo-Aristotelian *Questions of Mechanics* in Renaissance culture. *Studies in the Renaissance* 18: 65–104.

Ross, W. D. 1936. *Aristotle, Physics*. Oxford: Oxford University Press.

Schadewaldt, W. 1960a. Natur-Technik-Kunst. *Hellas und Hesperien: Gesammelte Schriften zur Antike und zur neueren Literatur*, 892–906. Zurich: Artemis-Verlag.

Schadewaldt, W. 1960b. Die Begriffe "Natur" und "Technik" bei den Griechen. *Hellas und Hesperien: Gesammelte Schriften zur Antike und zur neueren Literatur*, 907–919. Zurich: Artemis-Verlag.

Schneider, H. 1989. *Das Griechische Technikverständnis: Von den Epen Homers bis zu den Anfängen der technologischen Fachliteratur*. Impulse der Forschung 54. Darmstadt: Weidmannsche Buchhandlung.

Vernant, J.-P. 1957. "Remarques sur les formes et les limites de la pensée technique chez les Grecs." *Revue d'histoire des sciences* 10, 205–225. Reprinted in *Mythe et pensée chez les Grecs* (1971), vol. 2, 44–64. Paris: F. Maspero.

# 5

## Art, Nature, Alchemy, and Demons: The Case of the *Malleus Maleficarum* and Its Medieval Sources

William R. Newman

The attempt to equal or outdo nature was already a longstanding feature of alchemical literature by the time that the aurific art passed from the Islamic world to Western Europe in the High Middle Ages. It is perhaps obvious that alchemists would have held a vested interest in the claim that they could manufacture products that were identical in all respects to their natural exemplars. In maintaining this claim, alchemists typically employed a highly elastic concept of the natural, according to which human-induced change did not *ipso facto* lead to an artificial product. This blurring of the boundaries between the artificial and the natural, so reminiscent of the situation that contemporary bioengineering and synthetic organic chemistry encounter today, produced remarkable consequences over a broad array of disciplines in medieval and early modern Europe. In effect, the unequivocal claims of perfectly reduplicating natural products made by alchemists led others to view their art as a test case for the powers of technology in general.

Within a hundred years of alchemy's twelfth-century entrance into the medieval Latin world, the scholastics had fully appropriated the discipline, making it a subject of discussion in encyclopedias as diverse as those of Vincent of Beauvais and Arnoldus Saxo.[1] Of equal historical significance is the fact that the scholastics inaugurated a tradition whereby alchemy became the benchmark against which the power of all other arts could—and should—be measured. Since alchemists claimed—or could be seen as claiming—to have the power of transmuting *species* when they supposedly turned one metal into another, it was possible to view their art as verging on the creative power of God himself. This aspect of alchemy would soon come to provide a test case among theologians for the power of any beings who could employ arts—humans, angels, and more importantly, demons. It was a theological commonplace that demons could not perform genuine miracles that worked outside the realm of natural law—this was the province of God alone. Although demons were immensely clever and capable of transporting bodies at tremendous speeds, their powers stopped where God's had only begun—they could not, for example, create new beings ex nihilo.[2] But could demons transmute one species into another, as when Pharoah's magicians claimed to convert wooden staffs into snakes in Exodus 7? It was here that alchemy entered into the medieval discussion of demonic power and its limits.

If alchemists could really transmute species, then demons, being more powerful than humans, should also have this ability. On the other hand, if one could employ alchemy to show that art could not engage in species transmutation, then it should be impossible for demons as well, since their powers were limited by art just as those of humankind were. Such an argument promised important consequences, since traditional powers attributed to demonic magic in the realm of popular belief included that of changing humans into animals, and even inert materials into living ones. Beginning in the mid-thirteenth century, this use of alchemy to determine the limits of *ars* grew and ramified into a bewildering variety of different traditions. As the question of the power of art passed into new genres of discussion, its original raison d'être within theological texts often faded into the background, but the deracinated issue of alchemy as a test case for art lived on. In natural philosophy, the practice of determining the overall success of ars by reference to alchemy was still alive and well even in the century of Descartes and Newton, for the scholastic manuals of the time usually answered the question "Whether art effects certain works of nature?" by reference to the aurific art.[3] As we will see, however, the theological association between alchemy and demonic power was never entirely lost, and the issue occasionally reemerged in full force centuries after its original initiation.

The longevity of alchemy as a technological benchmark with consequences for demons finds no better illustration than the use made of that discipline by the two Dominican inquisitors Heinrich Kramer and Jakob Sprenger in their well-known witch-hunting manual of 1487, the *Malleus maleficarum*. Undoubtedly the most influential witch-hunting guide ever written, the *Malleus* went through at least 28 Latin printings between 1487 and 1669, having been composed, according to one unabashed modern fan, *sub specie aeternitatis*.[4] The *Malleus* has been given a central role in histories of the great witch hunt, from Joseph Hansen's influential early twentieth-century studies up to the recent book by Walter Stephens, *Demon Lovers*.[5] Given the unflinching gaze and single-minded dedication with which Kramer and Sprenger meet the evil eye of their malefic adversaries, it is perhaps surprising that the text starts with a rather unfocused beginning. For it is an indisputable, yet hardly noticed fact that the *Malleus maleficarum* begins with a denunciation of alchemy as well as witchcraft (figure 5.1). In response to the obligatory question whether it is heretical to deny the power of witches, the text launches into a series of *responsiones quod non*, or negative replies. One of these, found on the very first folio, contains the following thesis, tacitly borrowed from Thomas Aquinas's commentary on the *Sentences* of Peter Lombard: "Demons do not work except by art. But art cannot give a true form. Whence it is said in the chapter on minerals that the authors of alchemy should know that species cannot be transmuted. Therefore demons, also working by means of art, cannot induce real qualities of health or sickness. But if these [qualities] are real, they have some other hidden cause beyond the work of demons and sorcerors."[6]

One can easily grasp the general tenor of this argument. Beginning with the traditional premise that demons cannot work by genuinely supernatural means, since they lack the creative capacity of God, they must be limited to the realm of art—technology, as we would now say—just as humans are. But art cannot impose "a true form," as the alchemists should know from their failed attempts to convert base metals into gold. By extension, then, demons themselves and a fortiori witches, cannot really impose new forms on matter or change the species of things, any more than alchemists can.

But a preliminary understanding of this argument only leads to further questions. Why did Kramer and Sprenger choose to begin their famous exposé on witchcraft with this short yet unrelenting attack on alchemy, instead of using some other artisanal example? Did they perhaps view alchemists themselves as witches, or see the aurific art as a species or close relative of magic, in the way that some modern students of the subject have assumed?[7] And did they really want to dismiss alchemy as a sort of fraud, as the passage clearly suggests?

Despite their initial debunking of alchemy with its consequent limitation on demonic power, it is clear that the general intention of Kramer and Sprenger is to magnify the power of demons as much as they can. This is evident throughout the *Malleus maleficarum*, which attempts at every turn to erode the force of medieval texts restraining belief in witchcraft, above all the authoritative *Canon episcopi* of the early tenth century, which was incorporated into Gratian's *Decretum*. The *Canon episcopi* explicitly rejects the belief that sorcerers can change

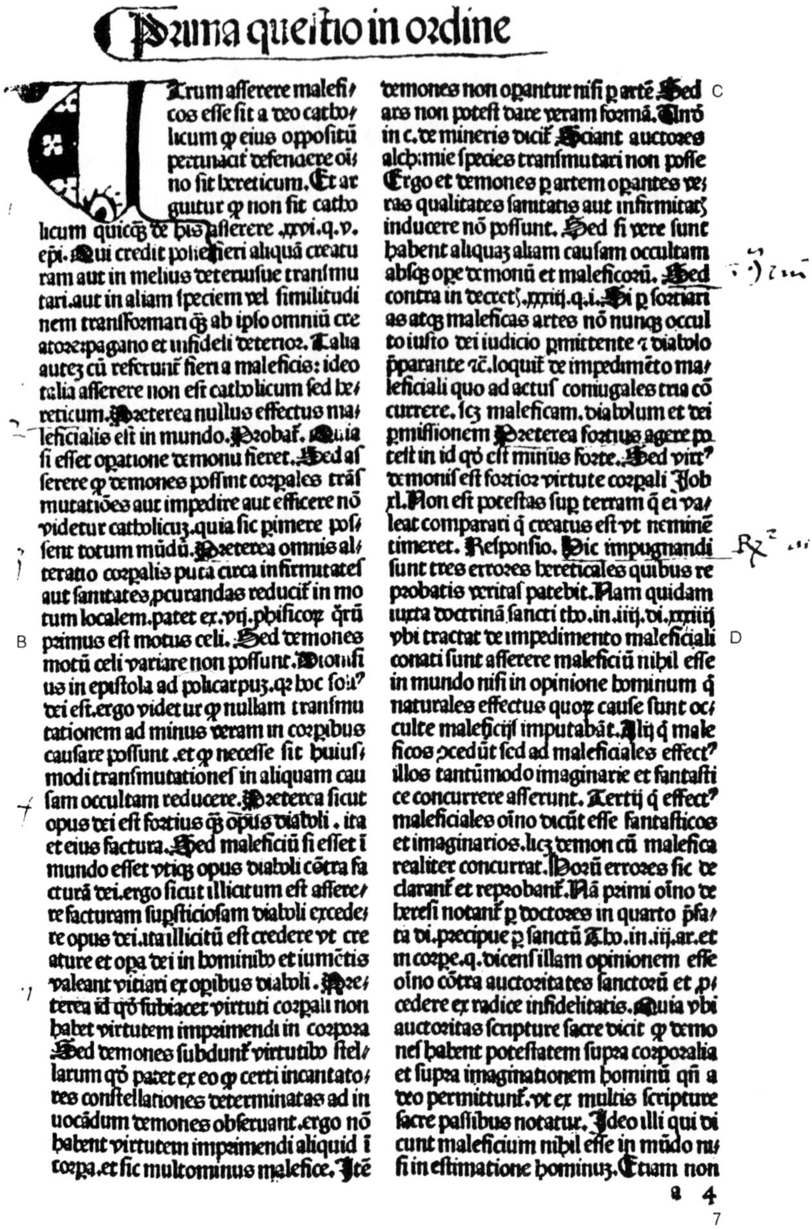

Prima questio in ordine

Utrum asserere maleficos esse sit adeo catholicum quod eius oppositum pertinaciter defendere omnino sit hereticum. Et arguitur quod non sit catholicum quicquid de his asserere. xxvi.q.v. epi. Qui credit posse fieri aliquam creaturam aut in melius deteriusue transmutari aut in aliam speciem vel similitudinem transformari quam ab ipso omnium creatore: pagano et infideli deterior. Talia autem cum referuntur fieri a maleficis: ideo talia asserere non est catholicum sed hereticum. Preterea nullus effectus maleficialis est in mundo. Probatur. Quia si esset operatione demonum fieret. Sed asserere quod demones possint corporales transmutationes aut impedire aut efficere non videtur catholicum. quia sic primere possent totum mundum. Preterea omnis alteratio corporalis puta circa infirmitates aut sanitates procurandas reducitur in motum localem. patet ex. vij. phisicorum quorum primus est motus celi. Sed demones motum celi variare non possunt. Dionisius in epistola ad policarpum. quia hoc solius dei est. ergo videtur quod nullam transmutationem ad minus veram in corporibus causare possunt. et quod necesse sit huiusmodi transmutationes in aliquam causam occultam reducere. Preterea sicut opus dei est fortius quam opus diaboli. ita et eius factura. Sed maleficium si esset in mundo esset utique opus diaboli contra facturam dei. ergo sicut illicitum est asserere facturam supersticiosam diaboli excedere opus dei. ita illicitum est credere ut creature et opera dei in hominibus et iumentis valeant viciari ex operibus diaboli. Preterea id quod subiacet virtuti corporali non habet virtutem imprimendi in corpora. Sed demones subduntur virtutibus stellarum quod patet ex eo quod certi incantatores constellationes determinatas ad inuocandum demones obseruant. ergo non habent virtutem imprimendi aliquid in corpora. et sic multominus malefice. Item demones non operantur nisi per artem. Sed ars non potest dare veram formam. Unde in c. de mineris dicitur Sciant auctores alchimie species transmutari non posse. Ergo et demones per artem operantes veras qualitates sanitatis aut infirmitatis inducere non possunt. Sed si vere sunt habent aliquam aliam causam occultam absque opere demonum et maleficorum. Sed contra in decretis. xxxiij.q.i. Si per sortiarias atque maleficas artes non numquam occulto iusto dei iudicio permittente et diabolo preparante etc. loquitur de impedimento maleficiali quo ad actus coniugales tria concurrere. scilicet maleficam. diabolum et dei permissionem. Preterea fortius agere potest in id quod est minus forte. Sed virtus demonis est fortior virtute corporali. Job xl. Non est potestas super terram que ei valeat comparari qui creatus est ut neminem timeret. Responsio. Hic impugnandi sunt tres errores hereticales quibus reprobatis veritas patebit. Nam quidam iuxta doctrinam sancti tho. in. iiij. di. xxxiiij ubi tractat de impedimento maleficiali conati sunt asserere maleficium nihil esse in mundo nisi in opinione hominum qui naturales effectus quorum cause sunt occulte maleficijs imputabant. Alij quod maleficos concedunt sed ad maleficiales effectus illos tantummodo imaginarie et fantastice concurrere asserunt. Tertij quod effectus maleficiales omnino dicunt esse fantasticos et imaginarios. licet demon cum malefica realiter concurrat. Horum errores sic declarantur et reprobantur. Nam primi omnino de heresi notantur per doctores in quarto prefata di. precipue per sanctum Tho. in. iij. ar. et in corpore. q. dicens illam opinionem esse omnino contra auctoritates sanctorum et procedere ex radice infidelitatis. Quia ubi auctoritas scripture sacre dicit quod demones habent potestatem supra corporalia et supra imaginationem hominum quando a deo permittuntur. ut ex multis scripture sacre passibus notatur. Ideo illi qui dicunt maleficium nihil esse in mundo nisi in estimatione hominum. Etiam non

a 4

7

**Figure 5.1**

First folio of the witch hunter's manual *Malleus maleficarum*, containing a version of Avicenna's famous anti-alchemical edict, the *Sciant artifices*, in the upper-right corner of the second column. From Heinrich Kramer and Jakob Sprenger, *Malleus maleficarum* (1487); p. 7 in the facsimile that appears in *Malleus maleficarum von Heinrich Institoris (alias Kramer) unter Mithelfe Jakob Sprengers Aufgrund der Dämonologischen Iradition Zusammengetellt*, ed. André Schnyder (Göppingen: Kümmerle Verlag, 1991).

their shape or species as heretical and even worse than the ignorance of the heathen. At the same time, the *Canon episcopi* also rejects the vaguer claim that "any creature can be changed or transformed to better or worse" (*posse aliquam creaturam aut in melius aut in deterius immutari*) by witchcraft.[8] This dismissal could obviously be taken to mean that witches are unable to work any real physical effects in the world, such as the inflicting of disease and death on their victims. It is clear that Kramer and Sprenger did in fact fear that the *Canon episcopi* would be read in that light, for they expend the greater part of their efforts in the first *Quaestio* of the *Malleus* in establishing that witches can indeed bring about physical diseases by means of demonic help.[9] The heavy weight that this issue exercised on the authors' minds appears clearly from the initial headings of the first *Quaestio* (all to be rebutted later)—there is no malefic effect in the world; all alterations of the body resulting in illness or health come from local motion due to the changes of the heavens, but demons have no power over the heavens; it seems that demons can cause no true change in bodies, and so forth. Kramer and Sprenger were eager to show that all of these propositions limiting demonic power—especially the power of causing sickness—were false.

The emphasis that the two Dominican authors put on proving the reality of demonic diseases can clearly be seen if we compare the passage that we quoted above in English to its original in Thomas Aquinas's *Sentence* commentary. Whereas Thomas is explicitly concerned with the alchemist's supposed ability to impart a new substantial form to matter, thereby converting a base metal into gold, Kramer and Sprenger emend "substantial form" (*forma substantialis*) to "true form" (*vera forma*). The reason for this change is not haphazard—it reflects the fact that the two Dominicans want to expand Thomas's discussion to include accidental forms like illness and health that do not change one substance (such as "man") into another substance (such as "wolf"), but merely work accidental change on a given substance (such as changing a healthy person into a sick person). I quote the two Latin passages below in parallel columns.

| *Malleus maleficarum*, quaestio 1 | Thomas Aquinas, *Sentence Commentary*, book 2, distinction 7 |
|---|---|
| Item demones non operantur nisi per artem. Sed ars non potest dare veram formam. | Praeterea, daemones non operantur nisi per modum artis. Sed ars non potest dare formam substantialem; |

| | |
|---|---|
| Unde in c. de mineris dicitur Sciant auctores alchimie species Transmutari non posse Ergo et Demones per artem operantes Veras qualitates sanitatis aut infirmitatis inducere non possunt.[11] | unde dicitur in cap. De mineris: "Sciant auctores alchimiae species transformari non posse." Ergo nec daemones formas substantiales inducere possunt.[10] |

Thomas's claim that the alchemists are unable to transmute species and impose new substantial forms now becomes a broader argument denying the ability of the alchemists to induce any "true forms" into matter. From this Kramer and Sprenger draw the explicit inference that such impotence on the part of alchemists will imply a similar failure on the part of demons trying to imprint the qualities of health or illness on the body. As one might expect, the broadening of Thomas's censure has the effect of turning it into something of a straw man, since it was much easier to uphold the thesis that witches cause disease (by changing accidents) than that witches change humans into animals (by transmuting substances). Belief in shape-changing was explicitly forbidden by the *Canon episcopi's* words that whoever believes that any creature can be "transformed into another species or likeness, except by the Creator himself who made everything and through whom all things were made, is beyond doubt an infidel."[12] Although Kramer and Sprenger seem here to equivocate on the issue whether demons and their minions can ever impose new substantial forms—by inducing spontaneous generation in lower, imperfect animals, for example—their general tendency when pushed is to avoid the stronger claim of specific transmutation in favor of the weaker one of accidental change.[13] When confronted with the apparent "fact" of specific transmutatuion brought on by demonic interventions, as in the case of lycanthropy, for example, Kramer and Sprenger could appeal to Satan's well-known ability to create illusions. The same explanatory move allowed them to explain numerous other seemingly marvelous phenomena, such as the apparent separation of the penis from a man's body.[14]

The revisionist spirit that drove Kramer and Sprenger to circumvent the inconvenient authority of the *Canon episcopi* also lay behind their approach to the initial Thomistic thesis debunking alchemy. At the end of the first *Quaestio*, the two authors return to the antialchemical argument, now explicitly invoking the authority of Thomas. I quote them as follows:

> It must be said according to Saint Thomas in 2, Distinction 7, in the solution of one argument where he determines the power of demons in their working, that although some substantial forms can be induced by art by the power of a natural agent, just as the form of fire is induced in wood by art, this cannot be done universally, because art cannot always find or conjoin proper active things to proper passive ones, although it can produce something similar. Thus the alchemists make something similar to gold as far as exterior accidents, but they do not make real gold. For the substantial form of gold does not come from the heat of the fire that alchemists use, but rather from the heat of the sun in a specific place where the mineral virtue flourishes. Therefore such [fake] gold does not operate according to the species [of real gold], and the same is true of [the alchemists'] other works.[15]

This passage is a close paraphrase of a section in book 2, distinction 7, of Thomas Aquinas's *Sentence* commentary. Thomas's own intention is fairly clear. First, he wants to limit the power of demons to the natural realm, which he does by comparing them to alchemists. The force of this comparison lies in the unspoken assumption that alchemy is a natural, rather than a supernatural, art. Second, he argues that even in the natural world some arts are ineffectual, since not every substantial form can be induced by humans into matter. The failure of alchemy, then, is used as an example to limit the power of demons and their followers, as it was in the antialchemical thesis with which the *Malleus maleficarum* began.

What is really interesting, however, is the way in which Kramer and Sprenger manage to wiggle out of Thomas's censure by turning his words against him. Immediately after the passage just quoted, they continue as follows: "Demons work by art according to malefic effects. Therefore they cannot induce any substantial or accidental form without the aid of another agent. And because we do not say that one can bring about *maleficium* by means of art [per artem *legitur pro* partem] without the aid of another agent, it follows that with such aid one can induce the true qualities of disease or of another effect."[16] It is important to note that Kramer and Sprenger are not responding directly to the Thomistic passage just cited, but rather to the initial Thomistic thesis with which they introduced the *Malleus*, the blanket injunction against the transmutation of species by alchemists. The conclusion there was, as we already stated: "Therefore demons, also working by means of art, cannot induce real qualities of health or sickness. But if these [qualities] are real, they have some other hidden cause beyond the work of demons and sorcerors."

Kramer and Sprenger's argument, then, is that demons cannot in fact induce forms by means of art alone. It does not follow, however, that demons absolutely cannot induce forms. This is what the two authors mean when they say, at the end of the first *Quaestio*, "because we do not say that one can bring about *maleficium* by means of art without the aid of another agent, it follows that with such aid one can induce the true qualities of disease or of another effect." In other words, as long as one can argue that the effects of witchcraft are not *purely* artificial, but rather products of art working on "another agent" supplied by nature, then the effects of witchcraft—and perhaps by extension even those of alchemy—can be genuine. This conclusion, of course, is strikingly different from what Thomas himself intended with regard to alchemy, for in his *Sentence* commentary he upheld the view that the aurific art was an unequivocal failure, and a useful example of the limitations placed by God on human and demonic power. Although Thomas certainly admitted that demons could produce marvelous effects by uniting hidden agencies to their properly receptive subjects, he wished to stress that this process had limits, as exemplified by the failure of alchemists to transmute metals.[17] In certain cases, such as attempted chryspopoeia, "art cannot find the appropriate actives and passives," Thomas insists.[18] Hence the art in question fails to meet its intended goal, the transmutation of species. But Kramer and Sprenger have taken a very different lesson from Thomas's conclusion. They read Thomas's limitation on demonic power as pertaining only to the use of purely artificial agents; by adding natural agents to natural patients, demons can act on matter and induce "true forms."[19]

We can see, then, that the conclusion of Kramer and Sprenger hinges on a distinction between an art that works purely "artificially," and an art that operates by working on or aiding nature. If demons try to work by the former route, they are doomed to failure, since unlike God, they cannot create or transmute purely by means of their own artifice or by a power that emanates from themselves, like the mirific word of God. If they choose the latter path, however, their harmful intentions can be realized in the physical world, since they can join actives to passives in the way that Thomas speaks of in his example of bringing fire to dry wood. Although Kramer and Sprenger do not go so far as to say that demons really can transmute species here, they make it quite clear that demons can operate on natural substances to bring them to deleterious ends.

Undoubtedly my analysis so far will raise more questions than it has solved. We are still left with the nagging query: Why did the two Dominicans introduce alchemy into their discussion, only to obviate the strictures on demonic power that a rejection of alchemy seemed to imply? To this general question we can now add a more nuanced one: Where do Kramer and Sprenger get the distinction between art operating alone and hence ineffectually, and art operating on natural agents to produce real natural effects?

## The Avicennian Rejection of Alchemy

The answer to both questions can be found by turning to the traditional disputation literature surrounding alchemy from the Middle Ages onward. As I have argued elsewhere, alchemy provided a compelling focus for the discussion of human artisanal power, since alchemists had an obvious and vested interest in proving that they could reproduce genuine natural products in the form of the precious metals.[20] This innate tendency on the part of alchemists to defend their products as natural received a further impetus from the fact that Islamic practitioners of alchemy had been the brunt of several powerful attacks coming from the most famous philosophers of the Arab-speaking world—Avicenna and Averroes. The more important of these attacks for the Latin West was that of Avicenna, since his words came to be mistakenly identified with those of Aristotle in the thirteenth century.

Avicenna's attack on alchemy forms a small part of an impressive treatment of geology and mineralogy found in his *Kitāb al-Shifā'*, or *Book of the Remedy*. This section of the *Shifā'* would be translated into Latin by Alfred of Sareshal at the beginning of the thirteenth century as the *Liber de congelatione et conglutinatione lapidum* (Book on the Congealment and Concretion of Stones). Alfred attached his translation of Avicenna to a manuscript version of Aristotle's *Meteorology Book IV* already prepared by Henricus Aristippus, leading to a widespread and mistaken view that the *Liber de congelatione*, with its denunciation of alchemy, was by Aristotle himself.

In the *Liber de congelatione* Avicenna frames two very powerful arguments against alchemy, which the Latins would refer to by the incipit *Sciant artifices alchimie* (Let the artificers of alchemy know . . .), the words that began his pronouncement against alchemy. We already saw a version of this phrase (*Sciant auctores alchimie* . . .) on the first folio of the *Malleus maleficarum*, where it had been borrowed indirectly from Avicenna via Thomas Aquinas. After admitting that artificers can fabricate clever simulations of natural products, Avicenna denies that alchemists can ever make genuinely natural products, for the following reasons:

> Art is weaker than nature and does not overtake it, however much it labors. Therefore let the artificers of alchemy know that the species of metals cannot be transmuted (*Quare sciant artifices alkimie species metallorum transmutari non posse*). But they can make similar things, and tint a red [metal] with yellow so that it seems gold, and tint a white one with the color that they want until it is very similar to gold or copper. They can also cleanse the impurities of lead, although it will always be lead. Even though it may seem silver, alien qualities will dominate in it, so that men err in this just as those who accept [artificial] salt and sal ammoniac err. I do not believe that it is possible to take away the specific difference by some technique because it is not due to such [accidents] that one complexion is converted into another, since these sensible things are not those by which species are transmuted; rather they are accidents and properties. For the differences of the metals are not known, and since the difference is not known, how will it be possible to know whether it is removed or not, or how it could be removed?[21]

The first of these sentences contains the universal proposition that art is inferior to nature and cannot therefore make a product that genuinely measures up to its natural exemplar. This idea is probably based on the ancient belief that all arts are learned by imitating nature. Avicenna has simply stated an implicit consequence, that the copy cannot equal its model.[22]

The second proposition in Avicenna's *De congelatione*, that the species of the metals cannot be transmuted, employs the Aristotelian opposition of species (*nau*$^{c}$) and genus (*jins*). Avicenna believes that the mere fact of belonging to a single genus (metallic substance) does not mean that the individual species (the metals taken individually) can be transmuted among themselves. He argues further in the text that our senses only allow us to perceive the accidents that superficially distinguish the metals, such as taste, color, and weight. The genuine species-determining characteristics of the metals are unknown to us, and lurk beneath the level of sense data. Since we cannot even perceive the real specific differences between the metals, how can we hope to transmute them?

If we now shift our gaze from Avicenna himself to the Latin world of the early thirteenth century, it is easy to see why his antialchemical pronouncement acquired a prestige out of all proportion to its length. In a world where Aristotle was referred to customarily as "the prince of the philosophers," or simply as "the philosopher," the integration of the *De congelatione* as a stowaway within the *Meteorology* added immense prestige to the Avicennian text. In practical terms, this Aristotelian rebaptizing meant that alchemy was an important and legitimate subject of discussion for commentators of Aristotle's *Meteorology*. At the same time, the Latin translation of the *De congelatione* did not reveal the full force of Avicenna's attack on alchemy, but terminated by suggesting that the artificial transmutation of metals might be possible if they were first reduced into their "prime matter," the undifferentiated material substrate of all things according to Aristotelian physics.[23] Hence the *Sciant artifices* came to be fair game for scholastic disputation both within and beyond the confines of alchemical texts themselves.[24] As we will now see, Avicenna's short pronouncement against alchemy soon entered a new and seemingly anomalous arena—the discussion of demons and their powers in the world of nature.

## Alchemy and the Power of Demons

We can observe the influence of Avicenna's *Sciant artifices* already in one of the earliest known treatments of alchemy by a university doctor, the commentary on Peter Lombard's *Sentences* written by Albertus Magnus, probably in the second half of the 1240s.[25] The *Sentences*, composed by Peter in the mid-twelfth century, comprise a four-volume collection of theological questions and answers, largely compiled from Saint Augustine, but reprising many other sources as well. Albert, like many writers in the thirteenth century and later, wrote an extensive commentary on the *Sentences*, expressing his views on a multitude of topics. As we will see, Albert is an early representative of the scholastic tradition of using alchemy to determine the power of demons that culminated, in a sense, in the *Malleus maleficarum* of Kramer and Sprenger. Here we must avoid the easy and modern habit of grouping such topics as magic and alchemy under a single, seemingly unproblematic rubric, such as "the occult sciences" or "the occult."[26] Albert definitely does not equate the two fields of alchemy and magic, and it is precisely their distinctness that allows him to draw meaningful comparisons between the two. For him, the claim of alchemy to transmute species represents the ultimate assertion of human power in the natural world. Alchemy is the high-water mark against which other arts—even the arts possessed by demons—must be measured. This view became a commonplace among scholastic authors that would last well into the seventeenth century.

To understand the points that Albert will make, we must first consider the passage from Peter Lombard that he is analyzing. His comments on alchemy form part of a gloss on book II, distinction 7 of the *Sentences.* In this section of the *Sentences* Peter, relying heavily on Augustine's *De trinitate,* had presented the position that "the magic arts work by means of the power and knowledge of the devil, which power and knowledge is granted to him by God."[27] What Peter has in mind is several passages from Exodus 7 and 8, where the magicians of Pharaoh are said to have made various animals, including serpents and frogs. While the Bible grants them this, Peter points out that the magicians had no power against the gnats that provided the third plague suffered by the Egyptians at Exodus 8:18. To him, this indicates that the power of the magician is extremely limited, and that the marvelous deeds of the magi are really only permitted to them insofar as God wills it. Otherwise, the demons would themselves be creators (*creatores*) like God himself, a possibility that Peter resolutely denies. Introducing Saint Augustine's concept of "seminal reasons," the *logoi spermatikoi* of the ancient Stoics, Peter suggests that demons merely gather together the otherwise separated and hidden "seeds" of things in order to produce their marvels.

In his commentary to this passage, Albert considers a number of ramifications of Peter's view, but the section that interests us asks "Whether demons can induce substantial forms in transmuted bodies?" In typically scholastic fashion, Albert first replies with a list of negative responses, the *responsiones quod non.* Beginning with earlier Christian commentary on Exodus 7, Albert proceeds to a discussion of the serpents that Pharaoh's magicians reportedly made from wooden staffs in their famous contest with Moses and Aaron. This passage from Exodus served as one of the paradigmatic witnesses of demonic power, since it was assumed, of course, that the Egyptian magicians could only work their sorcery with demonic help.[28] Since Albert is here presenting arguments against the claim that demons can induce a substantial form in matter, he argues first that the magicians' snakes were really just illusory, not transmuted substances. Among the arguments in favor of this view, Albert presents the following: "Likewise, art does not transmute a substantial form into [another substantial] form, because Aristotle says in *Meteorology 4* that 'the artificers of alchemy should know that species cannot be transmuted': therefore, demons cannot [transmute them], because they only work by means of art."[29] The thrust of this argument is that alchemy, the art that claims par excellence to work species transmutations, cannot really effectuate such radical change. If alchemists cannot do this, neither can demons: as a result, the marvelous changes wrought by demons must actually be illusions. It is very interesting to see that at this stage in his career, Albert still accepted that the *Sciant artifices*, which he here quotes verbatim, is a genuine statement of the Stagirite's. In his impressive study of mineralogy and alchemy, the *Liber mineralium*, written a few years later, Albert would explicitly reject this Aristotelian pedigree, and return the text to Avicenna.[30] There can be little doubt that Albert's attribution of the *Sciant artifices* to Aristotle encouraged his otherwise unlikely incorporation of it into a theological treatment of demons. What is remarkable about Albert's use of the *Sciant* is that he has omitted all reference to the metals, making Avicenna's dictum apply not only to them but to species in general.[31] The *Sciant* thereby acquires a universalist character that it otherwise lacked: it became a general statement about the limitation of art in the world of nature. And since demons were also thought to work by means of art, the *Sciant artifices* restricted their power just as it put limits on the power of humans.

After presenting a number of other arguments against the ability of demons to make genuine transmutations of species, Albert then passes to the other side of the issue. Following the typical method of scholasticism, he now produces a list of arguments in favor of demons' having actual power over physical substances. First, Albert recapitulates the Augustinian notion that all things on earth are generated from "seminal reasons" or hidden seeds that God imposed on matter during the Creation. When sorcerers perform their incantations, the demons respond by running off to collect these seeds in various parts of the world: "They suddenly (*subito*) bring together the seeds by which this is done, and thus, with God's permission, they lead forth new species of things from them." In this fashion, the ability of demons to perform marvels is preserved, without allowing them any supernatural power over the material world. All that they can do is join natural agents with natural patients, although their superior knowledge and speed allow them to do this more effectively than humans can. At this point, Albert reintroduces alchemy, now as a support for the power of demons. He begins by referring to a passage from Job (41:33), where the power of Leviathan is said to exceed that of other beings on earth: "Likewise, Job 41: there is no power on earth which can be compared to him. Therefore it seems that if the power of art worked in the transmutations of bodies, as in alchemy, that demons would be able to do this much more powerfully."[32] Albert's introduction of alchemy at this point makes an appeal to the chrysopoetic art that is entirely distinct from his earlier reference to the *Sciant artifices*. In the earlier passage, Albert used the *Sciant* in his negative arguments to show that demons worked by illusion only, since art cannot genuinely transmute species. Now, to the contrary, he takes it as a given that alchemy can indeed transmute species, and works outward from that point. If humans can actually transmute species, it follows that demons, who are much more powerful, can also do so. The implicit assumption behind this use of alchemy is absolutely clear. In terms of its claims, alchemy is the summum bonum of the human arts. As the apex of human artistry, alchemy serves as the high-water mark against which demonic power must be measured. This use of alchemy as a test case of the human ability to alter the natural world would have far-reaching consequences. Although Albert may or may not have been the first to use alchemy in this fashion in the tradition of the *Sentence* commentaries, his is an early example of a tradition that would continue to bud and ramify well into the seventeenth century.

After finishing with his list of reasons against and for the ability of demons to induce a substantial form in matter, Albert tries to resolve the issue to the best of his ability. In his solution to the question, he modestly admits that only God and the angels can know for a certainty whether demons have this power. Nonetheless, Albert asserts that the doctrines of churchly authority allow one to suppose that demons cannot induce a permanent substantial form into matter except in the case of beings that arise easily from putrefaction. To clarify his analysis, he argues that four types of transmutation are possible: first, where the ingredients of a mixture retain their identity and operation, while also working in unison to produce a new effect, as in the medicinal effect of theriac; second, the case where a body is dissolved into its components, as when fire resolves a body; third, the type of transmutation effected by alchemy; and fourth, the case where nature itself converts one substance into another, as when frogs and toads appear spontaneously, and without parents. Does Albert's inclusion of alchemy in this list mean that he believes in the power of the chrysopoetic art? His response is surprisingly thorough:

> The third [type of transmutation] occurs through the stripping off of properties, and the imposition of others through liquefaction, cibation, sublimation, and distillation, which the alchemists effect: and in this fashion by means of a quite well-known operation, bread, ink, and the like come into existence. I impute that [alchemists] do not give substantial forms, as Avicenna says in his alchemy: the sign of which is that one does not find the properties comprising the species in the things produced thus. For this reason, alchemical gold does not benefit the heart, and an alchemical sapphire does not cool off sexual ardor, or cure an affection of the windpipe [*arteriaca*]; nor does an alchemical carbuncle dispel a vaporous poison. And the test [*experimentum*] of all these things lies in the fact that alchemical gold is consumed more in the fire than the other, and also precious stones produced by alchemy; and likewise they do not last as long as the natural ones of that species. This is because they do not have the specific form [*species*], and so nature has denied them the virtues that are given with the specific form for the conservation of the same.[33]

In this extraordinary passage, Albert reveals a surprisingly negative attitude toward alchemy, given that his later *Liber mineralium* would provide a mechanism by which alchemists could indeed produce precious metals.[34] In his *Sentence* commentary, Albert evidently accepts the Avicennian position that alchemists cannot work real transmutation, but can only strip off transient accidents and replace them with equally superficial ones. It is for this reason that alchemical gold lacks the medical effect of strengthening the heart that medieval physicians granted to natural gold, and the artificial carbuncle and sapphire lack the marvelous powers ascribed to their natural counterparts.[35] But the real proof of their falsity lies in the inability of alchemical gold and precious stones to resist the dissolutive power of fire. Interestingly, this point emerges again in the *Liber mineralium*, where Albert says that he has tested alchemical gold and found it to decompose after six or seven firings.[36] In the later text he ascribes this to the shortcoming of alchemical practitioners, however, while in the *Sentence* commentary he apparently views it as a weakness of the art itself.

As we can see, then, Albertus Magnus was an early representative of a tradition that interjected alchemy into the discussion of demonic power. The reasons for this strange introduction are two. First, the *Sciant artifices* could be taken—and was taken by Albert—as a statement about the limits of art in general, not just alchemy. On the assumption that art cannot transmute species, it followed that demons could not really perform the wonders that were ascribed to them, at least not by art. Second, if one denied or ignored the *Sciant artifices*, then alchemy became the paragon of human artifice, since it actually claimed to replicate natural products rather than restricting itself to ersatz imitation. On this basis, alchemy became the benchmark against which all arts should be measured, including those of demons. If humans could transmute species, then so could demons, and more so. Albert's own position, as we saw above, is that alchemy cannot transmute species, at least not by the method of stripping off accidents and imposing new ones. Near the end of the question, however, he hints that there may be another way by which alchemy can really transmute metals by aiding nature—"art of itself cannot induce a form, as was said before, but it can help nature."[37] He may have in mind the same idea that he would describe later in the *Liber mineralium*—that proper alchemists act toward metals as doctors do toward their patients. The alchemist first cleans and purifies the old metal, just as a doctor employs emetics and diaphoretics to purge his patient. Then he strengthens the elemental and celestial powers in the metal's substance. As a result, the purged metal receives a new and better specific form from the virtues of the celestial bodies. The alchemist, then, has not really transmuted any species: he has only removed one specific form and prepared the way for another to be received.[38]

Albert's distinction between a purely artificial art that works by itself, and an art that aids nature, was also adopted by his student Thomas Aquinas. Thomas, who began lecturing on the *Sentences* in 1252, takes much the same position as Albert in his own commentary on book 2, distinction 7. Thomas begins his question by asking "whether demons can induce a true corporeal effect in corporeal matter?" Like Albert, he then proceeds to give a list of negative answers. A number of these are astrological, and have no concern for us. Alchemy soon appears, however, in the fifth *responsio quod non*, which is the direct source for the attack on alchemy found on folio one of the *Malleus maleficarum*: "Moreover, demons do not work except by the method of art. But art cannot give a substantial form, whence it is said in the chapter *on minerals*: 'the artificers of alchemy should know that species cannot be transmuted.' Therefore, neither can demons induce substantial forms."[39] Having introduced the *Sciant artifices* in exactly the same form as Albert, assuming it to be a pronouncement on the limits of art in general, Thomas then passes immediately to the strongest piece of contrary evidence—that on the authority of Exodus 7, Pharaoh's magicians really did convert their staffs into snakes. Thomas's solution to the problem is as follows.

Demons cannot act on matter by means of their minds alone, as God can. Despite the belief of Avicenna that matter automatically "obeys" separated substances like demons and angels, a view that the Persian philosopher expressed in his commentary on Aristotle's *De anima*, demons can only act on matter by means of art. They are limited to the application of agents to patients, just as humans are. Hence if a demon wishes to heat up a portion of matter, he cannot do it by means of his own power alone, but must subject the matter to fire. Now since Thomas takes the position that the demons must act by means of art, the issue of alchemy—the art of transmuting species par excellence—acquires considerable significance for him. He therefore considers it by drawing the analogy of fire applied to wood, which we already met at the end of the *Malleus maleficarum's* first *quaestio*:

> Art by its own power cannot confer a substantial form, but it can do this by means of a natural agent, as is clear in the following—that the form of fire is produced in logs through art. There are some substantial forms, however, which art cannot induce by any means, since it cannot find the proper active and passive subjects. Even in these art can produce a similitude, as when alchemists produce something similar to gold as to exterior accidents. But it is still not true gold, since the substantial form of gold is not [induced] by the heat of fire—which alchemists use—but by the heat of the sun in a determinate place where the mineral power flourishes. Hence such [alchemical] gold does not operate according to the specific form [of real gold], and the same is true for the other things that they [i.e., alchemists] make.[40]

Like Albert, Thomas clearly distinguishes between a purely artificial art that works "by its own power," and one that induces a substantial form by working on nature. Thomas's rejection of alchemy is similar to that of Albert except that the former introduces the concept of the *virtus loci*—the power of a specific place. His idea is that metals can only be generated by natural heat operating in the subterranean chambers where ores and metals come into being. It is a priori impossible for humans to make metals artificially, since they cannot erect their laboratories in the hidden subterranean depths where the "mineralizing power" operates with the aid of solar heat. Like Albert, however, Thomas is using alchemy to determine the limits of demonic power. Since humans cannot induce just any substantial form on matter, it follows that demons are subject to a similar limitation. Alchemy once again serves as the touchstone by which all arts, including those of Lucifer and his minions, are measured.

The tradition of using alchemy as a benchmark for the arts is also found in the early *Sentence* commentary of Saint Bonaventure, who composed his work between 1250 and 1253 (probably somewhat earlier than Thomas); he also takes the subject up in his treatment of book 2, distinction 7.[41] Like Albert and Thomas, Bonaventure asks "whether demons can induce true forms of things in matter?" His concern, again, is with the serpents and other animals seemingly produced by the magicians of Pharaoh. His third affirmative reason introduces alchemy into the discussion: "Likewise, the power of demons is greater than that of man through artifice; but men make the species of diverse metals by means of the art of alchemy: therefore demons can do this much more powerfully."[42] Although Bonaventure does not expend much more time on alchemy in his commentary, it is clear that he, unlike Albert and Thomas, finds its claims to be unproblematic. His acceptance is based on the fact that he distinguishes purely artificial actions from those where art and nature cooperate in a more thorough way than the other two writers maintain. In purely artificial things, Bonaventure insists, the agent imparts nothing to the patient, but either removes matter or changes its position, as appears in the case of a carving. Thus an agent cannot produce natural forms by its own power, unless the agent is pure act, as is God. Hence demons act only as *ministri* to nature, assistants rather than principal agents. Otherwise they would create things that differ from themselves in name and species, and the demons "would produce just as the Creator does, and thus they would be Creators."[43] Alchemists too administer agents to patients, in the same way as demons: they do not act in a purely artificial fashion, but employ their art to lead nature to an end that it would otherwise fail to attain. Once again employing the seminal reasons of Augustine, Bonaventure says that alchemists and demons do not produce their marvels by means of their own power, but by that of the "seeds" that they assemble and coax into their full maturity.[44]

## Conclusion

In conclusion, it is clear that the theologians of the thirteenth century initiated a tradition of discussing alchemy in the context of demonic power—not because alchemy was a form of magic, but because it represented the apex of the arts in its relationship to nature. To the writers whom we have considered, "magic" (*magia*) automatically meant the work of demons, which did not apply to alchemy as such, although demons, like humans, could certainly devote themselves to the transmutation of metals. The *Sentence* commentators found alchemy useful precisely because it was *not* in itself demonic, but an art known to humans—it could therefore be used as a yardstick to assess the things that demons could or could not do. This use of alchemy would live on in the *Sentence* commentaries of later writers, such as Richard of Middleton and Robert Kilwardby, who add little that we have not already discussed.[45] At the same time, however, the treatment of alchemy in relation to other arts provided by theological writers and by alchemists themselves spread out into different genres of literary production that I cannot discuss here, including vernacular poetry, treatises on painting and other artistic pursuits, and discussions on the possibility of manufacturing artificial humans and other forms of life.[46] As we have also seen from the example of Kramer and Sprenger's *Malleus maleficarum*, the discussion of alchemy begun in the *Sentence* commentaries of the thirteenth century had another repercussion as well. In the hands of the two Dominican inquisitors, alchemy became yet another tool for dismantling the limitations placed on demonic power by skeptical writers of the Middle Ages. Kramer and Sprenger, intent on aggrandizing the power of witches, weaken the Thomistic and Avicennian argument that limited the alchemists' ability to alter the nature of matter. By loosening the bonds of Avicenna's *Sciant artifices*, the witch hunters liberated their own diabolical quarry from the inability to impose "true forms" and hence wreak havoc on the world. As the beneficiaries of such gargantuan power, the witches clearly had to be destroyed, resulting in a call to action that we know all too well from the dismal history of the great witch hunt. Although the role of alchemy in this persecution was at most minor, it is testimony to the image of the aurific art as an exemplar of humankind's artisanal power in the natural world that Avicenna's debunking of alchemy is itself appropriated and in turn deflected in the *Malleus maleficarum*.

## Notes

1. Vincent of Beauvais, *Speculum doctrinale* (Douai: Baltazar Belierus, 1624), book 11, chapters 105–133, columns 1054–1072; Vincent of Beauvais, *Speculum naturale* (Douai: Baltazar Belierus, 1624), book 7, 426–449; Arnoldus Saxo, *Die Encyklopädie des Arnoldus Saxo*, ed. Emil Stange (Erfurt: Fr. Bartholomäus, 1906), 41–45.

2. On the widely discussed limits of demonic art, see for example the fifteenth-century Alfonso Spina, *Fortalitium fidei*, in Joseph Hansen, *Quellen und Untersuchungen zur Geschichte des Hexenwahns und der Hexenfolgung im Mittelalter* (Hildesheim: Olms, 1963; photoreproduction of Bonn, 1901), 148. See also Stuart Clark's excellent discussion of this issue in his *Thinking with Demons: The Idea of Witchcraft in Early Modern Europe* (Oxford: Clarendon Press, 1997), 161–172. As Clark points out (p. 165), paraphrasing the sixteenth-century writer on demonology Otto Casmann, "Devils were unable to produce any substantial form, create anything from nothing, make anything out of anything else, produce any effect they pleased from any cause or by any instrument, transform any natural thing into any other, or produce perfect living beings without seed." Early modern writers such as Spina and Casmann were beneficiaries of a long scholastic discussion where alchemy had originally played a quite significant role. For the Augustinian roots of Christian demonology, see also Hans Peter Broedel, *The* Malleus maleficarum *and the Construction of Witchcraft* (Manchester: Manchester University Press, 2003), especially 41–43.

3. See my book on alchemy and the art-nature debate, *Promethean Ambitions: Alchemy and the Quest to Perfect Nature* (Chicago: University of Chicago Press, 2004), chapter 2. See also Sister Mary Richard Reif, *Natural Philosophy in Some Early Seventeenth Century Scholastic Textbooks*, Ph.D. dissertation, St. Louis University, 1962, 238.

4. Montague Summers, trans., *Malleus maleficarum* (London: Pushkin Press, 1948, rpt. 1951), xvi.

5. Joseph Hansen, *Zauberwahn, Inquisition und Hexenprocess im Mittelalter und die Entstehung der grossen Hexenverfolgung* (Munich and Leipzig: R. Oldenbourg, 1900); Joseph Hansen, *Quellen und Untersuchungen zur Geschichte des Hexenwahns und der Hexenverfolgung im Mittelalter* (Bonn: C. Georgi, 1901); Walter Stephens, *Demon Lovers: Witchcraft, Sex, and the Crisis of Belief* (Chicago: University of Chicago Press, 2002).

6. Throughout this chapter, I have provided my own translations of the *Malleus maleficarum*. Although the new edition of the *Malleus* by Christopher S. Mackay is accompanied by an English translation, the translator has treated the scholastic vocabulary of the text in a way that is at times too free for the purposes of close analysis. See Henricus Institoris, O.P., and Jacobus Sprenger, O.P., *Malleus maleficarum* (Cambridge: Cambridge University Press, 2006). For the Latin, I have relied on the facsimile edition of the *editio princeps* of the *Malleus* published by André Schnyder. Any instances where I have taken Mackay's textual emendations into account are indicated in the notes. André Schnyder, *Malleus maleficarum von Heinrich Institoris (alias Kramer) unter Mithelfe Jakob Sprengers Aufgrund der Dämonologischen Tradition Zusammengestellt* (Göppingen: Kümmerle Verlag, 1991), 7: "Item demones non operantur nisi per artem. Sed ars non potest dare veram formam. Unde in c. de mineris dicitur Sciant auctores alchimie species transmutari non posse[.] Ergo et demones per artem operantes veras qualitates sanitatis aut infirmitatis inducere non possunt. Sed si vere sunt habent aliquam aliam causam occultam absque opere demonum et maleficorum." The first three sentences of the passage are taken more or less verbatim from Thomas Aquinas. See Thomas Aquinas, *Scriptum super libros sententiarum magistri Petri*

*Lombardi episcopi parisiensis*, ed. R. P. Mandonnet, O.P. (Paris: P. Lethielleux, 1929), vol. 2, distinction 7, question 3, articulus 1, 193–194.

7. See, for example, Brian P. Levack, *The Witch-Hunt in Early Modern Europe* (London: Longman, 1995), 7: "The most common forms of high magic are alchemy, which is the changing of base metals into precious ones, and divination (also known as conjuring), which is the use of various means to acquire secret or otherwise unknown knowledge"; Robin Briggs, *Witches and Neighbors: The Social and Cultural Context of European Witchcraft* (London: HarperCollins, 1996), 70. Briggs lumps "such pseudosciences as astrology, natural magic, and alchemy" together as parts of the supposed worldview rooted in the "seedbed of error" provided by contemporary natural philosophy.

8. Edward Peters, *The Magician, the Witch, and the Law* (Philadelphia: University of Pennsylvania Press, 1978), 72–73. An English translation of the *Canon episcopi* taken from Henry Charles Lea may be found in Alan C. Kors and Edward Peters, *Witchcraft in Europe 1100–1700: A Documentary History* (Philadelphia: University of Pennsylvania Press, 2001), 61–63. The Latin text of the *Canon episcopi* is found in *Corpus iuris canonici*, ed. Emil Friedberg (Graz: Akademische Druck, 1955), columns 1030–1031.

9. The *Canon episcopi's* transformation "into better or worse" is specifically referred to health and disease by Kramer and Sprenger at Schnyder, *Malleus*, 11: "Secundum etiam quod in melius deteriusve valeat transmutari intelligatur solumodo a deo authoritative et ad correctionem seu etiam punitionem. Sepius tamen ista ministerio demonum exercentur. Et sicut de primo dicitur. Dominus percutit et ipse medetur. Et ego occidam et ego vivere faciam. Ita de secundo dicitur immisionem per angelos malos ut supra tactum est."

10. Thomas Aquinas, *Scriptum super libros sententiarum magistri Petri Lombardi episcopi parisiensis*, ed. R. P. Mandonnet, O.P. (Paris: P. Lethielleux, 1929), vol. 2, distinction 7, question 3, articulus 1, 193–194.

11. André Schnyder, *Malleus maleficarum*, 7.

12. Kors and Peters, *Witchcraft in Europe*, 62–63.

13. André Schnyder, *Malleus maleficarum*, 11: "Hec tres partes si nude intelligantur sunt contra processum scripture et determinationem doctorum. Nam posse fieri aliquas creaturas a maleficis utpote vera animalia imperfecta. Inspiciatur sequens canon. Nec mirum post allegatum canon. Episcope quid Augustinus determinat de magis pharaonis qui virgas in serpentes verterunt inspiciatur glosa super illud Exo. vii. Vocavit pharao sapientes." Later in the text, they do explicitly deny that humans can transmute higher species, such as the more "perfect" animals that are not capable of undergoing spontaneous generation, as at Schnyder, *Malleus*, 119: "De primis loquitur canon et precipue de formali seu quidditativa transmutatione prout una substantia in aliam transmutatur. Cuiusmodi solus deus qui talium quidditatum creator existit facere potest."

14. Stephens, *Demon Lovers*, 289–321.

15. André Schnyder, *Malleus maleficarum*, 13: "Dicendum secundum sanctum thomam in ii. di. vii. in solutione unius argumenti. ubi declarat de virtute demonum in operando. licet quedam forme substantiales per artem induci possint virtute naturalis agentis ut quando forma ignis inducitur per artem in lignum hoc tamen non potest fieri universaliter eo quod ars non potest invenire semper seu coniungere propria activa propriis passivis potest tamen facere aliquid simile et sic alchimiste faciunt aliquid simile auro quantum ad accidentia exteriora sed tamen non faciunt verum aurum. Quia forma substantialis auri non est per calorem ignis quo utuntur alchimiste sed per calorem solis in loco determinato ubi viget virtus mineralis. et ideo tale aurum non habet operationem consequentem speciem et simile est de aliis eorum operationibus."

16. Schnyder, *Malleus maleficarum*, 13: "Demones operantur per artem circa effectus maleficiales et ideo absque amminiculo alterius agentis nullam formam substantialem vel accidentalem inducere possunt et quia non dicimus quod maleficia inferat partem absque amminiculo alterius agentis. Ideo etiam cum tali amminiculo potest veras qualitates egritudinis aut alterius passionis inducere." The troubling phrase "non dicimus quod maleficia inferat partem absque amminiculo alterius agentis" is clearly ungrammatical as it stands in the 1487 editio princeps (Schnyder, 13). I have consulted the 1574 Venice edition, *Malleus maleficarum in tres divisus partes* (Venetiis: Apud Io. Antonium Bertanum, 1574), 14, which alters "inferat" to "inferant," but leaves the problematic "partem," now acting as the object of "maleficia inferant." It is more sensible, in my view, to suppose that "partem" is a misprint or misreading of "ꝑartem." I note that Mackay has proposed the same emendation in his edition of the Latin text. See Mackay, *Malleus*, vol. 1, p. 229, note c.

17. Thomas Aquinas's desire to limit demonic power is well known. See Peters, *The Magician, the Witch, and the Law*, 95.

18. Thomas Aquinas, *Scriptum super libros sententiarum magistri Petri Lombardi episcopi parisiensis*, ed. R. P. Mandonnet, O.P., vol. 2, distinction 7, question 3, articulus 1, 1936: "Sed quaedam formae substantiales sunt quas nullo modo ars inducere potest, quia propria activa et passiva invenire non potest."

19. Kramer and Sprenger do not explicitly argue for the legitimacy of alchemical transmutation, presumably because such an argument could put them at loggerheads with the *Canon episcopi*'s injunction against the transmutation of "perfect" beings. In a later section of the *Malleus maleficarum*, they distinguish between "transmutatio substantialis" and "transmutatio accidentalis." The former, which would involve a quidditative transmutation of one substance into another, can only be performed by God. The second, as when leprosy is imposed on someone's face, can be induced by demons. See Schnyder, *Malleus maleficarum*, part 2, question 1, chapter 8, 119.

20. William R. Newman, "Technology and Alchemical Debate in the Late Middle Ages," *Isis*, 80(1989): 423–445. For a deeper treatment of this issue, see Newman, *Promethean Ambitions*, 34–114.

21. William Newman, *The "Summa Perfectionis" of Pseudo-Geber* (Leiden: Brill, 1991), 49–50. For the Arabic text and a translation therefrom, see Avicenna, *Avicen-*

*nae de congelatione et conglutinatione lapidum*, ed. and trans. E. J. Holmyard and D. C. Mandeville (Paris: Paul Geuthner, 1927), 85–86, 41–42.

22. The notion that art is unequal to nature was already a commonplace in antiquity. Examples abound in Cicero's *De natura deorum*, as at I, 92, II, 35, 57, and passim. For other examples of this ancient belief, see A. J. Close, "Commonplace Theories of Art and Nature in Classical Antiquity and in the Renaissance," *Journal of the History of Ideas* 30(1969): 467–486.

23. Newman, *Summa perfectionis*, 51: "Hec compositio in aliam mutari non poterit compositionem nisi forte in primam reducatur materiam, et sic in aliud quam prius erat permutetur."

24. Newman, "Technology and Alchemical Debate."

25. Fridericus Stegmüller, *Repertorium commentariorum in sententias petri lombardi* (Würzburg: Ferdinand Schöningh, 1947), vol. 1, 25.

26. For the problematic character of this modern habit, see William R. Newman and Anthony Grafton, "The Problematic Status of Astrology and Alchemy in Premodern Europe," in Newman and Grafton, *Secrets of Nature: Astrology and Alchemy in Early Modern Europe* (Cambridge, MA: MIT Press, 2001), 1–38.

27. Peter Lombard, *Sententiae in IV libris distinctae* (Quaracchi: Collegium S. Bonaventurae ad Claras Aquas, 1971), vol. 1, part 2, chapters 6–8, 362–364.

28. For the influence of Exodus 7 on medieval discussions of magic, see Valerie I. J. Flint, *The Rise of Magic in Early Modern Europe* (Princeton, NJ: Princeton University Press, 1991), 18–19, 29, 45, and passim.

29. Albertus Magnus, *Beati Alberti Magni, Ratisbonensis episcopi, ordinis praedicatorum, commentarii in II. et III. lib. sententiarum*, ed. Pierre Iammy (Lyon: Huguetan et al., 1651), vol. 15, 86: "5. Item, Ars non transmutat a forma substantiali in formam, quia dicit Arist. In 4. Metheo. Sciant artifices alchimiae species transmutari non posse: ergo nec daemones, quia ipsi non operantur nisi per modum artis."

30. Albertus Magnus, *Book of Minerals*, trans. Dorothy Wyckoff (Oxford: Clarendon Press, 1967), 170, 177.

31. The Arabic text of the *De congelatione* printed in E. J. Holmyard and D. C. Mandeville, *Avicennae de congelatione et conglutinatione lapidum* (Paris: Paul Geuthner, 1927), 41, 85, does not use the term for "metals" either, opening up the possibility that Albert had a manuscript of the *De congelatione* lacking the Latin word *metallorum*.

32. Albertus Magnus, *Commentarii in II. et III. lib. sententiarum*, vol. 15, 86: "4. Item Iob 41. Non est potestas super terram quae possit ei comparari: ergo videtur, quod si potestas artis operetur super corporum transmutationes, ut alchimia, quod daemones hoc multo magis facere praevaleant."

33. Albertus Magnus, *Commentarii in II. et III. lib. sententiarum*, vol. 15, 86–87: "Tertia est per expoliationem proprietatum, & dationem aliarum per liquefactionem &

cibationem & sublimationem & distillationem, quibus operantur alchimici: & hoc modo operatione satis nota fit panis, & attramentum, & huiusmodi. Et puto quod non dant formas substantiales, sicut dicit Avicen. in alchimia sua: cuius signum est: quia in talibus operatis non inveniuntur proprietates continentes speciem: unde aurum alchimicum non laetificat cor, & saphirus alchimicus non refrigerat ardorem, neque curat artheriacam & carbunculus alchimicus non fugat venenum vaporabile in aere. At omnium talium experimentum est in hoc quod aurum alchimicum consumitur plus in igne quam aliud, & similiter lapides alchimici: & iterum non durant ita diu sicut naturalia illius speciei. Et hoc ideo est: quia non habent species, & ideo negavit eis natura virtutes quae dantur cum speciebus ad conservationem specierum." The so-called "alchemy" of Avicenna refers to the pseudonymous *Epistola ad Hasen*, another text translated from Arabic. See Robert Halleux, "Albert le grand et l'alchimie," in *Revue des sciences philosophique et théologiques* 66(1982): 57–80.

34. Albertus Magnus, *Book of Minerals*, 177–179. See also Halleux, "Albert le grand et l'alchimie," 74–75.

35. For more on Albert's views regarding the powers of these precious stones, see Albertus Magnus, *Book of Minerals*, 77–78, 115–116.

36. Albertus Magnus, *Book of Minerals*, 179.

37. Albertus Magnus, *Commentarii in II. et III. lib.sententiarum*, vol. 15, 87: "Ad aliud dicendum, quod bene potest esse, quod ars de se non potest inducere formam, ut prius dictum est: sed potest iuuare naturam, & ita facit daemon."

38. Albertus Magnus, *Book of Minerals*, 178–179.

39. Thomas Aquinas, *Sancti Thomae Aquinatis commentum in secundum librum sententiarum*, in *Sancti Thomae Aquinatis opera omnia* (Parma: Petrus Fiaccadorus, 1856), vol. 6, 450: "Praeterea, daemones non operantur nisi per modum artis. Sed ars non potest dare formam substantialem; unde dicitur in cap. de numeris: sciant auctores alchimiae, species transformari non posse. Ergo nec daemones formas substantiales inducere possunt." The phrase "de numeris" must obviously be read as "de mineris."

40. Aquinas, *Commentum in secundum librum sententiarum*, vol. 6, 451: "Ad quintum dicendum, quod ars virtute sua non potest formam substantialem conferre, quod tamen potest virtute naturalis agentis; sicut patet in hoc quod per artem inducitur forma ignis in lignis. Sed quaedam formae substantiales sunt quas nullo modo ars inducere potest, quia propria activa et passiva invenire non potest, sed in his potest aliquid simile facere; sicut alchimistae faciunt aliquid simile auro quantum ad accidentia exteriora; sed tamen non faciunt verum aurum: quia forma substantialis auri non est per calorem ignis quo utuntur alchimistae, sed per calorem solis in loco determinato, ubi viget virtus mineralis: et ideo tale aurum non habet operationem consequentem speciem; et similiter in aliis quae eorum operatione fiunt."

41. *Bibliotheca sanctorum* (Rome: Istituto Ciovanni XXIII, 1963), vol. 3, 242, for the date of Bonaventure's *Sentence* commentary.

42. Saint Bonaventure, *Commentaria in quatuor libros sententiarum magistri*, in *Petri Lombardi Doctoris seraphici S. Bonaventurae opera omnia* (Quaracchi: Collegii S. Bonaventurae, 1885), vol. 2, 201.

43. Bonaventure, *Commentaria in quatuor libros sententiarum*, vol. 2, 202.

44. Bonaventure, *Commentaria in quatuor libros sententiarum*, vol. 2, 202: "1. 2. 3. Unde tres rationes primae verum concludunt, quoniam non probant, quod faciant *virtute sua*, sed virtute seminum adductorum."

45. Richardus de Mediavilla, *Clarissimi theologi magistri Ricardi de media villa seraphici ord. Min. convent. Super quatuor libros sententiarum* (Brixia: De consensu superiorum, 1591; rpt., Frankfurt am Main: Minerva, 1963), vol. 2, 99–100. Robert Kilwardby, *Quaestiones in librum secundum sententiarum*, ed. Gerhard Leibold (Munich: Verlag der Bayerischer Akademie der Wissenschaften, 1992), 133.

46. The influence of the alchemical art-nature debate on other genres forms the topic of Newman, *Promethean Ambitions*.

# 6

## Forms of Art in Jesuit Aristotelianism (with a Coda on Descartes)

Dennis Des Chene

The seventeenth century witnessed a great revision in the relations of art and nature. It was not uniform, or unopposed, or even perhaps inevitable. Things might have gone against Descartes, Boyle, and the other mechanical philosophers. The current of thought might have flowed the way of those who, like Leibniz, wanted to retain some part of the scholastic framework. As it is, the first phenomenon the historian must acknowledge is that of an apparently irreversible change in conceptions of human industry and natural agents, along with their effects.

That change is bound up with, intellectually and socially, the decline of Aristotelianism in the learned world of Europe. Other traditions had a part, notably the Neoplatonism with which Aristotelianism became tinged as it was reintroduced into the universities of Europe in the thirteenth century. Nevertheless, for many proponents of the new science, Aristotelianism was the ground against which they drew the new figures of knowledge.[1] The bulk of this chapter will be devoted to one prominent group of Aristotelians, the Jesuits. By virtue of their educative mission and the texts written in support of it, their version of Aristotelianism was often taken to represent the Schools generally. This is certainly true for Descartes and many of his followers; Boyle seems to have drawn on Suárez in his arguments against substantial form; Leibniz in his dissertation cites primarily Jesuits and their Scotist opponents. When "the Schools" came to designate, in the later seventeenth century, an undifferentiated mass of obsolete opinion, that mass was likely to be Jesuit in composition, especially in Catholic Europe.

The early texts of Jesuit Aristotelianism were commentaries. They included, as was customary, disputational questions on topics suggested by the text, and were thus constrained to follow the sometimes wayward thematic order of the works they commented on. In the early seventeenth century, commentaries gave way to the four-part *cursus*, which while retaining the disputational format, forsook Aristotle's order for a more systematic arrangement of topics. Among the questions on physics there was usually a question on the relations of nature and art. In its older form the question, based on a passage in *Physics* 2c2, was whether art imitates nature. So we find it in Toletus, the earliest of the texts discussed here. In later works, like that of the Coimbran commentary or Roderigo Arriaga's cursus, the question varies: the Coimbrans compare and contrast art and nature; Arriaga subsumes a question on the forms of artifacts under a section on the definition of nature, as does the Dominican John of St. Thomas. Honoré Fabri's *Physica*, unusual here as elsewhere, omits the question altogether. But the very late textbook of Antonio Mayr, published in 1729, includes a lengthy discussion of artificial form, alchemy, and magic. At present the most I can say about the history of the question is that in later works reflections on art no longer depended exclusively on the notion of imitation.

## The Character of Art

In *Physics* commentaries, questions or disputations on art and nature are appended to one of two passages in which Aristotle contrasts art and nature. "Art" here is taken broadly, in the sense of *technē*—any human craft or skill, the exercise of that craft, or its products. In the first passage, Aristotle, having argued that the form of a thing is to be included in its nature, turns to matter. That too must be included, he says, in its nature. The "sign" of this is that when a wooden bed, discarded and lying in earth, sends forth a shoot, it is not the bed qua shape but the bed qua wood that has done so. The word *bed* therefore comprises both a form and a matter; and as with artifacts, so too with the things of nature. Later, arguing that physics ought to give an account of both form and matter, Aristotle writes that "since art imitates the ingenuity of nature, so far as it is able, art is to the artifact what natural science is to physical things" (*Phys.* 2c2). Human art, when it imitates nature, and thereby expresses a practical grasp of it, *already* takes both form and matter into account. Physics ought to emulate art, and in its theoretical grasp include both form and matter.

Art—the skill of the maker—is to artifacts what natural science is to the things of nature. Since art is said to imitate nature, one might suppose that the understanding artists possess concerning the form and matter of what they make might be useful to physics. Bacon and Descartes certainly thought so. But in Aristotelian physics the utility of art in generating knowledge by analogy is strictly limited. Art is secondary, superficial, and subordinate.

**Secondary**

Art depends on the creative act of God and the generative acts of nature. God realizes in matter the exemplars or divine ideas of the forms of things. The act of creation thus resembles human art sufficiently that it may itself be called art. The active powers of nature, in generation, execute the divine plan, yielding individual substances composed of prime matter and substantial form, which is supported and embellished by suitable qualities. Art in turn then operates on the finished substances of nature—wood, metal, bone, and so forth. Unlike God, it cannot create from nothing; unlike nature, it cannot reduce an existing substance to prime matter and give it a new substantial form.

Art imitates nature both in operation and in the outward form of its products, which are doubly derivative: first from the forms of nature, and ultimately from the exemplars of those forms in the divine understanding:

> And as nature emulates divine art, so human art emulates nature, insofar as that can be done: for which reason divine art is said to be the exemplar of nature, and nature both the exemplum of a divine archetype and the exemplar of human art; and by some [philosophers] human art is usually called sometimes another nature, sometimes the ape of nature, since it fashions many things from nature, as pictures from shadow, temples from caves, sails from the flight of birds, oars from the fins of fish, and rudders from tails. (Coimbra *Physics*, 2c2q6; 1:215[115])

In imitating nature, human art imitates nature's imitation of itself. Imitation is universal. All natural agents emulate their Creator, but each of them also imitates those superior to it. The lifeless earth imitates the aspect of animals in its mountains and rivers. Sublunary things, in their cycles of generation and corruption, imitate the immutable periods of the heavens. The heavens themselves imitate "other more divine causes"—the intelligent substances that govern them.

It is not only the arts of depiction that imitate nature. Other arts—carpentry, tailoring, cobblery—though they do not imitate preexisting natural effects, imitate those that "ought to have preexisted" and strive to fashion them as nature would have (*Phys.* 2c2q6; 4:54v). The claws, fur, and fins that nature gave to animals, nature supplies to us by giving us an intellect capable of conceiving universally the forms of all those tools; so that human art, though dependent on nature for its materials, can attain to a more direct imitation of the divine mode of production.

**Superficial**

The products of human art are secondary in both matter and form. The forms they imitate, moreover, are not the substantial forms of things, but only their figures, their outward shapes. The species of nature are distinguished for our senses not by qualities like heat and cold, nor by quantity, but only by figure:

> And this was done according to the highest wisdom. Since man, the king of this world, for whose sake [the species of nature] were made, depends for his understanding on the senses, and since the senses do not perceive substances except through their accidents, clearly man would always be deceived and would be unable to distinguish the forms or substances of things, unless there were some sensible accident by which such substantial forms were designated; such are the figures and external forms of things. (Toletus *Phys.* 2q6, 4:54v)

Figures follow forms. They are the attendants of form, the indices of substance. Human art cannot lend to matter new substantial forms. To imitate the effects of natural substances on vision it can only exhibit the signs of form—that is, the shapes of things. Again this holds not only for the arts of depiction but also to some degree for the other arts. The rudder of a ship may look like a tail, but it is only by courtesy an organic part of the ship. Art stays at the surface: "As Plotinus observes . . . , and Averroes . . . , art directs itself primarily to the outside surface, because it affects most strongly the external, finished face [of things]. Nature, on the other hand, occupies itself much more with hidden workings" (Coimbra *Phys.* 2c1q6a2; 1:217[117]). Roderigo Arriaga, writing more than a generation later than the Coimbrans, holds that the forms of art consist merely in the locations [*ubicationes*] of preexisting substances. All we can change is the *ubi*, the "where," of various bits of stuff:

> This is more clearly understood in a wooden or stone statue: for precisely by the removal of superfluous or impeding parts . . . the remainder, without any new form, retain the the distances and expressive proportion of a man, and produce the statue; therefore no artificial form is produced that is distinct from the parts of wood so located and the absence of redundant parts. (*Phys.* disp6§2; 319)

The forms of artifacts, then, are simply displacements of their materials. In the Aristotelian world, unlike the Cartesian, mere displacement, whether by nature or by humans, cannot yield any new natural kind.

There are, of course, arts that do not operate merely by moving things around, like baking and cooking. In these, however, the human contribution amounts, as we will see, to the application of natural powers to suitable patients, as fire to dough. Nature produces whatever new forms are thereby generated; art merely provides the occasion.

**Subordinate**

Nature's powers are subordinate to God; human industry is subordinate both to nature and to God. The Coimbrans hold that "brings forth [things] from nothing: nature from the potential being: art from perfected being: God by creating, nature by generating, art by compounding or composing" (*Phys.* 2c1q5; 214[114]). Imitation, of course, already entails a kind of subordination: the act of the imitator is governed by the outward aspect of the thing imitated. In keeping with this, the universal imitation described earlier always is of the higher by the lower. There is, it must be admitted, a tension here. Humans are not, after all, inferior to the animals they imitate. But human *power*, the power proper to humans of realizing their conceptions in matter, is inferior to that of nature.

Since human industry is confined to rearrangements of natural materials and to furnishing occasions for natural powers to operate, human art is quite incapable of passing its own generative power on to its products:

> As Saint Thomas . . . says, from the forms induced by art the action of art does not follow as it does from forms brought forth by nature. A house does not give birth to another house, as a horse does another horse. The reason for this difference is that natural forms attain the same mode of being in the thing generated as in the progenitor, and thus the same power of acting issues from them; but artificial [forms] exist in one way in the mind and idea of the artisan, and in a different way in artifacts, as is evident. (*Phys.* 2c1q2a2; 216[116])

Natural forms, consequently, are "active [*actuosæ*] and as if alive." But the forms of art are "as if inert [*stolidæ*] and dead, having no effective force [*effectricem vim*]." They are nothing more than reshapings of things. But shape pertains to quantity, and quantity "of itself is idle [*ignava*] and inert; it is given by nature [to things] as if it were another matter, to sustain their accidents." Figure, which is a mere mode of quantity—one might say it was hardly a *thing* at all—inherits from quantity its passivity.

Art, in short, has no effects as such; and if the nature of a thing is, as Aristotle says, its principle of rest and motion, then artifacts have no natures.

## Art and the Work of Nature

The contrast could hardly be greater. Art deals only in the surfaces of complete substances, its forms barely deserve the name, and it is inert, more so even than the inanimate substances of nature. Nature, on the other hand, works from within, needs only prime matter for its material, and confers on its offspring the powers it itself possesses.

Yet there would seem to be instances in which art exceeds the limits thus set for it. The Coimbrans consider three cases: automata, magical figures and characters, and alchemy, to which they devote a special question.[2] The statues of Dædalus, the dove of Archytas, the animated stools of Apollonius of Tyana—a standard list—all seem miraculously to possess powers not unlike those of living things. Likewise the images and amulets produced by astrologers and other practitioners of natural magic seem even to exceed in their effects the powers of natural agents. And if the alchemists can generate, as they say, gold from base metal, then art will have managed not merely to relocate bits of stuff but to impose a new substantial form on prime matter.

But all this, save perhaps alchemy, is either fakery or can be ascribed to natural causes: "Neither art nor artificial form by its own power is capable of the work of nature" (Coimbra *Phys.* 2c1q7a2; 218[118]). Some of the reasons we have already seen. In the case of those arts that consist wholly in action, like dancing, the art itself is nothing more than a collection of precepts or rules in the mind of the actor. Those rules at most affect the manner of the action; they are not its efficient cause. The dancer leaps higher and more elegantly by art, but the force behind the leap issues from the dancer's soul, not from the art itself.

As for those arts that reshape things, they too at most modulate the actions of natural agents: "So also when something which was blunt is by art made sharp [*acutum*], it moves more quickly by benefit of that art, but not because the figure itself, blunt or sharp, concurs in the movement as an efficient cause, but rather the medium opposes itself more to a blunt body than to a sharp one" (Coimbra *Phys.* 2c1q7a2; 218[118]). This is true of figure generally, whether in art or nature. But art deals only in figure, and thus artifacts as such are never efficient causes. They only redirect the forces of nature in ways useful or pleasing to us.

Finally, medicine and other similar arts that consist in "bring[ing] active natures to bear on passive natures" do effect the works of nature. But it is clearly the active natures, and not the art itself, that produce the intended effects.

If witches and magicians seem to endow figures and characters with active powers, the actions they seem to bring about are sometimes due to "the industry of demons who at the sign [given by the witch] hasten by tacit or express agreement to play with the minds of men." The sign is only a visible manifestation of the witch's intention. In other cases ordinary natural causes are at work. The appearance of activity in artifacts is in that case owed to the concealment of those causes. So it is with automata, whose actions are brought about by "little machines hidden within," little machines that act in perfectly natural ways.

Concerning alchemy, the Coimbrans' conclusion is a bit of a surprise, considering all that has preceded it, not to mention their evident disdain for the art. "Even if," they write, "it is extremely difficult to produce true gold by chemical recipes, still it does not seem that one can judge it entirely impossible."

The arguments are a bit surprising as well. If art really is the "rival and imitatrix" of nature, then why should it not be capable of producing gold? An angel would know which ingredients to use and what temperament to give them, and when the influx of the stars would have the desired effect. The expertise required, though great, is finite. There is no reason to deny that humans can acquire it. It is not decisive, even, that in every previous claim to have made gold there was error or deceit. Human industry has discovered many things unknown to previous ages: gunpowder, the ship's compass, typography, the extraction of sugar from cane, flat glass.[3] Nor should we agree with Ægidius Romanus that gold has a native home outside of which it cannot be born. If gold has hitherto been made only under the earth, that is merely because only there have the requisite matter and the requisite agents been brought together. Gold has no homeland. The natural processes that produce it can occur, in principle, anywhere, even in the alchemist's den.

It is as if the Coimbrans had suddenly flung themselves forward into their own century from their customary place among the ancients, and for once felt the pressure placed on the old order of nature and art by new technologies. But the arguments, if not the examples, are taken from Albertus Magnus and Henry of Ghent, among others, to whom one could add Nicolas of Cusa. They are centuries old. The recognition here intimated, that human art can do more than imitate nature, was, though dominated by the trope of art as imitation, present within Aristotelianism itself.

The Coimbrans, however, immediately retrace their steps. They remind the reader that in fact no one has demonstrated the art of making gold. In every case the product was either not true gold or, if genuine, was surreptitiously introduced during the process. Alchemists therefore deserve their bad name. And if the day comes when gold is made by art, it will be by way of applying natural agents to suitable materials. Art itself will remain an inert bystander. Alchemy, like natural magic, tests the limits of art but does not exceed them.

## Cartesian Revisions

Surprisingly little, in a way, is required to transform the Aristotelian account of art and nature into a Cartesian account. Descartes no more attributes active powers to artifacts than do the Aristotelians. But for accidents of terminology, Arriaga's description of the actions of art could be his: all change in the material world consists in displacements of bits of stuff; a fortiori so do the changes wrought by art. Natural magic and the reputed feats of alchemy are to be referred, as by the Coimbrans, if not to deceit then to entirely natural causes.

The Archimedean point by which the whole of Aristotelian nature was moved, and the order of art and nature overturned, is the identification of matter with extension. Cartesian extension is, if we neglect fine distinctions the Aristotelians found it necessary to introduce here, the same as Aristotelian continuous quantity. We have seen that neither quantity nor its modes have any active powers. The identification of body with extension and of the qualities of bodies with the modes of extension (I set aside here Descartes' more complicated treatment of sensible qualities) therefore eliminates all at once the forms, active powers, and qualities of Aristotelian physics.

Descartes' commitment to what for brevity's sake I will call mechanism was fixed early in his career, in his collaboration with Isaac Beeckman if not before. At first, it would seem, Descartes simply set aside the doctrines of the Schools, doctrines he would have learned at the Jesuit *collége* of La Fleche. The flowering of his opposition to Aristotelianism is first made evident in *The World* and *Man*, the extant parts of a *Treatise on Light* on which Descartes labored in the early 1630s.

In those texts, comparisons between the works of nature and those of art are ubiquitous. As is well known, the "world" of *The World* is not the real world but a simulation, and the men of *Man* are not real humans but statues made by God purposely to resemble actual humans as much as possible, short of providing them with souls. The slogan "art imitates nature" is reversed: now nature imitates art. The forms of art provide a *model of intelligibility* for the science of nature. This is true even if our understanding of artificial forms is itself less than exact. Descartes' man is a mostly hydraulic device, operated by the animal spirits; his hydrodynamics was hardly more than a collection of intuitions, and hardly up to the task of describing the complex flows required in his physiology.[4] Nevertheless the body-as-fountain or the body-as-pipe-organ provide models for understanding, in the sense both of offering sources for comparison and of establishing an ideal of intelligibility.

We have seen that in Aristotelian natural philosophy artificial forms, being merely superficial and devoid of activity, are of limited use in understanding natural change. At most they may help in explaining the modulation of the effects of natural agents whose powers must be explained by other means. Suárez likens the organs of speech to pipes. But what is at issue is not the motive power that drives air through the throat and mouth. It is only how the movement of the air is altered so as to yield articulate sounds. The active powers of animals, and the souls from which they emanate, cannot be understood by comparing animals to machines. An Aristotelian physiology instead classifies the operations and powers of the soul and describes the dispositions in the body required for those powers to operate normally.

If, on the other hand, bodies are regions of space, and therefore have no powers, then all that remains is to explain the modulation, so to speak, of the divine force by which all things, artificial and natural alike, are moved. The artifact then becomes *eligible* to serve as a source of comparisons, which will in turn tell us all we need to know about the causes of natural change. The artifact is, no doubt, better known to us than the body, which is compared to it in part because the artifact is of our own making. But the convertibility of maker's knowledge—the skill to produce artifacts, which as we have seen is for the Aristotelian a superficial sort of knowledge—into knowledge of nature rests on a prior decision that has converted all of nature into surfaces.

What had seemed a limitation on art—that it deals only in displacements—now turns out to be no limitation at all. Nature is itself merely displacements, merely art. The Coimbrans treat the Platonic notion of divine art with some caution, aware no doubt of its latent charge. For if nature really were divine art, then natural things ought to lack active powers. Secondary causes would have no efficacy, and God alone would act in the world. Divine art must therefore remain only a metaphor. By it philosophers meant to express their intuition that "the variety of dissimilar things, from which the universe is constituted, come together in a single order and are contained in [a single] composition," as in an artwork (2c1q5; 1:214). "Nature is art" is, to use a later terminology, an aesthetic, not a scientific judgment. Science tells us that God, unlike human agents, gives to his creatures active powers and the capacity to transmit the art within them to others.

The subordination of art to nature is not altogether overturned in Cartesian philosophy. But the difference between human and what is now literally divine art is no longer the all-or-nothing presence or absence of generative powers. It is instead the difference between the finite and the indefinitely large, a difference in number and intricacy of parts. Human art is only accidentally, not essentially, subordinate to nature. If we had the intellects of angels, we could manufacture anything in nature, even a living human body. Or again, as Descartes argues, if we had lived in a world in which we built ourselves various animal-like machines, and then were transported to this world, it would not occur to us to regard the animals here as other than machines.

The barrier between art and nature is thus displaced. Art is, one might say, that which is actually made in accordance with our desires; nature is that which is not, or which is only potentially so. It is not surprising that in the *Dioptrique* a new entity, a fusion of art and nature, should make an appearance. Descartes, having laid out his theory of light and the functions of the eye, in the seventh part, turns his attention to the practical question of improving vision.

The best way to do this is to augment the size of the images on the back of the eye. That will be accomplished if one can arrange that rays from the object should converge as far from the back of the eye as possible. Suppose we affix to the cornea a tube of water whose front surface has the same shape as the cornea. Rays from the object will then converge after passing through that surface, much farther from the base of the eye than before. "Vision will occur," Descartes writes, "in the same manner as if Nature had made the eye longer than it is"—as long as the tube, in fact. Moreover the natural pupil of the eye will become "not only useless but even deleterious, insofar as it excludes, by its smallness, rays that could otherwise proceed toward the edges of the back of the eye" (*Dioptrique* 7, AT 6:156–157).

Nature's own eye is superseded by a new organ, part natural, part artifact, deprived of its pupil and having a long tube attached to it. The only arguments against this arrangement are that there is "great discomfort" in putting water against our eye and that the curvature of the cornea cannot be preceisely known. It is not that there is anything *unnatural* in the arrangement. Only its awkwardness, its unsuitability to our overall purposes, rules it out. Nature no longer supplies a norm; utility takes its place.

Even for Descartes, I should note, this is not the whole story. The Sixth *Meditation* and the *Passions of the Soul* rely on norms supplied by nature to the human being by way of its feelings, whose purpose or *usus* is to promote the conservation of the mind-body union. That is a species of utility, but unlike the utility of machines, it is not according to our fancy; the union supplies a *stable* norm for action by revealing to us, through sensation and passion, God's intentions for us in this world. Joy and sadness, pain and pleasure, are genuine, though fallible, guides to the maintenance of the union, and thus of the body on whose integrity the union depends. The body is God's *machine*, and not just an aggregate of corpuscles. Though the multitude and intricacy of its parts offer some testimony to the intelligence of its maker, it is God's machine primarily because we understand him to have made it to serve our ends.

Descartes' position was among the most radical in the seventeenth century, equaled among major philosophers only by Spinoza. The enormous movement of the later seventeenth and throughout the eighteenth centuries to put the new natural philosophy in the service of theology, a movement that consisted in discovering ever nicer instances of divine design, accepted the analogy proposed in Descartes' *Man*. But unlike Descartes it took that analogy seriously. To impute art to nature was not a heuristic device, to be *supposed* at the beginning of the philosophy of nature and then set aside; *discovering* the art in nature was the aim of philosophy, which thereby remained the handmaiden of a higher science.

## Notes

1. See Mercer 1993 and Menn 1998 for surveys of late scholasticism and its relation to seventeenth-century philosophy.

2. For an extensive and informative survey of Jesuit texts on alchemy, see Matton 1998, which includes some seventy pages of extracts. According to Matton, the Coimbrans' question draws heavily on the earlier work of Benedictus Pereira (Benito Pereyra); see Pereira 1576.

3. The list of inventions may be taken from Jean Bodin's *Methodus ad facilem historiarum cognitionem* (1566). On Bodin, see Anthony Grafton's contribution, chapter 8, this volume.

4. See Sutton 2000 for a discussion of the role of fluids in Descartes' psychology, and Gaukroger 2002 for his work in hydrostatics and fluid mechanics.

## References

Arriaga, Rodericus de. 1632. *Cursus philosophicus*. Antwerp: Balthasar Moretus. (Ten subsequent editions, including four in Paris, through 1669. Facs. microfilm, *Manuscripta* list 84 reel 7.)

Bodin, Jean. 1951. *Methodus ad facilem historiarum cognitionem*. 1572. In Jean Bodin, *Œuvres philosophiques*, ed. Pierre Mesnard, 104–269. Paris: PUF. (Mesnard's text is the second edition of 1572.)

Coimbra [Collegium Conimbricensis]. 1594. *Commentarii Collegii Conimbricensis . . . in octo libros physicorum Aristotelis*. Coimbra. (Facsimile repr. Hildesheim: Olms, 1984. Note: In this edition the page numbers 331 to 352 are used twice. Pages from the second series will be marked with an asterisk.)

Descartes, René. 1964–1991. *Œuvres de Descartes*. Ed. Charles Adam and Paul Tannéry. Nouvelle présentation. Paris: Vrin.

Gaukroger, Stephen. 2002. *Descartes' System of Natural Philosophy*. Cambridge: Cambridge University Press.

Long, Pamela O. 2001. *Openness, Secrecy, Authorship: Technical Arts and the Culture of Knowledge from Antiquity to the Renaissance*. Baltimore: Johns Hopkins University Press.

Matton, Sylvain. 1998. Les théologiens de la Compagnie de Jésus et l'alchimie. In Frank Greiner, ed., *Aspects de la tradition alchimique au XVIIe siècle*, 383–501. Milan: Arché. (Textes et Travaux de *Chrysopoeia*, 4.)

Menn, Stephen. 1998. The intellectual setting. In Daniel Garber and Michael Ayers, eds., *The Cambridge History of Seventeenth-Century Philosophy*, 33–86. Cambridge: Cambridge University Press.

Mercer, Christia. 1993. The vitality and importance of early modern Aristotelianism. In Tom Sorell, ed., *The Rise of Modern Philosophy: The Tension between the New and Traditional Philosophies from Machiavelli to Leibniz*, 33–67. Oxford: Oxford University Press.

Pererius, Benedictus. 1576. *De communibus omnium rerum naturalium principiis et affectionibus libri quindecim*. Rome: Venturinus Franciscus.

Sutton, John. 2000. The body and the brain. In Stephen Gaukroger, John Schuster, and John Sutton, eds., *Descartes' Natural Philosophy*, 697–722. London: Routledge.

Toletus, Franciscus. 1985. *Opera omnia philosophia*. Introduction by Wilhelm Risse. Hildesheim: Georg Olms. (Facs. repr. of Köln 1615–1616 ed. in 5v.) The *Physica*, vol. 4, was first published in 1572.

Vérin, Hélène. 1993. *La gloire des ingénieurs: L'intelligence technique du XVIe au XVIIe siècle*. L'évolution de l'humanité. Paris: Albin Michel.

# 7

## The Artificial and the Natural: Arcimboldo and the Origins of Still Life

Thomas DaCosta Kaufmann

*Quanto naturale et artificioso!* How natural and artificial! This is the reaction to a painting by one of the interlocutors in the treatise *Il Figino overo del fine della pittura*, published by the Milanese cleric Gregorio Comanini in Mantua in the year 1591.[1] Conjoining two key terms in Renaissance art theory, Comanini indicates that the perennial question of the relation of human art to nature's creation provided an important topic for discussion of the visual arts in the early modern period. Comanini's treatise on painting addresses this issue chiefly in reference to artistic imitation, and he derives his concepts from ancient definitions, especially from an idiosyncratic interpretation of Plato. While Comanini refers to Plato and to other ancient authors, including Aristotle, Pliny, and Virgil, he also adduces contemporaneous paintings as examples. The art of Giuseppe Arcimboldo (or Arcimboldi; 1526–1593) is conspicuous among them.

*Il Figino* includes poems by Comanini on Arcimboldo's paintings *Flora* and *Vertumnus*. His poems elucidate the identity of *Vertumnus* as a disguised portrait of the emperor Rudolf II, and its meaning as such. Comanini also discusses several other similar pictures by the artist. These are further examples of Arcimboldo's composite heads made out of fruits, flowers, animals, and other utensils relating to the subject depicted, in the way that Flora is constituted out of flowers. Finally, Comanini also mentions a color cembalo (harpsichord) designed by Arcimboldo.[2]

*Il Figino* is in fact only one of many contemporaneous texts that comment on Arcimboldo, many of which originated in Arcimboldo's immediate circle of collaborators and friends. Many such texts were written in Milan, where the artist was born, where he retained contacts and visited during his stay in Central Europe, and to which he returned in 1587. At that time Comanini lived in the same house in Milan as Arcimboldo did; copies of his poems on *Flora* and *Vertumnus* (figure 7.1) had initially been used to accompany these pictures, along with other similar poems on the artist by other Lombard literati, when they

**Figure 7.1**
*Vertumnus*, Giuseppe Arcimboldo. Courtesy of Skoklosters Slott, Balsta, Sweden.

were sent with Arcimboldo's approval and collaboration from Milan to the artist's patron, Emperor Rudolf II, in Prague. Other references to Arcimboldo and poems on his work abound in Milanese publications from the end of the sixteenth century and the beginning of the seventeenth.[3] Before this time, poems on his paintings and other descriptions of his work had already been penned at the imperial court, chiefly in Vienna and Prague, where Arcimboldo worked from 1562 to 1587. These include poems by Giovanni Battista Fonteo (Fontana), Arcimboldo's quondam collaborator in the execution and design of court festivals, which had been presented to Rudolf II's father Maximilian II in order to explicate the first versions of series of paintings of the *Four Seasons* and *Four Elements* by Arcimboldo when they were given to the emperor.[4]

Comanini and some other sources have already been utilized in addressing this issue of the artificial versus the natural in Arcimboldo's art. Following a well-established line of critical discussion linking Arcimboldo with Mannerism, Giancarlo Maiorino has elaborated an interpretation that relates Arcimboldo and Comanini to "the portrait of eccentricity" and the "Mannerist grotesque" as expressions of the fantastic. Maiorino views Arcimboldo's and Comanini's terms of reference as those of Mannerist parody and paradox.[5]

Many other terms have also been used to describe Arcimboldo's works. His pictures have for instance been called hybrids of art and nature.[6] They have also been considered in relation to a number of other contexts in which these realms of art and nature meet: the concept of the marvelous or the wonderful, the monstrous in art and nature, the *Kunstkammer* (early collections of works of art and wonders of nature), and objets d'art such as mounted Seychelles nuts that include both the natural and the artificial.[7]

The present paper revisits Comanini's text along with other poems and documents, as well as paintings (or references to them) that have been recently rediscovered. The continuing recovery of poems and archival references relating to the artist's work, including some texts that have not yet been fully analyzed, among them some probably by Arcimboldo himself, and also the appearance of a previously unknown painting, allow for a further review of the issue of the artificial versus the natural as it relates to Arcimboldo's art. These discoveries provide insights into the genesis of his inventions, and lead to further reconsideration of the ways Arcimboldo's paintings can be associated with the invention, or reinvention, of still-life painting.

Still-life painting was known in antiquity. Elements of still life appear in fresco painting from the fourteenth century. Still-life details are also to be found in panel painting from the fifteenth century, and independently in *intarsie* from that epoch. While perhaps already made in the fifteenth century, independent easel paintings of still life survive only from the sixteenth century, when they were painted both north and south of the Alps. By the early seventeenth century a distinctive genre of painting had come into being. There are many sources for the Renaissance recovery of still life. Among them, as art historians have long recognized, and as will be reiterated here, was the inspiration of ancient texts, including in Arcimboldo's instance, several that dealt with the contest of art with nature.[8]

Arcimboldo has long been mentioned in connection with still life. Yet the significance of this association, and of the artist's possible role in the invention of still life, has until recently been largely discounted.[9] One problem is that the genre of still life and its origins are usually connected with the antithesis of "Mannerism," namely with "realism" in art, and with the representatives of what is called "naturalism" in painting, as this notion has been treated in the literature of art since the seventeenth century, including Caravaggio. Since Arcimboldo has however most often been connected with Mannerism and fantasy, he has thus seemed remote from the naturalist current.[10] While it has been recognized that Arcimboldo may have painted works that might be taken as some of the first extant paintings of still life, this observation has remained unexamined.

Recently however Giacomo Berra has presented fuller arguments that link Arcimboldo with the development of still-life painting in Italy.[11] Arcimboldo may also be associated with nature painting and the study of natural history at the imperial court in Vienna and Prague, not just in Italy. And his involvement with the invention of still life has even broader implications for an understanding of his art, as it does for a more general consideration of the art-nature debate.

Arcimboldo's contribution to the reinvention of still-life painting may be related to his engagement with issues central to considerations of artistic imitation. It derives in part from his response to the question of the relation of the artificial to the natural, which results not only in fantastic inventions, or displays of artifice, as has been noted in the past. In other words, Arcimboldo's response to the art-nature debate is much more than merely a matter of mannered excess or eccentricity.

This causes a reconsideration of responses to Arcimboldo's work, which, as Comanini's *Il Figino* indicates, has since the artist's own time primarily evoked ideas of artifice. Comanini in fact directly invoked the concept of the artificial specifically in reference to a painting by Arcimboldo. Comanini called attention to the artifice exhibited in his picture of a head composed of animals, in which various creatures were used to correspond to the parts of the face, the elephant for the cheek, the wolf for the eye, and so forth. Another of Arcimboldo's Milanese contemporaries, Gian Paolo Lomazzo, reacted similarly, when he said that all of Arcimboldo's pictures of composite heads were done with the "height of artifice."[12]

Comanini related his understanding of artifice to his discussion of artistic imitation. He linked *sprezzatura artificiosa*—*sprezzatura* may be understood in the sense of artful artlessness discussed by Baldassare Castiglione in the *Cortegiano*—directly with *imitazione fantastica.* As Maiorino suggests, in this manner of thinking, the artfully natural thus becomes the spectacularly artificial.[13] Comanini defines *imitazione fantastica* with terms he takes from Plato's discussion of imitative arts, the *mimetikai technai*, in book 10 of the *Republic* and from the *Sophist*, both of which dialogues he cites, and defends painting as a mimetic art. As Erwin Panofsky recognized long ago, Comanini's defense however alters Plato's definitions, particularly that of fantastic imitation.[14] According to Comanini's understanding of Plato, icastic imitation is that which imitates things that are, and fantastic imitation is that which makes (*finge*) things that are not existent. Comanini says "that painter therefore who will imitate something formed by nature, as would be a man, a beast, mountain, sea, a plain and other similar things would make an icastic imitation; but he who would imitate his whim, and something never designed by any one else, at least that he might know, would make a fantastic imitation."[15] He adds, consequently, "that painter makes a fantastic imitation, who depicts something from his whim and his own invention, which would not have its being outside of his own intellect."[16] As he also says, the icastic is the imitation of things that are in nature, and the fantastic of things that only have their being in the intellect of the imitator.[17]

While Comanini gives a portrait as an example of icastic imitation, Arcimboldo's paintings of Flora and Vertumnus (figure 7.1) are employed to exemplify fantastic imitation. Arcimboldo is specifically called an artificer of fantastic imitation (*artefice di fantastica imitazione*).

His works are examples of his *capriccio* and *invenzione*. It is for such inventions that Comanini calls the painter an *ingegnosissimo pittor fantastico*, a most ingenious fantastic painter. Comanini says that these pictures are products of his power of imagination, his *virtu fantastica*, which in him is most powerful (*gagliardissima*): this power has the task to receive species from the exterior sense in the common sense, to retain them, and to compose them together.[18]

Comanini elaborates these notions in other terms that frequently appear in Renaissance discussions of the imaginary.[19] *Il Figino* introduces several other topics of the fantastic to which Arcimboldo's paintings might be compared. His composite heads might be compared to various imaginary creatures, such as the chimera, the mythical being composed of the foreparts of a lion, the body of a goat, and a serpent's tail. They might be compared to the monster described by Horace at the beginning of the *Ars Poetica*: this is the creature painted with a human head on a horse's body, whose limbs are covered with feathers, like a beautiful woman on top, and a fish tail below.[20] This is of course a commonplace reference for the imaginary, and for painters and poets to create what they may imagine. Comanini compares Arcimboldo's creations directly to the images found in dreams.[21] He says that the ministers of dreams must be his familiars. Only their arts seem to be surpassed by his, because "he knows how to make the arts and the transformations that they effect. So that he makes more things with effect than they do, transforming animals and birds and serpents and trunks and flowers and fruits and fish and plants and leaves and ears of grain and straw and grapes into men and into men's vestments, and into women and into women's adornments."[22]

Comanini's discussion of Arcimboldo is exceptionally extensive and idiosyncratic in its particular philosophical orientation, but its descriptions correspond to the paintings of the elements and the seasons that are items pertaining to them (e.g., *Spring* out of flowers). Arcimboldo's heads are composed of individual elements that taken on their own might actually be found in nature. Fruits, flowers, fish, animals, birds, and the like abound in his work. But these elements of nature are combined in his art in ways that could not be found in nature itself, hence they are examples of fantastic artifice. Comanini also indicates why many authors' reactions to Arcimboldo's transformations of the natural might themselves be considered artificial. One of the characters in his treatise describes his own poem on the painting of Vertumnus as having much artifice together with marvelous charm.[23] This may also provide one reason why so many responses to his paintings take the form of poems.

In any event, Comanini's terms of discussion, as well as his manner of expressing them in poems, were echoed in other contemporaneous descriptions of the painter's art. For example, in his poems of the 1560s Fonteo referred to Arcimboldo's pictures of the *Seasons* and *Elements* as *chimerae*. His notes to his poem also called them *grilli*. Although this term has various meanings, Fonteo says that Arcimboldo "meant by *grillo* an image of painting such as the theme is supposed to express, that is, an image made up of instruments and objects pertaining to the object itself." He exemplifies this with a boor made up of farmers' tools, and a cook with his utensils.[24]

The vocabulary of the imaginary, of the whim, the invention, the bizarre, was regularly applied to Arcimboldo, also in works that were not themselves poetic in form. In his biography of Arcimboldo, Paolo Morigia (Morigi) for instance called his composite heads *invenzioni* and *bizarrie*.[25] In his writings Lomazzo also refers to Arcimboldo's *invenzioni* and *capricci*. Lomazzo says, too, that he was always prepared to do something *capricciosa* for his patron.[26]

Arcimboldo's composite heads and the language used to describe them recall the Renaissance discussion of another composite, the grotesque. Lomazzo himself defined the grotesque as *confuse* of *diverse cose*. The grotesque was also regarded as a characteristic product of the *fantasia* understood as the inventive imagination, inventive in the power to produce something not found in nature. Hence A. F. Doni describes grotesques as *fantasie*, *sogni*, and *chimere*, all words that were used to describe Arcimboldo's paintings. Doni also talks about *macchie* in this context. *Macchia*, meaning a spot, referring to an expression of an idea, as in an artist's idea made in a sketch, was also a term employed by Arcimboldo in a letter describing a form of grotesque he designed. It also seems appropriate that Lomazzo referred to Arcimboldo in his *Rime*, which he defined as being in the manner of grotesques.[27]

The interpretation has thus lasted of Arcimboldo as the creator of fantastic, bizarre images that are the height of the artificial and are comparable to the grotesque. In the eighteenth century, Arcimboldo continued to be regarded as an extravagant painter of bizarre pictorial thoughts and his paintings were also called caprices. After a period in which he seems to have received little critical attention, he was "rediscovered" in the twentieth century chiefly in the context of Dada and surrealism, when his supposedly fantastic images were seized on as the precursors of these forms. Much twentieth-century criticism treated the artist in similar fashion, as the creator of *dipinti ghiribizzosi*, as the prince of "pictorial caprices," and in general as a painter of amusing, ludic, images. This last interpretation also stems from a tradition of reception dating back to Arcimboldo's contemporaries. Comanini called one of his images a joke, and spoke of them as ridiculous; Lomazzo suggested that his paintings of the *Seasons* might be used to inspire paintings in inns; *grilli* is also a term used to mean joking images, as in Pliny.[28]

Maiorino's account of Arcimboldo as an artist who produced a "panegyric to artifice" and as the creator of the "Mannerist grotesque" is thus but the most extended recent expression of a long critical tradition. Maiorino follows the tendency of many twentieth-century authors to link Arcimboldo with surrealism. He picks up the interpretation of rhetorical elements, the play of pictorial artifice in Arcimboldo's work that was emphasized in an interpretation of Arcimboldo's rhetoric by Roland Barthes.[29] Maiorino's notion of Arcimboldo's work as a form of parody was also expressed earlier by Paul Wescher.[30] Arcimboldo's stress on the ludic is also well established, as is his emphasis on the extravagance that leads him to transcend the rules of art.[31]

But there is much more to Arcimboldo's art. While Comanini and other sources say that one response to Arcimboldo may be to regard his paintings as ridiculous and to laugh at them, they also say a more informed reaction is called for. Comanini and other commentators tell us that Arcimboldo's pictures convey serious messages. Comanini himself refers to these paintings as allegories, which work in the form of hieroglyphs. Like Socrates, they hide meaning beneath a seemingly risible exterior. Comanini describes *Vertumnus* in this manner, and Fonteo's lengthy poem gives an account of the series of the *Four Seasons* and *Four Elements*. Fonteo allows for an explication of these pictures as imperial allegories and for an understanding of *Vertumnus* as the culmination of these series.[32]

There is another side to Arcimboldo's art that, while noted by some scholars, has not been thoroughly examined even by those who have discussed its hybrid character. The interpretation of Arcimboldo's paintings as involving fantastic imitation does not fully cover Comanini's description of how they were made. While he gives license to the interpretation of Arcimboldo's paintings as products of fantasy, he also says that his painting of a head composed of animals sent some years before by the emperor to the king of Spain was fully taken from nature, because the emperor had let him study this painting from nature, allowing him to see live all the animals that constituted the head.[33]

As in many other instances, Comanini's account can be shown to have a basis in fact. Documents discovered in Spain indicate that a version of this painting, a representation of *Earth*, was indeed sent to Spain along with other depictions of the four elements and four seasons probably in 1580, some years before the period in which *Il Figino* was set; this painting was hung with a group of pictures described as *cuadri di capriccio*.[34]

Moreover, although Comanini says that one should be astonished at the human artifice in this work,[35] and the aspect of the wondrous has been noted in Arcimboldo's oeuvre, the naturalistic basis of this work should be emphasized, since its existence can also be verified. Several creatures that appear in the head of the existing version of *Earth* (figure 7.2), among them a cheetah, leopard, and antelope (figure 7.3), are also found in a compendium of nature studies that belonged to Rudolf II.[36] The original studies, whatever their exact location at present, were probably done by Arcimboldo himself. Moreover, a representation of the antelope identical to that in the Vienna compendium, and in the head of *Earth*, is to be found in the collection of nature studies that once belonged to the Bolognese naturalist Ulisse Aldrovandi, now in the university library in Bologna. A number of pages that differ from others in Aldrovandi's volumes are pasted into his volumes labeled *Tavole degli animali*, and are comparable to the studies now in Vienna. These seem identical to some of the sheets described in a letter, also in Aldrovandi's collections, sent to him in 1585 from Prague by Franciscus de Paduanis (Padovano), where it is said that he had obtained many images of birds and quadrupeds delineated with colors from the life (*ad vivum*) by Arcimboldo, the imperial painter.[37] This coincides with Comanini's claim that he painted the creatures in the head *dal naturale*.

The letter from De Paduanis (Padovano) may also give credence to another statement by Comanini about Arcimboldo's compositional techniques: Comanini's remark occurs in the same context in which he makes observations about the painting of a head composed of animals. He says there that every fruit or flower in the paintings of *Vertumnus* and *Flora* was taken from nature and imitated with the greatest diligence possible.[38] De Paduanis (Padovano) says that he obtained a *lilium persicum*—described in Carolus Clusius as being observed in Hungary—with the intention of having it depicted by the artist, but the plant faded before he could have Arcimboldo fulfill his promise to paint it.[39]

**Figure 7.2**
*Earth*, Giuseppe Arcimboldo. Private Collection, Vienna, Austria.

**Figure 7.3**
"Study of Antelope, and of Goat," Giuseppe Arcimboldo. Courtesy Österreichische Nationalbibliothek, Vienna.

Although Arcimboldo did not make this particular image, De Paduis (Padovano) thus indicates the artist was known to make similar studies.

The Vienna volume from Rudolf II's collection, which contains animal studies attributable to Arcimboldo, provides further evidence indicating that studies of flowers were indeed known and collected at the imperial court, where Arcimboldo worked. It contains a number of oil sketches of flowers, and some of animals, by Ludger Tom Ring the Younger. These studies of flowers were then used in larger oil paintings by Tom Ring. Some of his studies are even bound in reasonably close proximity in the volume to animal studies resembling Arcimboldo's.[40] Similar studies from nature could well have been made at the imperial court. Certainly Arcimboldo's animal studies were also used by later artists: the so-called *Museum of Rudolf II*, another set of manuscripts with oil paintings on paper and vellum depicting naturalia, contains other copies of the drawings (including the antelope studies) attributed to Arcimboldo.[41]

Two more of Arcimboldo's paintings from the series *The Four Elements* in addition to *Earth* are also noteworthy in this regard. His head composed of sea creatures, representing *Water* (figure 7.4), resembles the many fish studies that also are found in a Vienna codex (cod. min. 42). Likewise, his head made of birds, standing for *Air*, can be compared to bird studies in Viennese codices. Similar studies of fish and birds were also found, and made, at the imperial court (see figure 7.5).

In any case, the making of studies from nature that are meant to portray creatures from life (or to be more specific, so as not to expand further on this artistic and epistemological problem, to represent them as if they were taken from life) can be best related, to use Comanini's terms, to the form of imitation that he exemplifies by portraiture: they are, after all, also meant to be portraits of living things. To follow Comanini's definition, they can therefore be regarded as products of icastic imitation.

In the end, despite its usefulness, Comanini's tract is thus not to be taken as a totally logical guide or adequate source for a full interpretation of Arcimboldo's work. But Comanini's logic is not the issue at stake, for his treatise has its own theoretical objective and approach to the question of the artistic imitation of nature. Comanini uses Arcimboldo for his own purposes. The point is that a one-sided emphasis on artifice and fantasy does not take into account the full range of Arcimboldo's art and its implications.

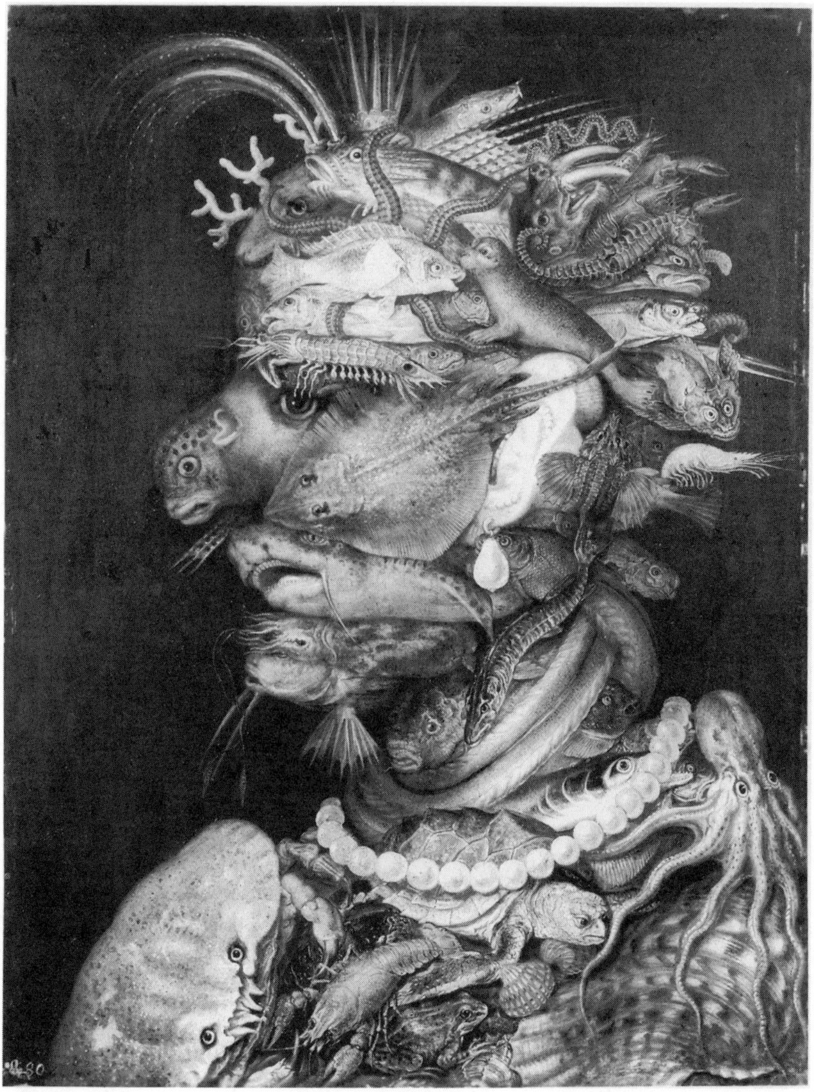

**Figure 7.4**
*Water*, Giuseppe Arcimboldo. Courtesy Kunsthistorisches Museum, Vienna.

**Figure 7.5**
"Fish Studies," Giacomo Ligozzi. Courtesy Österreichische Nationalbibliothek, Vienna.

Arcimboldo's nature studies confirm that the artist may be linked not just with the fantastic, but with the naturalistic, in this instance also meaning what at the time was called natural philosophy, or history. Comanini says that a painter was to have had a sort of universal knowledge, and suggests that Arcimboldo did; for this reason and others, he may be considered the Habsburgs' Leonardo. Besides designing festivals and masquerades, Arcimboldo invented ciphers and was involved in managing rivers.[42] His drawings for a grotesque decoration actually presented designs for sericulture.[43] Drawings for the 1571 festival probably reveal his knowledge of astronomy.[44]

There is also much evidence for Arcimboldo's knowledge of natural history. Comanini himself points to some of this display of knowledge, when he indicates that the composition of Arcimboldo's head made out of animals corresponds to a description in book 8 of Pliny. This aspect of natural historical theory, relying on the system of natural correspondences, is physiognomy, by which the function of parts of the body are paralleled to or exemplified by animals.[45] Moreover, Arcimboldo's supposedly bizarre *grilli* may not only be connected to the Leo nardesque caricatural tradition in Lombardy, but they may also be related to the physiognomic investigation that is also involved in Leonardo's so-called caricature heads.[46] Another aspect of Arcimboldo's method of composition may involve phytognomic theory, on which authors such as Giovanni Battista della Porta, whose work was known in Prague, also wrote.[47] It is possible that the comparison of parts of the body to plants was operative in Arcimboldo's method of composition as well.

However Arcimboldo's paintings may have been composed, their various elements do indicate that he examined the world of nature. Comanini suggests that he observed it both closely and widely. Though not discussed by Comanini, the head of *Water* (figure 7.4) contains at least sixty-two separate species of aquatic creatures, including vertebrates as well as invertebrates. In addition to fish, mollusks, echinoderms, platyhelminths, annelids, cnidarians (i.e., coral), amphibians, reptiles, and sea mammals are found represented in his painting of *Water*.[48]

For these reasons, and more, Arcimboldo was probably regarded as something of an expert on natural history. In 1582 he was sent by Rudolf II to Germany to acquire rare creatures from the Americas.[49] He seems to have enjoyed a broad reputation for this sort of expertise, as is suggested by his having been called on to work for the naturalist Aldrovandi.

The association with Aldrovandi's investigations is informative, because it indicates that while Arcimboldo's studies from nature may have served him for the composition of his supposedly capricious inventions, that was not their only, and perhaps not even their original function. Like other nature studies, his paintings could be employed as substitutes or representations of rare or wondrous plants and animals. This places him in a role comparable, it has been noted, to the many artists, Jacopo Ligozzi most prominent among them, who supplied studies of plants, snakes, and other creatures to the Bolognese naturalist.[50]

Rudolf II also owned two books, one containing studies of birds, and one of acquatic creatures, mainly fish, by Ligozzi (figure 7.5).[51] They were part of the large collection of nature studies brought together at the imperial court. These painted studies complement the imperial gardens and menageries used for the study of nature as well as for enjoyment in Prague. Comanini in fact suggests that they were employed for nature study, when he says that Arcimboldo made observations of animals in them. They complement the collections of rare specimens of naturalia found in the Kunstkammer as well as elsewhere in the imperial collections (the emperor's lions and the like). Painted depictions of naturalia also may be related to the studies carried out by the many important scholars of natural history who were active at the imperial court, beginning with Pier Andrea Mattioli, Carolus Clusius, and Rembert Dodoens during the reign of Maximilian II, and continuing with Anselmus Boethius de Boodt at the time of Rudolf II.[52]

Numerous important nature studies by deceased as well as active artists and contemporaries were collected in the imperial library and Kunstkammer. The groups of nature studies owned by Rudolf II begin with older works by Albrecht Dürer and Simon Marmion. In addition to Ligozzi and Arcimboldo, Giorgio Liberale, who had collaborated in the illustration of the herbal published by Mattioli in Prague, made splendid nature studies for the Archduke Ferdinand; these also ended up in the imperial collections. The imperial court painters Hans Hoffmann, Joris Hoefnagel (figure 7.6), Jacob Hoefnagel, Daniel Fröschl, and Dirck de Quade van Ravesteyn all made nature studies for the emperor. Joris Hoefnagel's four volumes corresponding to the four elements, which present the creatures of the natural world in emblems, may be related to the ways Arcimboldo similarly allegorized nature: Hoefnagel made compilations of nature studies, and Arcimboldo compressed these into individual heads. The Netherlandish painters Hans Verhagen, Hans Bol, and most significantly, Jacques de Gheyn all made splendid sheets or books that Rudolf II owned.[53]

Hence, while the Italian aspect of Arcimboldo's activity, and his later career, have been emphasized in recent scholarship, his nature studies, which Comanini correctly informs us were carried out in Germany (in the broader sense that Austria and Bohemia might be regarded as part of Germany), can also more directly, and with as much consequence, be associated with the imperial court. Vienna, and later Prague, through the artists active at the court, the scholars attached to them, and the works delivered to the collections, were centers for the study of nature. Lee Hendrix thus rightly suggests that Arcimboldo may be considered the "first major producer of natural history illustration employed directly by Rudolf II and as a figure of larger importance for Rudolfine nature illustration as it would develop."[54]

The approach informing the imperial artists' painting of studies of nature may also have had consequences for more than natural history. Hendrix relates the collections of nature painting in Prague to the emergence of independent paintings of animals in easel paintings, seen for

**Figure 7.6**
"Miniature with Nature Studies in Still-life-like Composition," Joris (Georg) Hoefnagel. Oxford, Ashmolean Museum.

example in the oeuvre of the imperial painter Roelant Savery. An inventory reference suggests that Arcimboldo may also have made such a painting of a cat eating fish.[55] Nature studies may with more certainty be related to the impetus involved in the origins of another genre of painting, namely still life.

As noted, a number of paintings by Arcimboldo have long been associated with still life. These are heads composed, for example, of meats or vegetables (figure 7.7). When these heads are inverted 180 degrees, still-life compositions are revealed: in the one instance a painting with hands holding a platter, and in the other, a bowl with vegetables.[56]

**Figure 7.7**
*Reversible Head with Still-life of Bowl of Root Vegetables*, Giuseppe Arcimboldo. Courtesy Cremona, Museo Civico a la Ponzone.

It may even be that Comanini's failure to mention these particular kinds of paintings by Arcimboldo, or any other independent still-life paintings in Lombardy, may have been in part responsible for the relative neglect of these paintings. Comanini's silence is noteworthy in that Lombardy, and in particular Milan, was a site for the origins of still life. Some of the first still-life paintings showing fruits were painted there by artists such as Ambrogio Figino, Fede Galizia, and Vicenzo Campi. Caravaggio, whose basket with fruit in the Ambrosiana is the best known example of this game, stems directly from this tradition.

As Giacomo Berra has repeatedly argued, Arcimboldo may be connected with this tradition as well. Arcimboldo's pictures could be related to the work of several of these artists, not just to that of Caravaggio, who lived relatively near Arcimboldo in Milan and whose paintings, Berra argues, share many similarities with the naturalistic thrust of Arcimboldo's work.[57] Figino also painted still lifes, and Figino is the eponymous character who supplies the name for Comanini's treatise. A number of other poems deal with both Figino and Arcimboldo together. Arcimboldo knew Figino, and Arcimboldo was involved in the acquisition of a painting by that artist for Rudolf II.[58] This is a painting with some natural details. Arcimboldo was also involved in mediating the court's acquisition of a work by Fede Galizia, also imaginable as a still life, and it can now be demonstrated that Galizia made fruit still lifes.[59] It might be added that several paintings by the Campi, with still-life elements, were also in the imperial Kunstkammer.

The appearance of a painting on the art market completes the connection between Arcimboldo and the other Lombard still-life painters (figure 7.8). This is an invertible head composed of fruit, which, when reversed, shows a basket containing fruit. This painting can be firmly attributed to Arcimboldo for several reasons. Its support, linden wood, suggests a subalpine origin. (In southern Germany lindenwood is most commonly used for paintings and sculpture.) The painting's provenance can be traced to Sweden, whence many of Arcimboldo's paintings, including versions of *Flora* and *Vertumnus*, were taken after the sack of Prague by Swedish troops in 1648. It may correspond to listings of a number of paintings mentioned in the inventories of the imperial, and later Swedish, collections. Its execution resembles that of Arcimboldo's later work. Most important, its features, when read as a composed head, closely correspond to those of *Vertumnus*. For instance, a berry and a nut form the eyes, grapes make up the ears or hair, a pear represents the nose, and so forth.[60]

**Figure 7.8**
*Reversible Head with Still-life of Bowl of Fruits*, Giuseppe Arcimboldo. Courtesy French & Company, New York.

**Figure 7.8**
(continued)

When inverted and read as a still life, Arcimboldo's painting resembles these pictures, which it even probably antedates. Most important, Arcimboldo's paintings of reversible heads, while seeming to be humorous or ridiculous heads, may not have been read, or hung, as composite heads, but as still lifes. A document describes yet another such reversible head, probably by Arcimboldo, and allows for this reading. Among other things, Italian nobleman Ottavio Landi reported to Adam von Dietrichstein on a painting he saw on the occasion of a February 1573 visit of Elector August of Saxony to Maximilian II in Vienna. In the tradition of diplomatic visits, in which princes were shown parts of collections, August was shown *pitture et altre cose*. Among them was a painting probably by Arcimboldo, not only because the description corresponds to those of other works by the artist, but because a series of the *Seasons*, like those then visible in Vienna, were made for the Saxon duke in 1573. The painting was a face of Doctor Zasius, which was formed of documents, and of flowers, some dried. It seemed to be a flower vase when shown one way, but when reversed was a ridiculous face.[61] The primary viewing perspective is suggested by the description, which speaks of the *vaso* being shown in its *essere* and then being turned (*voltato*).

The appearance of this painting in the imperial collections in Vienna in 1573 points, however, not to Lombardy as the place of origin of the painting, but to the imperial court. If the painting is considered a still life, this place of origin is especially likely. The imperial court was the home or recipient of many of the first examples of the still-life genre. Joris Hoefnagel produced several independent miniatures with flowers in vases. The book of nature studies including rare flowers by De Gheyn, owned by Rudolf II, was in Prague, where they were used for marginalia in a prayerbook in Munich, once associated with Duke-Elector Maximilian I of Bavaria, but now known to have been made in Prague.[62] These may have also been part of the inspiration for De Gheyn's own independent paintings of floral still life, which were executed soon after 1600. And the first extant independent floral still-life paintings by a Dutch artist were done by the imperial painter Savery either just before or after he arrived in Prague; many examples of the type were painted by him there.[63]

The 1573 reference antedates all of these examples, however, as it does the Lombard examples: all the northern and Italian examples mentioned so far date from the 1590s at the earliest. By providing a terminus ad quem for this painting, this reference allows a dating close to that for

Arcimboldo's original creation of the *Seasons*, 1563, and *Elements*, 1566. It also means that this picture is one of the earliest paintings of still life made by an Italian-born artist.

The existence of the still life/reversible head of Doctor Zasius also places Arcimboldo in proximity to some of the earliest German still lifes, some of which, or whose preparatory studies, he may well have seen at the imperial court. Still lifes depicting flowers in vases and bearing the date 1562 exist by Ludger Tom Ring the Younger. Tom Ring also painted pictures with other sorts of still lifes in 1565. Studies for still-life details related to other paintings by Tom Ring, including a kitchen scene (formerly Berlin, now lost) dated 1562, also exist. These include studies of individual fruits and flowers, and of fruits on plates. Especially worthy of note are two oil sketches, one with flowers in a vase, and one with flowers in a basket. These are pasted onto pages in the Vienna codex, which, as remarked, includes nature studies attributable to Arcimboldo.[64] It is thus possible that Arcimboldo was familiar with these studies, or the paintings made after them.[65] If works by Tom Ring were already available in Vienna in the 1560s or thereafter, they may have even inspired Arcimboldo.

In any case, some recently rediscovered poems suggest another source for Arcimboldo's painting of nature, and bring us back to the issue of the contest of art and nature. Two of these are by Bernardino Baldini. One of them, reads (in translation):

> On the Painter Giuseppe Arcimboldo
> The Divine goddess is able to form men from human members, and she clothes fields with leaves and flowers; but she has not learned how to weave human limbs with fronds, and verdant leaves; in this the unique ability of Arcimboldo has been able to succeed, in which he has surpassed the work of nature.[66]

Berra calls this praise of the ludic activity of the painter, but while the description of play may be allowable, once again it is a form of serious play that is in effect a form of contest with nature.[67] Here the reception of Arcimboldo's art indicates that it engages in debate over another conception of artistic imitation, which had also been articulated in antiquity by Aristotle, Cicero, and Seneca, among other authors. Art is said to imitate not merely what nature has produced, but to be like nature in that it may be compared to nature in its process of creation.[68] Arcimboldo's imagination does not merely create that which might be, or is not found in nature, in the form of art. His work may be considered a form of artifice, but this art employs a creative power that contradicts a point of view that had been represented by such authors as Cicero (*De Natura Deorum* 1:92, 2:35, 2:57–58, 2:82ff.) and Seneca (*Epistulae Morales* 90). Against the current of their arguments, Arcimboldo's art is said to demonstrate the capacity to equal and even surpass the power of nature.[69]

A phrase appearing in another poem by Baldini, written on Figino and Arcimboldo, indicates that Baldini also held notions of the mimetic qualities of art, and that, also in keeping with ancient thought, Arcimboldo's art was a form of emulative competition. In his poem, which is found in a compendium on Milanese figures, Baldini refers specifically to Arcimboldo's paintings of *Flora* and of *Vertumnus.* The painter is described as sparkling even more "Parrhasiusly" among artists (*Talis parrasior micat Arcimboldius inter/Artifices*).[70] This turn of phrase alludes to the well-known story, told by Pliny (*Natural History* 35:65–66), of how the ancient painter Parrhasius entered into a competition with Zeuxis. Zeuxis produced a picture of grapes so successfully that birds flew up to it, but Parrhasius produced such a realistic picture of a curtain that Zeuxis, although proud of the verdict of the birds, requested that the curtain should be drawn. Whereupon Zeuxis yielded up the prize, because while he had deceived birds, Parrhasius had deceived him, an artist (*artificem*). Arcimboldo's work is again thus considered one of artifice, and the poem calls him an artificer, evoking Pliny's language, but the reference to Parrhasius invokes the power of art to deceive nature itself, as well as to deceive an artist. It thus invokes the idea of competition. Comanini also compares Arcimboldo favorably to both Zeuxis and Parrhasius with reference to the same story.[71]

Two more poems reinforce the thesis that conceptions of the art-nature competition were current in Arcimboldo's circles. These poems appear in a booklet prepared by Giovanni Filippo Gherardini, Arcimboldo's friend and another of his housemates, which accompanied the paintings of *Flora* and *Vertumnus* when they were sent to Milan. One is a madrigal praising Arcimboldo's painting of *Vertumnus*,[72] and the other is a sonnet praising his paintings in general.[73] These poems are signed with the monogram "G. A. da Milano." Revising his earlier opinion, Berra states that no other Milanese *litterato* can be identified with this monogram, and it is reasonable to identify Arcimboldo, who signs his name similarly elsewhere, as the author. Although the attribution might seem questionable, because it would have the artist praising his own work, something similar seems to have happened in Comanini's writing:

Comanini has a character in his dialogue praise his own poem, a version of which was indeed included in the booklet containing the two poems in question. And just as the humanist Fonteo was a draftsman, Arcimboldo may have been a poet.[74] I have also recently found a poem of self-praise by Figino. In any case, as previously noted, the poems were sent with the artist's knowledge and approval.

The first poem again says that although the image of *Vertumnus* might seem to elicit reactions of stupor and laughter, and merely be a bunch of flowers and fruits, there is more to it: it is the product of the *ingegno* of Arcimboldo, which jousts on equal terms with nature. The poem next alludes to the divine theft of the great son of Japeth. This is Prometheus, who stole fire and created humankind.[75] Arcimboldo, the creator of works like *Vertumnus*, is thus compared to Prometheus the creator. His fashioning of a type of human being from natural elements in art thus competes with the creation of humans in nature.

The second poem applauds Arcimboldo's powers of mimesis. It says that painters are praised for representing just one object, but Arcimboldo does more, since he succeeds in representing many figures, piecing them together from parts. Hence he does even more than Apelles, the famed ancient painter. The viewer of his work stands astonished, not knowing if what is created is the work of art or of nature. Arcimboldo thus surpasses even the ancients in his ability to replicate what is seen in nature.

These poems specifically refer to Arcimboldo's compositions of heads made up of parts taken from nature, but there are ways they help us interpret his approach to the painting of nature in terms of the invention of still life. First, Figino, the subject of one of the poems, is a painter of still life, and praised for his mimetic abilities: given the range of his art, and the association with Arcimboldo, still life is most likely to have inspired this remark. Second, the topos of the contest between Parrhasius and Zeuxis was used during the Renaissance specifically to allude to the painting of still life. This topos, and a related story (*Natural History* 35:67) about a deceptive image made by Zeuxis showing a child carrying grapes, were known by the imperial court painters, as they were by Caravaggio, and indeed by many other sixteenth-century artists, who allude to them in their own drawings and paintings. Artists, including those in imperial service, seem to be playing with these topoi when they paint children holding clusters of grapes, the subject of Zeuxis's painting, for example.[76] Third, groups of grapes, the source of the illusion, are also recalled in Arcimboldo's painting of *Vertumnus*. Fourth, the *Vertumnus* painting must have been closely associated in Arcimboldo's mind with what we would call his painting of a fruit still life, for the newly discovered reversible head, which when turned around becomes such a still life, possesses features very similar to those of *Vertumnus* (pear for a nose, apple and peach for cheeks, and so forth). It therefore seems likely that Arcimboldo's ideas about *Vertumnus* were also connected in his mind with the reversible head that becomes a still life.

The issues involved in the art-nature debate as it was discussed in Arcimboldo's circles seem consequently to help us understand some of the thinking behind the genesis of his paintings or composite heads, and, moreover, why Arcimboldo's paintings of still life take the form of reversible heads. Seen one way, like the head composed of fruits that resembles *Vertumnus*, they rival nature in creating a new form of creature. Turned around, they rival nature in replicating its forms. One image thus sets both forms of imitation, of competitive emulation, into play.

Furthermore, viewed one way, Arcimboldo's art involves a naturalistic approach to the creation of a fantastic artificial head. Viewed another way, this process is simply reversed. As Celeste Brusati has argued, still life is often viewed as the summa of realistic depiction, which from a certain point of view it is, but it is also a product of descriptive or illusionistic artifice.[77] In carefully representing nature, in the form of fruits, Arcimboldo is also supremely artful, as the comparison with Parrhasius suggests.

Arcimboldo's paintings may be regarded as caprices of nature as much as they are caprices of art.[78] They engage in icastic and fantastic imitation simultaneously. They also evoke both mimetic and productive conceptions of imitation.[79] Arcimboldo's paradoxical but not parodistic employment of the approach of *serio-ludere*, his creation of serious jokes, employ all these seeming contradictions together in one image. Hence his reversible pictures can make of the seemingly fantastic composite head, the height of artifice, a compelling vision of nature that also constitutes one of the origins of still life.[80]

## Notes

1. As printed in the modern standard edition (Barocchi 1962, vol. 3, 254), to which reference is made in this chapter.

The present chapter represents an annotated and slightly reworked version of the paper presented at the conference "The Artificial and the Natural," MIT, Cambridge, MA, May 18, 2001. Various versions of that paper have been given subsequently at other venues. The chapter as presented here represents an initial statement of the author's views of the subject on which much research has continued: the results will be forthcoming in a book tentatively titled *Giuseppe Arcimboldo: Serious Jokes, Art, Science and the Origins of Still Life Painting* (Chicago: University of Chicago Press). Aside from bringing references up to date, I have chosen to retain the form of the paper largely as given, in order not to overburden the text.

2. For discussions of Comanini's treatise and its relation to Arcimboldo's paintings, see the commentary in Barocchi 1964; Maiorino 1991; Kaufmann 1987; Berra 1988, 11–13, passim; Kaufmann 1993, 100–103, 126–129, passim; Berra 1996b; Comanini 2001.

3. In addition to the references (including Comanini) compiled and annotated in Falchetta 1987, see for these texts, and Arcimboldo's involvement with them, Berra 1988, 1996b.

4. The text of Fonteo's poems is given in Kaufmann 1993, appendix 2, 197–205; for the interpretation based on them see the same source, 103–128. For other texts and documents relating to activities of Arcimboldo during this period see Kaufmann 1978.

5. Maiorino 1991.

6. Lorraine Daston and Katherine Park have called them "characteristic hybrids of art and nature." This phrase occurs in a draft of their *Wonders and the Order of Nature* (1998), which they generously allowed me to consult, but it does not seem to appear in the book as finally published.

7. See for example Kenseth 1991, 43; Kaufmann 1997; Schlosser 1978, 173–176; Kemp 1995, 186.

8. There is a large literature on the subject of still-life painting. Useful general works, making the points referred to here, include the classic study by Charles Sterling (1952); the important exhibition catalog *Stilleben in Europa* (1979); the survey by Norbert Schneider (1998); Sybille Ebert-Schifferer's (1999) historical study; and the compendium of texts, commentaries, and interpretations brought together by Eberhard König and Christian Schön (1996).

9. See in particular Longhi 1950; Sterling 1952, 38; *Stilleben in Europa*, 1979, 154, 155; Spike 1983, 13; Maiorino 1991, 34.

10. For a critique of this interpretative framework and an effort to reconcile this apparent contradiction, see Kaufmann 1982, 1997b. These essays have now been reprinted in Kaufmann 2004, 33–89, with further commentary, 455–458.

11. Berra 1996a, 2000.

12. Comanini 2001, 265f.; Lomazzo 1973, vol. 1, 362: "Come meravigliose in somma sono tutti gl'altri quadri da lui fatti con sommo artificio."

13. Maiorino 1991, 19f.

14. Panofsky [1924] 1968, 212–215.

15. Comanini, *Figino*, in Barocchi 1962, 256: "Quel pittore adunque, il quale imiterà cosa formata dalla natura, come sarebbe uomo, fiera, monte, mare, piano et altre simili, farà imitazione icastica; ma quegli che dipingerà un suo capriccio non più disegnato da alcun altro, almeno che egli sappia, farà imitazione fantastica."

Except where otherwise noted all translations are my own.

16. Comanini, *Figino*, in Barocchi 1962, 256: "Voi dite quel pittore fare imitazion fantastica, il qual dipinge cosa di capriccio e d'invenzion sua, e che non abbia l'essere fuori del proprio intelletto."

17. Comanini, *Figino*, in Barocchi 1962, 256; also 274: "L'icastica e imitazione di cose che sono in natura, e la fantastica di cose che hanno solamente l'essere nell'intelletto dell'imitante."

18. Comanini, *Figino*, in Barocchi 1962, 270: "L'ufficio della quale e di ricevere le specie apportate dagli esteriori sensi al senso commune, e di ritenerle, et ancora di comporle insieme."

19. For a discussion of Arcimboldo's work in relation to this topic, including further references, see Kaufmann 1987; Berra 1988, 11–13; Berra 1999, 391–392.

20. For the relation of Arcimboldo's paintings to chimeras, also see Kaufmann 1993, 107, and in general for these topoi, Kaufmann 1987.

21. This may also be related to a discussion in Lomazzo: for Arcimboldo's paintings and dream images, see Kaufmann 1993, 102, 162; for their relation to other sorts of visions, see Berra 1998.

22. Comanini, *Figino*, in Barocchi 1962, 270: "Poiché egli sa fare l'arti e le trasformazioni che eglino fanno. Anzi, fa di vantaggio piú cose che non fanno essi, trasformando egli animali et uccelli e serpenti e bronchi e fiori e frutti e peschi et erbe e foglie e spiche e pagli et uve in uomini et in vestimenti d'uomini, in donne et in ornamenti di donne."

23. Comanini, *Figino*, in Barocchi 1962, 265: "Maravigliosa vagezza congiunta con molto artificio."

24. Kaufmann 1993, 107, with related notes. For more on *grilli* see Bredekamp 1989, and in relation to Arcimboldo, Berra 1998.

25. Morigia 1592, 566.

26. Lomazzo 1973, vol. 1, 363.

27. This paragraph summarizes the discussion in Kaufmann 1993, 154–162, with further references. For Arcimboldo and the grotesque, also see Berra 1998. While Morel (1997, 12–13) allows for the relation of Arcimboldo's paintings to *grilli* and grotesques, he distinguishes his paintings from both of them, and from their method of composition.

28. For full references see Kaufmann 1993, 100–103, with notes, summarized here.

29. Barthes 1978.

30. Paul Wescher, review of Geiger 1954; Legrand and Sluys 1955, in *Burlington Magazine* 98 (1958): 29.

31. The ludic comes into play, as it were, in considering Arcimboldo's paintings as jokes; for a full discussion of this notion, see Kaufmann 1990.

32. Kaufmann 1990, 1993, 103ff.

These themes will be developed more fully in my book. For the context of this discussion, also see Waddington 2004, especially 117–132 for Arcimboldo.

33. Comanini 2001, 266–267: "Lasciamo che no v'ha testa la quale dall'Arcimboldo non sia stata tratta del naturale, percioche l'Imperadore gliene diede la commodità, facendogli veder vivi tutti i sopradetti animali."

34. Rudolf 1995, 167, 183–184, 211.

35. Comanini, *Figino*, in Barocchi, ed. 267: "Vedete pure l'artificio d'uomo et stupitene."

36. This observation pertaining to the Österreichische Bibliothek, Handschriftensammlung, Cod. Min. 42, was first made by Thea Vignau-Wilberg, "Le 'Museum' de l'empereur Rodolphe II et le Cabinet des arts et curiosités," in Vignau-Wilberg et al. 1990, 40–41, 61. These and other similar nature studies by Arcimboldo (or possible identifications and uses thereof) are discussed in Conigliello 1992; Staudinger in Staudinger and Irblich 1996, 233, 260; Hendrix 1997, 157–171; Kaufmann 1997a, 37, 49 n.; Kaufmann 1998, 171, 175 n.; Berra 2000, 70, 71, 83 n.

37. The reference to Arcimboldo in Aldrovandi's *Nachlass* in the University Library, Bologna, and the identification of sheets by him there was first made by Staudinger, and confirmed by Kaufmann (1997a) as well as by Berra (2000). On p. 70, figure 10, Berra illustrates one of the sheets attributable to Arcimboldo in Aldrovandi's volumes; on p. 83, n. 41, he gives the full citation of the notation by Aldrovandi relating to Arcimboldo.

38. Comanini 2001, 265–266: "Fate stima, che non c'é frutto o pur fiore, che non sieno cavati dal natural et imitati con quella maggior diligenza che possibil sia."

39. See Berra 2000, 70, and Aldrovandi's text on 83 n. 41.

Through the continuing research of Anne-Marie Jordan Gschwendt, some of the actual creatures after which Arcimboldo made his nature studies can now be determined. I identify many more drawings and discuss the further significance of Arcimboldo's nature studies in my forthcoming book.

40. Vienna, Österreichische Nationalbibliothek, Cod. Min. 42. For these images and related ones, see Lorenz 1996, 2; *Katalog und Werkverzeichnis*, cat. no. 178–193, pp. 633–638; also Segal, Koreny 1985, 240–247, cat. 88–90. The question of the composition of the volume, its date, the acquisition of the sheets in it, and, most

important for the present question, their attribution, including the assignment of sheets to Arcimboldo, all need further comment. I consider these issues, as well as the discovery of many more nature studies of various kinds in various places, in my forthcoming book.

41. See Vignau, "Le 'Museum,'" in Vignau-Wilberg 1990, 40, 41, and cited by Kaufmann 1997a, 1998.

42. On this point, see Kaufmann 1998. Berra (1998) also draws out connections with Leonardo, but not in reference to these points; Berra's point of reference is reformulated in relation to the article by Kaufmann cited here, in Berra 2000, 69–70.

43. See most completely Kaufmann 1993, 171–193.

44. See Kaufmann 1998, 170, and figure 3 on 173.

45. See Kaufmann 1998, 172. These interests can now be demonstrated to have been continued in Leonardesque circles in Milan, by among others the artist Figino. This issue is discussed further in my forthcoming book.

46. Berra (1998) develops the argument about the relationship to Leonardesque (and Lombard) caricatures, but without reference to the physiognomic tradition. I deal with this and other issues in my forthcoming book.

47. Kaufmann 1998, 172.

48. This point is established in *The Arcimboldo Effect*; see Kaufmann 1987, 96–97.

49. According to a 1582 document cited in Geiger 1954, 125.

50. On Aldrovandi's relations with artists, see in general Olmi 1992.

51. For Ligozzi's studies for Rudolf II, see Conigliello 1991.

52. While reference has continued to be made to this aspect of Prague court culture, the subject remains to be studied as a whole. In the meantime the best general introduction is still found through the references in Evans 1973. But Evans does not devote a separate chapter to natural history or astronomy, and does not emphasize this aspect, or astronomy, at court; see Kaufmann 1993 for other approaches.

Arcimboldo's connection with these figures, and his importance for the development of nature studies, is a theme of my forthcoming book.

53. For a good overview of these studies, see Hendrix 1997.

54. Hendrix 1997, 158.

55. See Kaufmann 1988, 172.

56. See Kaufmann 1988, 170, cat. no. 2.18 and 2.19. Although the second of these pictures has long been accepted as Arcimboldo's work, there has been reluctance to bring either into the context of the development of still life. See however for the

first work Cavalli-Björkman 1994–1995; Berra 1996a, 2000. For the second picture, see most recently Bayer 2004, 162, no. 56.

57. See Berra 1996a, 2000.

58. See Kaufmann 1990, 75–80.

59. See Berra 1989; for a fruit still life by Galizia see Segal 1998, and more recently *Vincenzo Campi*, 222–223, cat. no. 44, entry by Berra.

60. For more detailed information on this painting, see the entry in Berra 2000, 212–213, cat. no. 39, where the image is however reversed. Berra cites my unpublished "Comments on a Reversible Anthropmorphic Portrait of a Man Composed of Fruit by Giuseppe Arcimboldo." The disagreement on the dating of this picture between myself and Berra is slight, and moot for the argument of the present chapter.

61. Rudolf 1995, 166, 166 n. 3: "Et tra l'altre un retratto del Dottor Zasio Iddio gli perdoni, fatto tutto di scritture, di cedoli, di polize, di lettere, et di memoriali; et di un naso [*recte*: vaso] di fiori diversi la metà contrafatti netti, et la metà secchi: che stando il [*recte*: vaso] in suo essere rappresentava una vaghesta mirabili di fiori, voltato il naso al rovescio mostrava una faccia incredibilmente ridicola."

I discuss the significance of the occasion on which this picture was seen in Kaufmann 2007.

62. See Hopper 1988; Vignau-Wilberg 1986.

63. See Kaufmann 1988, 228ff.

64. For these paintings, see Lorenz 1996, vol. 2, cat. no. 178–196, pp. 633–642.

65. This question depends on the determination of when the Tom Ring material came into the imperial collection. Cod. Min. 42 contains materials of various dates, and was bound later than many of the dates shown on the sheets it contains. As Manfred Staudinger has observed to me (personal communication), the fact that the sheets were pasted onto the pages in the codex suggests that it results from two different compilations. It is entirely possible that the Tom Ring studies were in the imperial collection at an early date, regardless of when they were placed in the book.

I provide more discussion of the relation of Arcimboldo to the visual sources of his depiction of nature in my forthcoming book.

66. Berra 1996b, 58:

> DE IOSEPHO ARCIMBOLDIO PICTORE
> Ex hominum membris homines formare/creatrix /Diva potest, folijs, floreq[ue] vestit agros;/ Frondibus humanos, herbisq[ue] virentibus artus/ Haud eadem simili texere docta modo./ Hoc Arcimboldi potuit praestare facultas/ Unica naturae qua superavit opus.

67. Berra 1996b; for the serious joke and Arcimboldo, see Kaufmann 1990.

68. This and other concepts of imitation, and the contest of art and nature, are discussed in other chapters in this book; see especially chapter 3, by Francis Wolff.

69. It may even be that Arcimboldo's representation of the god of transformations, Vertumnus, who is equated with the god of the elements as well, and his series of representations of the four elements also engage with Cicero's comments (*De Natura Deorum* 84) related to the cycle of the elements, and the ability of art to compete with them in its own cycles.

70. The full text of this poem was first published in Kaufmann 1990, 75. I discuss other aspects of the poems in the volume in which this was collected, and Arcimboldo's relation to poetry and to humanism, in my forthcoming book.

71. Comanini, *Figino*, in Barocchi, ed cit, 260 says of Arcimboldo's brush, "Ch'avanza/Pur quel di Zeusi, o quello/Di chi gli fe' l'inganno/Del sottil vel dipinto," as pointed out by Konečný (1988, 154 n. 21).

72. Cited in full in Berra 1996b, 58:

> IL MEDESIMO VERTUNNO/ a riguardanti./ Mad. Di G. A. da Milano
> A che tanto stupore./ E poi lieti ridete?/ certo, che solo un mucchio mi credete/ Di frutti, e di verdura. / Son Vertunno, e mi pregio esser fattura/ De l/ Arcimboldo, di sì alto ingegno,/ Che ben giostra di par con la Natura./ ma che gli manca? Giove il facci degno/ Di quel furto divin con grato ciglio./ Che di Giapeto fece il magno figlio.

73. Cited in Berra 1996b, and previously in Berra 1988, 20:

> ALL/ARCIMBOLDO./ Son. Di G. A. da Milano
> Se di lode, e di fama, assai è degno/ Pittor, che ben ritragga un solo oggetto./ Di quanti ogn'hor ne forma a suo diletto,/ Il supremo rettor del trino Regno:/ o più d'Apelle, e chi giunge al tuo segno?/ Che tanti in un ne aduni, e con effetto/ un viso human ne fingi sì perfetto,/ Che l''ammira ciascun più bell'ingegno?/ Tu di libri, di frondi, frutti, e fiori,/ D'animal vari, e di vari stromenti,/ E d'altre cose ancor, che non descrivo/ Con giuste linee espresse, e suoi colori/ Il viso altrui si ben ne rappresenti/ Che 'n dubbio stassi, qual sia il finto, o 'l vivo.

74. Berra (1996b, 59, n. 22) casts doubt on the possibility of this activity, misstating my reasons for the attribution of a drawing (now in the Berlin Kunstbibliothek) to Fonteo, and ignoring the evidence for this attribution. As stated in Kaufmann 1978, 59, where the monogram is copied, the drawing clearly bears a monogram that can easily be deciphered as "Fonteo."

There is much more to be said about Fonteo's associations, and the importance of Arcimboldo's relation to him as well as to other humanists; these themes are pursued in my forthcoming book.

75. Ovid (*Metamorphoses* 1:82) refers to Prometheus as *satus Iapeto.*

76. Konečný 1988. Konečný, 154–155, no. 21, also refers to the treatment of Arcimboldo as being superior to that of Zeuxis and Parrhasios.

77. Brusati 1997.

78. For this argument, see Kaufmann 1997a.

79. For more on these ideas, and on the Renaissance concept of the art-nature competition, see other chapters in this book, including that by Anthony Grafton (chapter 8).

80. After finishing the initial version of this chapter, I discovered the locus classicus for reversible paintings. The significance of this discovery in relation to humanism, philosophy, and Arcimboldo's still lifes is discussed in my forthcoming book.

## References


Barocchi, Paolo, ed. 1962. *Trattati d'arte del Cinquecento fra Manierismo e Controriforma.* Bari: Giuseppe Laterza e Figli.

Barthes, Roland. 1978. *Arcimboldo.* Parma: Franco Maria Ricci.

Bayer, Andrea, ed. 2004. *Painters of Reality: The Legacy of Leonardo and Caravaggio in Lombardy.* Exhibition catalog, Metropolitan Museum of Art, New York. New Haven, CT: Yale University Press.

Berra, Giacomo. 1988. Allegoria e mitologia nella pittura dell'Arcimboldi: La 'Flora e il "Vertunno" nei versi di un libretto sconosciuto di rime. *Acme* 41 (2): 11–39.

Berra, Giacomo. 1989. Alcune puntualizzazioni sulla pittrice Fede Galizia attraverso le testimonianze del letterato Gherardo Borgogni. *Paragone* 40 (469): 14–29.

Berra, Giacomo. 1996a. Arcimboldi e Caravaggio: "Diligenza" e "patienza" nella natura morta arcaica. *Paragone* 8–10: 108–161.

Berra, Giacomo. 1996b. Un 'Autoritratto cartaceo di Giuseppe Arcimboldi. *Arte Lombarda* 116 (1): 53–62.

Berra, Giacomo. 1998. Arcimboldi: Le teste "caricate" leonardesche; Le "grillerie" dell'Accademia della Val di Blenio. In *Rabisch. Il grottesco nell'arte del Cinquecento. L'Accademia della Val di Blenio Lomazzo e l'ambiente milanese,* 57–67. Exhibition catalog. Lugano, Milan: Skira.

Berra, Giacomo. 1999. Immagini casuali, figure nascoste e natura antropomorfa nell'immaginario artistico rinascimentale. *Mitteilungen des Kunsthistorischen Institutes in Florenz* 43 (2–3): 358–419.

Berra, Giacomo. 2000. Arcimboldi, Vincenzo Campi, Figino, Fede Galizia, Caravaggio; congiunture sulla nasciata della natura morta in Lombardia. In Franco Palliaga, ed., *Vincenzo Campi: Scene del quotidiano,* 60–85. Exhibition catalog. Cremona, Milan: Skira.

Bredekamp, Horst. 1989. Grillenfänge von Michelangelo bis Goethe. *Marburger Jahrbuch für Kunstwissenschaft* 22: 169–180.

Brusati, Celeste. 1997. Natural artifice and material values in Dutch still life. In Wayne Franits, ed., *Looking at Seventeenth-Century Dutch Art: Realism Reconsidered,* 144–157. Cambridge: Cambridge University Press.

Cavalli, Björkman Görel. 1994–1995. A reversible fantasy portrait by Giuseppe Arcimboldo. *Art Bulletin of Nationalmuseum Stockholm* 1–2: 15–17.

Comanini, Gregorio. 2001. *The Figino, or On the Purpose of Painting: Art Theory in the Late Renaissance.* Trans. and ed. Ann Doyle-Anderson and Giancarlo Maiorino. Toronto: University of Toronto Press.

Conigliello, Lucilla. 1991. Pesci, crostacei e un'iguana per l'imperatore Rodolfo II. *Paragone* 42 (493–495): 22–29.

Conigliello, Lucilla. 1992. L'altra faccia di Arcimboldo. *Paragone* 43 (509–511): 44–50.

Daston, Lorraine, and Katherine Park. 1998. *Wonders and the Order of Nature.* New York: Zone Books.

Evans, R. J. W. 1973. *Rudolf II and His World: A Study in Intellectual History 1576–1612.* Oxford: Oxford University Press.

Falchetta, Pietro, ed. 1987. Anthology of sixteenth-century texts. In *The Arcimboldo Effect*, 164–188. Exhibition catalog. Venice, Milan: Bompiani.

Geiger, Benno. 1954. *I dipinti ghiribizzosi di Giuseppe Arcimboldi: Pittore illusionista del Cinquecento (1527–1593).* Florence: Vallecchi.

Hendrix, Lee. 1997. Natural history illustration at the court of Rudolf II. In *Rudolf II and Prague*, 157–171. Exhibition catalog. London, Milan, Prague, Thames and Hudson, Skira, Prague Castle Administration, 96–106.

Hopper, Florence. 1988. Jacques de Gheyn II and Rudolf II's collection of nature drawings. In *Prag um 1600: Beiträge zur Kunst und Kultur am Hofe Rudolfs II.* Freren: Luca Verlag.

Kaufmann, Thomas DaCosta. 1978. *Variations on the Imperial Theme in the Age of Maximilian II and Rudolf II.* Outstanding Dissertations in the Fine Arts. New York: Garland.

Kaufmann, Thomas DaCosta. 1982. The eloquent artist: Towards an understanding of the stylistics of painting at the Court of Rudolf II. *Leids Kunsthistorisch Jaarboek* 1: 119–148.

Kaufmann, Thomas DaCosta. 1987. The allegories and their meaning. In *The Arcimboldo Effect*, 89–108. Exhibition catalog. Venice, Milan: Bompiani.

Kaufmann, Thomas DaCosta. 1988. *The School of Prague: Painting at the Court of Rudolf II.* Chicago: University of Chicago Press.

Kaufmann, Thomas DaCosta. 1990. Arcimboldo's serious jokes: "Mysterious but long meaning." In Karl-Ludwig Selig and Elizabeth Sears, eds., *The Verbal and the Visual: Essays in Honor of William Sebastian Heckscher*, 59–86. New York: Italica Press.

Kaufmann, Thomas DaCosta. 1993. *The Mastery of Nature: Aspects of Art, Science, and Humanism in the Renaissance.* Princeton, NJ: Princeton University Press.

Kaufmann, Thomas DaCosta. 1997a. Caprices of art and nature: Arcimboldo and the monstrous. In Ekkehard Mai and Joachim Rees, eds., *Kunstform Capriccio: Von der Groteske zur Spieltheorie der Moderne*, 33–51. Cologne: Verlag der Buchhandlung Walther König.

Kaufmann, Thomas DaCosta. 1997b. Perspectives on Prague: Rudolfine stylistics reviewed. In *Rudolf II and Prague*, 96–106. Exhibition catalog. London, Milan, Prague: Thames and Hudson, Skira, Prague Castle Administration.

Kaufmann, Thomas DaCosta. 1998. Giuseppe Arcimboldo, the Habsburgs' Leonardo. In Lubomír Konečný, Beket Bukovinská, and Ivan Muchka, eds., *Rudolf II, Prague and the World: Papers from the International Conference Prague 2–4 September, 1997*, 169–176. Prague: Artefactum.

Kaufmann, Thomas DaCosta. 2004. *The Eloquent Artist: Essays on Art, Art Theory and Architecture, Sixteenth to Nineteenth Century*. London: Pindar Press.

Kaufmann, Thomas DaCosta. 2007. Arcimboldo and the Elector of Saxony. In Sybille Ebert-Schifferer, ed. *Scambio culturale con il nemico religioso: Italia e Sassonia attorno al 1600*, 27–36. Rome: Sylvana.

Kemp, Martin. 1995. 'Wrought by no artist's hand': The natural, the artifical, the exotic, and the scientific in some artifacts from the Renaissance. In Claire Farago, ed., *Reframing the Renaissance: Visual Culture in Europe and Latin America*. New Haven, CT: Yale University Press.

Kenseth, Joyce, ed. 1991. *The Age of the Marvelous*. Exhibition catalog. Hanover, NH: Hood Museum of Art, Dartmouth College.

König, Eberhard, and Christian Schön, eds. 1996. *Stilleben*. Geschichte der klassischen Bildgattungen in Quellentexten und Kommentaren, 5. Cologne: Dietrich Reimer.

Konečný, Lubomír. 1988. Zeuxis in Prague: Some thoughts on Hans von Aachen. In *Prag um 1600: Beiträge zur Kunst und Kultur am Hofe Rudolfs II*, 147–155. Freren: Luca Verlag.

Koreny, Fritz. 1985. *Albrecht Dürer und die Tier- und Pflanzenstudien der Renaissance*. Munich: Prestel.

Legrand, F. C., and F. Sluys. 1955. *Giuseppe Arcimboldo et les Arcimboldesques*. Brussels: Le Nef de Paris.

Lomazzo, Gian Paolo. 1973. Idea del Tempio della Pittura. In Roberto Paolo Ciardi, ed., *Scritti sulle Arti*. Florence: Marchi and Bertolli.

Longhi, Roberto. 1950. Un momento importante nella storia della "natura morta." *Paragone* 1 (1): 34–39.

Lorenz, Angelika, ed. 1996. *Die Maler Tom Ring*. Exhibition catalog. Münster: Westfälisches Landesmuseum für Kunst und Kulturgeschichte.

Maiorino, Giancarlo. 1991. *The Portrait of Eccentricity: Arcimboldo and the Mannerist Grotesque*. University Park, PA: Pennsylvania State University Press.

Morigia, Paolo. 1592. *Historia dell'antichità di Milano*. Milan: Giovanni'Antonio de gli Antonij.

Olmi, Giuseppe. 1992. *L'inventario del mondo: Catalogazione della natura e luoghi del sapere nella prima età moderna*. Bologna: Società editrice Il Mulino.

Panofsky, Erwin. [1924] 1968. *Idea: A Concept in Art Theory*. Trans. Joseph J. S. Peake. Columbia: University of South Carolina Press. (1st ed. Leipzig, 1924.)

Rudolf, Karl. 1995. Die Kunstbestrebungen Kaiser Maximilians II. im Spannungsfeld zwischen Madrid und Wien: Untersuchungen zu den Sammlungen der österreichischen und spanischen Habsburger im 16. Jahrhundert. *Jahrbuch der kunsthistorischen Sammlungen in Wien* 55: 165–256.

Schlosser, Julius von. 1978. *Die Kunst- und Wunderkammern der Spätrenaissance: Ein Beitrag zur Geschichte des Sammelwesens*. 2nd ed. Braunschweig: Klinkhardt & Biermann.

Schneider, Norbert. 1998. *Still Life: Still Life Painting in the Early Modern Period*. Trans. Hugo Beyer. Cologne: Taschen.

Segal, Sam. 1998. An early still life by Fede Galizia. *Burlington Magazine* 140 (1140): 164–171.

Spike, John. 1983. *Italian Still Life Paintings from Three Centuries*. Exhibition catalog. New York: Centro Di.

Staudinger, Manfred, and Eva Irblich. 1996. Naturstudien Kaiser Rudolfs II. (1576–1612). In *Thesaurus Austriacus: Europas Glanz im Spiegel der Buchkunst*. 229–286. Exhibition catalog. Vienna: Österreichische Nationalbibliothek.

Sterling, Charles. 1952. *La Nature morte de l'antiquité à nos jours*. Paris: Éditions Pierre Tisné.

*Stilleben in Europa*. 1979. Exhibition catalog. Münster: Westfälisches Landesmuseum für Kunst und Kulturgeschichte.

Vignau-Wilberg, Thea. 1986. Die Randilluminationen und Initialen. In *Das Gebetbuch Kurfürst Maximilians I. Von Bayern*, 65–118. Frankfurt am Main: Fischer, and Stuttgart: Müller & Schindler.

Vignau-Wilberg, Thea, et al., eds. 1990. *Le Bestiaire de Rodolphe II. Codex 129 et 130 de la Biblioteque Nationale de Vienna*. Paris: Citadelle-Mazenod.

Waddington, Raymond B. 2004. *Aretino's Satyr: Sexuality, Satire, and Self-Projection in Sixteenth-Century Literature and Art*, 117–132. Toronto: University of Toronto Press.

# 8

## Renaissance Histories of Art and Nature

Anthony Grafton

In *The City of the Sun*, Tommaso Campanella envisioned the ideal state as a visual encyclopedia of nature and culture. In the Solarians' circular city on a hill, seven walls surround an enormous central temple. Covered with maps, diagrams, and specimens, the walls survey the natural world, realm by realm. Each of them provides full information, not just about the properties of the things or beings portrayed on it and the specimens in bottles in its niches, but also on their relations to the cosmos as a whole. The outer wall of the third circuit, for example, shows "all manner of fish to be found in river, lake or ocean; their particular qualities; the way they live, breed, develop; their use; their correspondence to celestial and earthly things, to the arts, and to nature."[1] Outward marks—that of the bishop fish, for instance—serve as signatures, clues to inner natures. They reveal each fish's place in a web of natural sympathies and antipathies that runs from the stars to the earth, and from the fish themselves to the bodily organs they resemble.

Campanella's fish stories—like those about stones and animals—involve not only cosmic correspondences, but human agents. For over the centuries, in and outside the City of the Sun, human effort has uncovered natural virtues that would otherwise have remained hidden. The Solarians record the history of this collaboration between humanity and nature with special care and passion. Accordingly, the inner wall of the sixth circuit displays the mechanical arts "together with their inventors, their diverse forms, and their diverse uses in different parts of the world," and its outer wall commemorates not only the founders of laws, but also the inventors of weapons—including the Chinese creators of printing and firearms.[2]

These visual histories of crafts do more than celebrate the works of great individuals. The Solarians' children, exposed to the series of powerful images from birth, "come to know all the sciences pictorially before they are ten years old."[3] No wonder that, unlike Europeans, they nurse no prejudices against the mechanical arts: "They laugh at us because we consider craftsmen ignoble and assign nobility to those who are ignorant of every craft and live in idleness."[4] Indeed, they require candidates for their highest official position, that of the Sun, to have mastered all the mechanical arts and their histories. A material history of the relations between art and nature, in other words, was to play a vital role in the utopia—a society with no idle members, no beggars, and no nobility—that Campanella hoped to create from the urban and rural poor of south Italy after he overthrew their Spanish rulers.[5]

At the other end of late Renaissance Europe, Francis Bacon nourished similar hopes. He did not represent his *New Atlantis* as a single great museum of natural history. But Salomon's House, its central institution, closely resembled one, as many interpreters have pointed out. And it too includes—along with its parks, orchards, "places for breed and generation of those kinds of worms and flies which are of special use," and "houses of deceits of the senses"—"two very long and fair galleries." One of these contains "patterns and samples of all manner of the more rare and excellent Inventions," the other "the statua's of all principal inventors." The member of Salomon's House who explains the institution to Bacon's narrator emphasizes that these twin galleries of invention are vital to its imaginative—and material—economy. His colleagues take special care to establish the identity of older inventors "by more certain tradition" than Europeans have, and to provide the creators of new mechanical arts with splendid brass, marble, jasper, quartz, or cedar statues of their own and "a liberal and honourable reward" as well.[6]

When Campanella and Bacon juxtaposed natural knowledge with the fine and mechanical arts, as many interpreters have pointed out, they built on rich precedents, written and material. The courts and cities of sixteenth-century Europe harbored *Kunst- und Wunderkammern*, where the marvels of nature flanked the most spectacular products of human industry. Samuel Quiccheberg surveyed many collections, most notably the spectacular one of the rulers of Bavaria, before he offered his plan for displaying "artificial and miraculous things" in a "theater" in 1565. Like the *City of the Sun* and *Salomon's House*, Quiccheberg's theater sets out an inventory of nature and culture: pictures of rare animals and fish, skeletons of men and animals, seeds, fruits, metals, precious stones, and every imaginable product of human industry, from clothing to scalpels to weapons of war—as well as maps, city views, paintings, and prints.[7]

Bacon and Campanella drew on many models, including the working establishments of Giambattista della Porta, John Dee, and Cornelis Drebbel.[8] But it seems clear that both men had the sixteenth-century Kunst- und Wunderkammer primarily in mind as they sketched plans for what later generations, resolutely anachronistic, would see as modern laboratories and scientific states.[9] Their visions of the ideal site for study of nature resembled those of traditionalists like Jean Bodin, whose natural philosophy was a mosaic drawn for the most part from books—but who imagined the Venetian house where seven wise men met to discuss the religions of the world as a "pantotheca" divided into 1,296 "capsulae," stuffed with specimens of metals, plants, fossils, and animals, as well as images of creatures too numerous or large to be accommodated, from insects to the rhinoceros.[10]

In one respect, Bacon's and Campanella's state museums seem to break with long-standing cultural traditions. Both men portray nature as changing, gradually, because of human intervention. The Solarians use art, eugenics, and astrologically timed sex to improve the human stock of their city. After watching young men and women wrestle naked, the elders pair them off in ways determined by their physical characteristics and temperaments: "Tall handsome women are not matched with any but tall brave men, while fat women are matched with thin men and thin women with fat ones, so as to avoid extremes in their offspring." Before intercourse, the young digest their dinners and pray—and the women examine "fine statues of illustrious men." Only then do they meet, at an astrologically determined time. The results are clear: this careful eugenic regime and the constant exercise required of girls has eliminated ugly women from the city—as opposed to Europe, where women's idleness "makes them pale, fragile, and short and creates a need for artificial coloring, high heels, and beauty care."[11]

The denizens of Salomon's House seek "the knowledge of Causes, and secret motions of things; and the enlarging of the bounds of Human Empire to the effecting of all things possible." And they too can alter nature in crucial respects—for example, by changing the normal course of human life. The artificial metals that they create in their deep caves, for instance, cure otherwise incurable diseases, and prolong the lives of "some Hermits that choose to live there, well accommodated of all things necessary, and indeed live very long."[12] Similarly, their subtle breads nourish chosen individuals so well that they live long lives without taking any other nourishment. They can even create serpents, worm, flies, and fish that become "perfect Creatures, like beasts or birds, and have sexes and do propagate."[13] In their belief that humanity's relation with nature could be altered in a uniform way, could be steadily improved, both men seem to look forward to Descartes and other later writers, who would make the possibility of material improvement in the human condition one of the most powerful slogans of the New Philosophy.

Some years ago, Horst Bredekamp connected these assertions to the Kunst- und Wunderkammer. Both collections and images of them, he pointed out, often juxtaposed stones, plants, and animals with statues—sometimes partially completed ones, half hewn from native rock. They often included automata—creatures crafted from inanimate materials to imitate the movements and qualities of living beings. And as Bredekamp and others have shown, these objects and devices spoke volumes, at least to informed viewers. Lorraine Daston, Katherine Park, Martin Kemp, and others have taught us how many objects in Renaissance collections deliberately challenged the onlooker to determine where nature left off and art began. And Thomas Kaufmann has revealed that Arcimboldo and other artists favored by the makers of Kunst- und Wunderkammern were fascinated by the historical development of arts and crafts.

Bredekamp brought his argument to a characteristically provocative close. Natural historians in the fifteenth and sixteenth centuries, following Pliny, saw natural history as a synchronic, not a diachronic, discipline. They mapped the variations of what they saw as a stable natural world, not its transformation over time. Scholars believed that art could help and even perfect nature—but only by drawing on nature's own resources, and within the narrow limits those resources imposed.[14] But artists and collectors came to see nature as changeable and human effort as a force constantly acting on it. And they arranged collections and devised images to tell a visual story about the direction in which nature moved—a story they literally could not have told in words, since they lacked both terminology and textual models. Two generations after Kunst- und Wunderkammern, botanical gardens, and grottos came into their full, variegated flower, Campanella and Bacon finally found the words for an argument long made by assemblages of objects and in pictures. What seem the most radical segments of their works thus express a view of nature that had long been literally embodied in specific sites of intellectual and artistic inquiry. More recently, Markus Popplow has argued, in a similar vein, that philosophers and humanists did not articulate a coherent vision of technological progress until they saw the one that engineers expressed, with dazzling theatricality, in the "Theaters of Machines" of the decades around 1600: the printed collections of designs, too general and inaccurate to serve as working drawings, and sometimes impossible to carry out at all, that advertised the prowess of inventors like Agostino Ramelli, Jacques Besson, and Salomon de Caus.[15]

These analyses have permanently altered our understanding of the great manifestos of *Science in Utopia.* They form a prism, read through which older works take on a radically new look. Quiccheberg, for example, made clear that his "theater" should include both real "miraculous and rare animals, such as rare birds, insects, fish, shells and the like" and, next to them, "sculpted animals of metal, plaster, clay, and any artificial material, which art makes appear to be alive."[16] He also called for small-scale models of machines for drawing water, sawing wood, and pulling ships, so that "the examples of these little machines or structures might make it possible to create other, larger ones in the proper way, and gradually to invent better ones."[17] Quiccheberg nowhere argues explicitly that people change nature in a directed, consequential way. But the implication seems clear when his work is read in light of Bredekamp's argument.

Both Campanella and Bacon, however, cast their nets very widely. Much though the Kunst- und Wunderkammern mattered to them, other sources, visual and textual, also helped them frame their visions of nature in directed motion. Consider, for example, the textual world of natural history as it developed in the sixteenth century. Even a dedicated humanist who worked in the field, like Marcello Virgilio Adriani, boasted of his efforts to consult "doctors expert about herbs, who work in the fields," and to offer "eyewitness testimony about plants," though he had to import them expensively from other countries.[18] Adriani insisted that "nature's power extends widely, and its majesty is great; hence it can produce many things in different forms, and grant others what it has denied us and our world."[19] Just as crops improved when many worked together to plow and burn over fallow fields, so medicine would improve when many collaborated to bring textual knowledge and observation together—a program carried out to a large extent, by Mattioli, whose edition of Dioscorides became a compendium of recent botanical research.[20] The textual histories of nature that Campanella wished to transform into frescos and cabinets already transmitted parts of the message that his imagined city was to preach with its very stones.

Other texts also helped to stimulate the utopian imagination. Bacon, for example, began to imagine his ideal research center, as Kaufmann and others have shown, in the 1590s. By 1608, in a set of notes on what he called "a place to command wytts and pennes," he expressed his disapproval of the denizens of existing institutions like Eton, Winchester, Trinity, and St. John's. "*It must be the postnati,*" he wrote, not established scholars, who would create a new way of studying nature. And they would do so, ideally, in "a college for Inventors." This, in turn, must have "2. Galeries wth Statuas for Inventors past and spaces and bases for Inventors to come And a Library and an Inginary"—not just a material history of the previous development of the mechanical arts, in other words, but one designed to follow their growth and change over the years.[21]

Just before this passage, Bacon made clear the sort of subjects that the inhabitants of his new college would pursue. On the one hand, they would compile from ancient and modern sources, "wth Judgmt and without credulity," a "History of Marvailes, Historia naturae errantis aut variantis"—or, as he put it elsewhere, "Hystorie of all sorts for matters strange in nature told in serie temporum heare and there inter caetera." On the other hand, they would "procure an History mechanique," which would lay out "the experimts and observations of all Mechanical Arts." This would include materials, instruments or engines, their uses, and "the woork it self and all the processe thereof," as well as "observacions, Axiomes, directions."[22]

Between the history of marvels and that of inventions Bacon cited one near-contemporary source: "Pancarolus, de reb. Memorabilibus."[23] And this reference suggests something of the wider range of sources on which he drew. The Italian jurisconsult Guido Panciroli had published, some decades before Bacon wrote, a short treatise in Italian on the lost inventions of the ancients and the new ones of the moderns. In 1603, just before Bacon wrote, Heinrich Salmuth republished this in a Latin translation, with an exhaustive commentary. This compendium made clear that the ancients had created wonders that the moderns could not match, such as Egyptian obelisks and pyramids. To forget this, Salmuth warned, was to imitate the Greeks, whose ignorance of ancient history the Egyptians had ridiculed, calling Solon himself a mere child when he revealed that he did not know the history of Atlantis. Knowledge of the history of inventions, in short, was urgent, and required careful reading of ancient texts. But it also required elaborate study of the postclassical world, which had devised such remarkable innovations as Greek fire and the compass, and had used the latter to discover the New World.

Panciroli and Salmuth made clear, finally, that culture always mediated the ways humanity exploited nature. The ancients, Panciroli argued, had had sugar, but had used it only for medicine. In recent times, however, new ways of making and purifying sugar had produced immense quantities of it. And human ingenuity had produced a world of things made up of this ancient substance, a new sort of sweet-tasting Kunst- und Wunderkammer: "Nowadays this art has reached such a pitch of subtlety, that rhubarb, pine nuts, pistachios, cinnamon and other species are candied in sugar, and thus preserved as if they were still fresh. Very pretty figures and little images are also fashioned from sugar, and fruits of all kinds are represented, so that they seem natural and wild."[24] Salmuth developed the point more gloomily, pointing out that the new availability of sugar had produced addicts, who overheated their blood, developed a perpetual thirst, and contracted disgusting black marks on their teeth by their immoderate eating of sugar and "sugary confections."[25] Nature, evidently, could jump through any hoop that human inventiveness held up for it—including the apparent transmutation of metals, which both Panciroli and Salmuth described at length in their entry on alchemy. Bacon envisioned his collection of material on the arts as both a compilation—rather like a commonplace book, like the work of Panciroli and Salmuth—and a museum, in which statues would commemorate inventions and empty pedestals would stimulate others to emulate them. And he thus revealed his belief that he could draw on existing verbal and material sources to frame his vision of human agency on nature.

Existing literary and artistic traditions offered Bacon and Campanella rich resources to work from. Pliny himself made clear, for example, that the fine arts had developed over time. A long series of anecdotes, repeated ad nauseam and beyond by Renaissance writers, transmuted the histories of painting and sculpture into a series of problems that artists had posed and solved, first in antiquity and then, after the arts revived, in the fourteenth and fifteenth centuries. Artists did not need to be erudite in order to frame their own work in these terms. When Lorenzo Ghiberti examined an ancient statue of a hermaphrodite, for example, he called special attention to the way "one of the legs was stretched out and the large toe had caught the cloth, and the pulling of the cloth was shown with wondrous skill."[26] Similarly, he pointed that when he fashioned the Gates of Paradise for the Florentine Baptistery, he "strove to imitate nature as closely as I could, and with all the perspective I could produce." He had set all his "histories" in frames, for instance, "because the eye from a distance measures and interprets the scenes in such a way that they appear round."[27] From the beginning of the fifteenth century, in other words, the fine arts provided perfect examples of a human pursuit in which continued inventiveness revealed many—if not limitless—new possibilities.

Leon Battista Alberti took this argument a step further in his treatise *De statua.* This little work began, like Ghiberti's autobiographical *Commentaries*, with a history of the art of sculpture. But Alberti, here as elsewhere, told an idiosyncratic and revealing story. The first likenesses, he explained, had been images made by chance: "certain outlines," observed in tree trunks, clods of earth, and other inanimate objects, "in which, with slight alterations, something very similar to the real faces of Nature was represented." This view was not completely unprecedented. In a classic article H. W. Janson collected other passages, ancient and modern, which referred to nature's ability to produce likenesses.[28] In fact, many believed that natural forces could stamp an exact likeness on stones. Writing around 1260, the Dominican philosopher Albertus Magnus noted that the stars, in certain conditions, might impress a recognizable image on "precious stones and certain marbles"—as they had impressed the image of a king's head on a cameo that he saw in the shrine of the Three Kings in Cologne Cathedral.[29]

Alberti, however, had a different point in mind. He described the natural images of faces as approximate and imperfect. Sculptors, he argued, "began by diligently observing and studying such things, "to try to see whether they could not add, take away or otherwise supply whatever seemed lacking to effect and complete the true likeness." Correction and competition, new creations and critical discussions, ensued. "By emending and refining the lines and surfaces" produced by nature, sculptors "achieved their intention and at the same time experienced pleasure in doing so."[30] Sculpture, in other words, crystallized into an art only because humans collaborated with and corrected nature, and one another as well. Alberti meant his own treatment of the making of statues—as he made clear—to improve the art still more, by giving sculptors a "firm method" that would enable them to avoid mistakes.

Sculpture, Alberti claimed, needed only the means that nature itself provided in order to provide better likenesses than unaided nature could: "Just as in a tree-trunk or clod of earth Nature's suggestions made men feel it possible to create something similar to her products, so in Nature herself there lies to hand something which provides you with a method and certain, exact means whereby you may with application achieve the highest excellence in this art."[31] When the sculptor studied the nature of likeness or resemblance, when he mastered the mathematical techniques that made it possible to measure and record human bodies in all their postures, he was only learning to apply "the convenient and necessary means . . . that Nature offers to sculptors to execute their work perfectly."[32] Human artistry could imitate *natura naturata* more elegantly than *natura naturans* could.

Yet Alberti assigned the artist tasks even more complex and demanding than straightforward imitation of the natural world. In his treatise of the mid-1430s *On Painting*, he told the story of how the painter Zeuxis went about his task:

> The idea of beauty, which the most expert have difficulty in discerning, eludes the ignorant. Zeuxis, the most eminent, learned and skilled painter of all, when about to paint a panel to be publicly dedicated in the temple of Lucina at Croton, did not set about his work trusting rashly in his own talent as all painters do now; but, because he believed that all the things he desired to achieve beauty not only could not be found by his own intuition, but were not to be discovered even in Nature in one body alone, he chose from all the youth of the city five outstandingly beautiful girls, so that he might represent in his painting whatever feature of feminine beauty was most praiseworthy in each of them.[33]

In this case, Alberti took Zeuxis's efforts as directed toward perfect representation: "He acted wisely, for to painters with no model before them to follow, who strive by the light of their own talent alone to capture the qualities of beauty, it easily happens that they do not by their own efforts achieve the beauty they seek or ought to create; they simply fall into bad habits of painting, which they have great difficulty in relinquishing even if they wish."[34] Yet even here Alberti did not suggest that created beauty could prove superior to natural beauty.

In *De statua*, Alberti cited the same passage to a radically different effect, to explain why he had composed his own table of ideal human measurements: "I proceeded accordingly to measure and record in writing, not simply the beauty found in this or that body, but, as far as possible, that perfect beauty distributed by Nature, as it were in fixed proportion, among many bodies; and in doing this I imitated the artist at Croton who, when making the likeness of a goddess, chose all remarkable and elegant beauties of form from several of the most handsome maidens and translated them into his work."[35] Intensive study of nature and resemblance should yield the ability to create not only the likeness of an individual—such as Alberti's own famous self-portrait plaquette, with its unforgettable beaky profile—but also that of a type, one that embodied an ideal of beauty not found in any individual man or woman.

Over time, as Erwin Panofsky showed in *Idea*, artists and writers on the arts followed the Alberti of *De statua* in asserting that the artist must produce a higher beauty than nature could. Lodovico Dolce, for example, wanted the painter "not only to copy nature but also to surpass it ... to show ... in a single body all that perfection of beauty, that nature hardly chooses to reveal in a thousand." Raphael complained, in a famous letter directed to (and perhaps written by) Baldesar Castiglione, that he could not find the beautiful models on whom Zeuxis had relied, while in any case Castiglione was not available to help him make a selection. "Since there are so few beautiful women and so few sound judges," he concluded, "I make use of a certain idea that comes into my head." Giorgio Vasari argued in his *Lives* that the artist could form his ideas only from experience: "Design is the imitation of the most beautiful things in Nature in all forms."[36] But he also argued that the greatest ancient artists had surpassed nature, and that Michelangelo had surpassed them.[37] As art developed, for these writers, it became not nature's ape, but nature's rival.

As early as 1557, Julius Caesar Scaliger made clear, in his famed attack on Girolamo Cardano, what such a doctrine might imply. The human mind, Cardano had argued, delights in false things, because they provoke wonder. Accordingly, only those leeser intellects that were capable of amazement and delight could experience the pleasure of the fake: those of boys and fools, not those of wise, mature men. Scaliger accepted this argument, but also qualified it. Even wise men loved Homer's inventions, he pointed out, and kings loved fools. His argument was simple. The human mind, being infinite by nature, loves new things—especially those, like "pictures of monsters," that "surpass the common boundaries of truth." The wise man praises pictures, "even though he hardly fails to realize that they are feigned." Moreover, "the wise man prefers a pretty image to one that resembles a particular natural being. For art surpasses nature in this respect. The symmetry [of the human form] has undergone many forms of corruption since the first man. But nothing prevents the artist from raising, lowering, adding, removing, twisting or pointing. In fact, this is my view: with two exceptions, that of the first man and that of the true man and true God, no human body was ever as skillfully made by nature as they are perfectly framed nowadays by the learned hands of craftsmen."[38] The artist, in Scaliger's view, approached divinity. He created forms so beautiful that they almost seemed exempt from the general corruption of nature introduced by the Fall. In the later sixteenth and early seventeenth centuries, a number of theorists of art drew the practical implications of Scaliger's thesis. Giovanni Paolo Lomazzo, for example, noted that the artist making a portrait of a woman "should use extreme diligence to achieve beauty, using art, so far as possible, to remove the errors of nature."[39] The most remarkable implication of Scaliger's view, however, was perhaps the thesis it implied about the direction of time's arrow: that high art could reverse the degenerative course of human and natural affairs.

The distinction between fine and mechanical arts, as Bredekamp points out in his book, took shape long after the early modern period. In the fifteenth and sixteenth centuries, many artificers now remembered above all as painters, sculptors, or architects earned their living by what now looks like mere craft—as Jean Fouquet did when he made gunpowder and cannon for the Estensi.[40] When Leonardo promised Ludovico Sforza that he could make movable bridges, tanks, mortars, and mines, before he mentioned sculpture, he was promising not much more than Brunelleschi had offered his civic patrons almost a century before.[41]

Most of the humanists who discussed the arts also paid close attention to the crafts and their results. The "Anonymous Life" of Alberti mentions that he interrogated skilled craftsmen about their techniques. He himself treated the sheer size of Brunelleschi's dome for the Florentine Cathedral, "vast enough to cover the entire Tuscan population with its shadow," as the clearest evidence that nature had not, as he himself once feared, "grown old and weary."[42] The power of engineering, Alberti now realized, as well as the artistry of Florentine sculptors and painters, showed that Lucretius was wrong to see nature as degenerating. Brunelleschi outdid the ancients not because his building was more handsome than theirs but because his skills and knowledge enabled him to create an "artificio" that even the Greeks or Romans could not have pulled off. In this case too, Alberti portrayed relations between human effort and nature as moving in a clear direction, not because art surpassed the beauty of nature, but because nature herself was still producing "ingegni" on a grand scale, and they in turn exploited—and revealed—her hidden powers.

Not long after Alberti wrote his letter to Brunelleschi, the supreme humanist Lorenzo Valla confronted a related set of problems. The consummate master of classical Latin, Valla recognized that modern ingenuity had created devices unknown to the ancients: the bell (*campana*), the clock (*horologium*), and the compass (*pyxis nautica*), among others. Accordingly, even a peerless classicist like Valla had to use the proper modern terms for them. Valla clearly took more than a casual interest in these matters. In discussing the clock, for example, he made clear that he knew the ancients had had sundials, hourglasses, and water clocks. Even so, he insisted on the novelty of the modern device: "I mean the one that is truly a clock [*horologium*], in which one detects not only the numbers of their hours, but, so to speak, their speech. . . . It does not only set out the hour to the eye, but also announces it to the ears of those far away and those at home, using the bell mounted on top of it to distinguish the numbers. Nothing could be more useful or more pleasant."[43] The great iron escapement clocks that rang the hours in Italian cities were something genuinely new in the world. "And it is certainly necessary," Valla wrote, "that the learned decide by what names we should refer to things that were invented not very long ago." Like Alberti, Valla took the existence of such devices as evidence of something fundamental about the order of things: "The minds of mortals are not yet exhausted." Like Alberti too, he argued that in some respects at least, the moderns "approached quite near to the high competence of the ancients."[44] Valla himself did not publish this text—but his friend Giovanni Tortelli inserted most of it into his famous lexicon, the *De orthographia*.[45] And the Florentine humanist Giannozzo Manetti, who may well have discussed these matters with Valla at Naples, cited the achievements of engineers and architects as powerful evidence for the dignity of man in the oration on the subject that he wrote for Valla's patron, Alfonso of Aragon, in 1452. Manetti cited the achievements of outstanding modern individuals, like Brunelleschi and Ghiberti; traced the history of navigation, which had opened up the whole world; and celebrated the architects and builders who had transformed the world itself into a human creation: "These creations are ours—that is human—since they are clearly the work of men: all the houses, all the towns, all the cities, and finally all the structures in the world, which are so great and so excellent, that they may well seem, because of their excellence, the work of angels rather than men. The paintings are ours, the sculptures are ours, the arts are ours, the sciences are ours."[46] By the middle of the fifteenth century, in short, the history of crafts was well established as an area of interest to scholars as well as artists—and one comparable, in its dignity, to the history of philosophy itself.

The craftsman's effective intelligence, moreover, had come to serve as model of the divine mind at work. Marsilio Ficino interpreted the speaking statues of the ancient Egyptians not as automata but as stone sculptures animated by the proper incantations. He distinguished sharply between "the work of craft," which he defined as "the mind of the artist in disjunct matter," and "the work of nature," which he defined as "the mind of nature in conjunct matter": "The order of a work of nature ... is more like the order in the art of nature than the order of a human artifact is like the art of man. This is to the degree that matter is closer to nature than to man and nature has greater sway over matter than man does."[47]

But Ficino also treated the work of especially skilled craftsmen as the best analogy for that of the creator of the universe itself: "We saw recently in Florence a small cabinet made by a German craftsman in which statues of different animals were all connected to, and kept in balance by, a single ball. When the ball moved, they moved too, but in different ways: some ran to the right, others to the left, upwards or downwards, some that were sitting stood up, others that were standing fell down, some crowned others, and they in turn wounded others." One single ball produced all these movements, he reflected, and the sounds of horns and trumpets and the songs of birds at the same time. "Thus God through His own being... has only to nod His head and everything which depends on him trembles."[48] No wonder that Ficino recommended contemplation of the mechanical clock in the Palazzo della Signoria as a way to gain deep understanding of the cosmos.[49]

In the course of the later fifteenth and sixteenth centuries, finally, three separate developments dramatized and focused attention on the history of the arts. Artists from Piero di Cosimo on gave precise and powerful visual form to the primitive life evoked in different ways by Lucretius and Vitruvius. Fra Giocondo, Cesare Cesariano, and others, for example, offered visual commentaries on Vitruvius in the form of woodcuts. These images highlighted the ways the practical arts developed. Vitruvius described, following older sources, how civilization began when humans had learned to use and control the fires made by branches rubbing against each other in the forest (*De Architectura*, 2.1.2). The discovery of fire's uses stimulated the development of language and housing, and coexistence and collaboration led to the continual improvement of the material condition of humankind. Sixteenth-century illustrators depicted the "life of the first humans" in detail, in ways that showed increasing precision in following Vitruvius's lead. Fra Giocondo, for example, imagined a group of clothed men and women sitting by a fire, equipped with vases (in the background, incongruously, he placed a Gothic city). He thus celebrated the crafts as essential to civic life, but suppressed what Vitruvius had said about their early development. Cesariano followed the text more closely, using the arts as a marker for the development of society. He depicted one group of humans, naked, as fleeing a fire in a clearing. The members of another group, some of them already wearing clothes, tended their common hearth and practiced such crafts as basket weaving. A single woodcut thus became a conjectural history of the origins of civilization.[50]

The exploration of Africa and the discovery of the New World confronted travelers, scholars, and illustrators with the practical problem of representing non-European peoples and the more refined one of describing and assessing their level of cultivation. Occasionally someone met this with an open mind. Dürer, who examined jewelry, clothing, and weapons from the New World at Brussels in 1520, described them as "wonderful works of art" and "marvelled at the subtle *ingenia* of men in foreign lands."[51] But most writers and artists saw primitive artifacts through a scrim of assumptions. Feather headdresses and facial jewelry, for example, came to be associated with societies that had no government, did not maintain European-style regulation over marriage, and ate human flesh. Royal entries, drawings, watercolors, and woodcuts—especially the famous work of Jacques Le Moyne and John White—popularized vivid images of tattooed men and women bearing primitive arms. The artists labeled their figures sometimes as Indians, sometimes as Picts or Germans. The latter step, though it rested on the ancient notion that time and space are convertible in ethnography, had a radical implication: it suggested that Europeans themselves had once gone naked, or largely so, painted their bodies, and wielded simple, primitive tools and weapons.[52] These images reached a vast public, especially in the engravings and captions of Theodore de Bry's *America*. Well before Bacon and Campanella, scholars and artists had outlined the early segments of a "history mechanique"—one that clearly bore the motto "antiquitas mundi juventus saeculi."

The more recent segments of the mechanical history also took on sharper outlines. From the fourteenth century on, engineers like Giovanni Dondi began to highlight the novelty of their work more and more dramatically. Brunelleschi took extreme measures to protect his intellectual property. He carved model machines out of turnips, which would rot away after his workmen had used them and before his enemies could steal his ideas. And he warned the Sienese engineer Mariano

Taccola never to reveal his new ideas to the public, since doing so meant courting insult or plagiarism—or both.[53]

Other engineers, however, adopted a different strategy. Dondi, Taccola, Georg Kyeser, Roberto Valturio, and others produced massive, splendidly illustrated compendiums of their inventions. With these vast manuscripts they staked a double claim, to membership in the community of the learned and to authorship of new and valuable devices.[54] Up-to-date patrons like the Estensi recognized how much these claims meant to engineers, and offered especially gifted ones not only salaries and other privileges, but also protection from competitors. In the sixteenth century, splendidly printed and illustrated books of inventions and the programmatic drawings of *Nova reperta* by Jan van der Straet, later engraved by Johannes Galle, defined a canon of modern inventions as especially dramatic and significant.

By the middle of the sixteenth century, some scholars synthesized these developments and gave them forceful expression.[55] Louis Le Roy argued as early as 1551 that early humans had lived in a barbarous state, mired in the "hard primitivism" described by the historian Diodorus Siculus and others.[56] In his synthetic work of 1575 *De la vicissitude ou variété des choses en l'univers*, he also celebrated in phosphorescent rhetoric the invention of printing, the compass, and gunpowder: "antiquity," he stated baldly, "had nothing comparable to these three." No optimist, Le Roy recorded the appearance of syphilis and the degradation of morality among the "marvels" of his age. But he also predicted that "what is now hidden will come to light with time, and our successors will be amazed that we were ignorant of them."[57]

Le Roy's contemporary Jean Bodin went further. In his *Methodus ad facilem historiarum cognitionem* of 1566, he used the same three canonical inventions, printing, the compass, and the gun, as well as modern alchemy and many others, to refute in formal terms the myth of the golden age. The earliest ancients had been little more than savages, he argued; they had not even understood that they should assess the death penalty for theft. And even the greatest ancient achievements in technology and warfare paled before the discoveries and inventions of the moderns. The ancients' catapults, Bodin scornfully pointed out, "might seem little more than toys if compared with modern cannon." The ancients had not known the "divine use" of the magnet for purposes of navigation. And printing "could easily compete on its own with all the inventions of all the ancients." More startling discoveries about the "hidden secrets of nature," moreover, would be made in future years. For nature, Bodin insisted, "holds inexhaustible treasures of knowledge, which can never be exhausted."[58] Like Le Roy, Bodin expected the progress of the arts to cease at times, as the vicissitudes of the cosmos required, and he assumed that humans could only learn what nature would teach. Nonetheless, he and Le Roy made clear that, over time, the progressive work of human ingenuity had radically changed the human relation to the natural world.

At times, finally, a still more radical attitude appeared—for example, in the world of what Francesco Zorzi and Henry Cornelius Agrippa christened "mathematical magic," the sort of magic whose practitioners operated by mixed mathematics and claimed that they could create automata, burning mirrors, submarines, and flying machines. Agrippa argued in both his *De occulta philosophia* and his oration *On the Vanity of the Arts and Sciences* that the insatiable desire of humankind to build had led architects to attack and alter the face of nature itself: "Thorowe whiche insatiable desire and studie of building, it is come to passe, that there is no measure or ende appointed herein: for this cause are hilles cut away, Valleys filled up, Mountaines made plaine, stoanes perced thorowe, and the rockes of the sea discouered, the entrailes of the earthe digged, the riuers turned from their course, seas ioined to seas, lakes consumed, marshes dried vp, armes of the sea barred out, the bottomes of the sea searched out, new Ilandes made, and againe other restoared to the maine lande. All which thinges, and more than these, albeit thei repugne against nature, yet oftentimes haue broughte verie greate commoditie to all the worlde." And he made similar claims about the power of mathematical magic to alter the face of the world—for example, by raising great stones, as the ancient builders of complexes like Stonehenge had done.[59] Agrippa, as is well known, derived part of his optimism about the human ability to work on the cosmos from the vast range of magical, alchemical, and cabalistic works that he read, excerpted, and rewrote.[60] In this case, however, he drew on a different source. An experienced military and mining engineer who had studied the Italian illustrated editions of Vitruvius, Agrippa learned from one of the most articulate Italian writers on the mechanical arts, Alberti himself, that architects could "cut down cliffs, cut through mountains, fill valleys, confine lakes and seas, dry out swamps, build ships, change the direction of rivers" and by doing so "make all the provinces of the world accessible."[61] Generations of mathematical practitioners repeated Agrippa's claims, adding new stories—like the one, apparently created by Ramus, about the mechanical fly and eagle devised by Joannes Regiomontanus to welcome the Holy Roman Emperor to Nuremberg—but never radically changing his central thesis.[62] As late as the middle of the seventeenth century, Gaspar Schott used much the same set of anecdotes Agrippa had—as well as descriptions of his own marvelous machines—to confirm the thesis, founded on Agrippa's fusion of magical and technological traditions, that magic of a particular sort was actually more powerful than nature: "This form of magic not only aids, and perfects nature, as the other [mathematical disciplines] do, but clearly overcomes her." The dazzling mechanical contrivances of the magicians, Schott insisted, involved the enhancement of natural powers and the imitation of every form of motion. He defined their art as "less a discipline that contemplates nature, than one that rules over her."[63] Here—and in other texts where traditions crossed and sparks flew—the idea found expression that man's arts could not only provided increasing control over the natural world, but actually transform it in radical ways.

Even those who set less store by human agency sometimes arrived at similar conclusions. Cardano and Campanella thought that all the inventions of their time—including such "stupendous" ones as the canonical three—had been caused by the action of the stars, which affected everything, if in different ways. "Hence," as Campanella put it, "the constellation that drew infectious vapors from Luther's cadaver drew fragrant exhalations of virtue from the Jesuits of that period and from Hernando Cortés."[64] But they no longer saw the heavens as stable and unchanging, since comets and the new star in Cassiopeia appeared there. They interpreted the radically new technologies of their day, accordingly, as evidence not only of what man could do to nature, but also of an impending transformation of the entire natural and human order. "The conviction grows," Cardano wrote, "that as a result of these discoveries, the fine arts will be neglected and but lightly esteemed, and certainties will be exchanged for uncertainties. These things may be true sometime or other, but meanwhile we shall rejoice as in a flower-filled meadow. For what is more amazing than pyrotechnics? or than the fiery bolts man has invented so much more destructive than the lightning of the gods?"[65] The Christian world, Campanella predicted, would profit from the great changes that impended, "but first the world will be uprooted and cleansed, and then it will be replanted and rebuilt."[66] Meantime the rise of new technologies confirmed other signs that the world was upside down—for example, the rise of women rulers in many parts of Europe, and the growth of effeminacy among men, which made them call each other "Your Lordship" and resort, at least in Fez and Morocco, to male brothels.[67] The transformation of the arts thus accompanied—even if it did not cause—a transformation in the order of nature.

No one made this general point more clearly than the Jesuit Giovanni Botero, author of a pioneering study of the reasons why cities failed or flourished, which naturally examined the roles of the arts and the crafts. In a chapter of his treatise *The Reason of State*, which developed the arguments of his earlier work on cities, Botero explicitly raised the question of which provided more value to humankind, art or nature: "Since art is the rival of nature I must consider which is of more importance to make a state great and populous, the fertility of the soil or the industry of man." The answer was clear: "Without hesitation I shall say industry." Consider, Botero noted, what human ingenuity could make of the simplest materials: "Wool is a simple and crude product of nature: but in how many marvelous and various ways is it transformed by art! . . . Silk is another simple fruit of nature: but what variety of beautiful clothes are fashioned from it by art! Art contrives that the excrement of a vile worm becomes the admiration of princes and the delight of queens." Industry supported far more people, Botero argued, than rents based on natural resources. Even the richest sources of natural wealth could not compete with human effort: "Such is the power of industry that no mine of silver or gold in New Spain or in Peru can compare with it." "Nature," Botero concluded, "gives a form to the raw materials and human industry imposes upon this natural composition an infinite variety of artificial forms: thus nature is to the craftsman what raw material is to the natural agent." All intelligent rulers would do their best to foster the industry and trade that alone could support large populations and build strong states, as natural resources on their own never could.[68]

Conjectural histories of the arts, like material Kunst- und Wunderkammern, expressed powerful and innovative views about the relation between human inventiveness and the natural world. The verbal and the visual interfered with one another constantly as these texts and images took shape. Perfectionist optimism warred with fear and pessimism when scholars drew out their implications. Yet all of their authors treated the history of nature as, in substantial part, a history of culture, and many of them portrayed this as moving in a clear forward direction. The City of the Sun and Salomon's House are porticos that have many entrances and many exits. Some of the paths that led to them began in craftsmen's shops, others in museums, others in botanical gardens. But others had their origins in the overstuffed libraries where scholars desperately scanned their texts in the partly reasonable hope that reading books and studying pictures could help them to master a rapidly changing material universe.

## Notes

1. Tommaso Campanella, *The City of the Sun: A Poetical Dialogue*, ed. and trans. Daniel Donno (Berkeley: University of California Press, 1981), 34–35: "Nel di fuora tutte maniere di pesci di fiumi, lachi e mari, e le virtù loro, e 'l modo di vivere, di generarsi e allevarsi, e a che servono, e le somiglianze c'hanno con le cose celesti e terrestri e dell'arte e della natura; sì che me stupii, quando trovai pesce vescovo e catena e chiodo e stella, appunto come son queste cose tra noi."

2. Campanella, *City of the Sun*, 34–37: "Nel sesto, dentro vi sono tutte l'arte mecchaniche, e l'inventori loro, e li diversi modi, come s'usano in diverse regioni del mondo."

3. Campanella, *City of the Sun*, 36–37: "E li figliuoli, senza fastidio, giocando, si trovan saper tutte le scienze istoricamente prima che abbin dieci anni."

4. Campanella, *City of the Sun*, 42–43: "Onde si ridono di noi che gli artefici appellamo ignobili, e diciamo nobili quelli, che null'arte imparano e stanno oziosi e tengono in ozio e lascivia tanti servitori con roina della republica."

5. Gisela Bock, *Thomas Campanella* (Tübingen: Niemeyer, 1974); John Headley, *Tommaso Campanella and the Transformation of the World* (Princeton, NJ: Princeton University Press, 1997); Germana Ernst, *Tommaso Campanella: Il libro e il corpo della natura* (Rome and Bari: Laterza, 2002), chapter 3.

6. Francis Bacon, *New Atlantis*, in *Works*, ed. J. Spedding, R. L. Ellis, and D. D. Heath, 14 vols. (London: Longman, 1857–1884; rpt. Stuttgart and Bad Cannstatt: Frommann-Holzboog, 1963), vol. 3, 156–166.

7. Samuel Quiccheberg, *Inscriptiones vel tituli theatri amplissimi* (Munich: Berg, 1565); new ed. by Harriet Roth, *Der Anfang der Museumslehre in Deutschland: Das Traktat "Inscriptiones vel Tituli Theatri Amplissimi" von Samuel Quiccheberg* (Berlin: Akademie, 2000).

8. Rosalie Colie, "Cornelis Drebbel and Salomon de Caus: Two Jacobean Models for Salomon's House," *Huntington Library Quarterly* 18 (1954): 245–260; Rosalie Colie, *"Some Thankfulnesse to Constantine": A Study of English Influence upon the Early Works of Constantijn Huygens* (The Hague: Nijhoff, 1956); William Sherman, *John Dee: The Politics of Reading and Writing in the English Renaissance* (Amherst: University of Massachusetts Press, 1995); Lorraine Daston and Katharine Park, *Wonders and the Order of Nature, 1150–1750* (New York: Zone, 1998); Anthony Grafton, "Where Was Salomon's House? Ecclesiastical History and the Intellectual Origins of Bacon's New Atlantis," *Die europäische Gelehrtenrepublik im Zeitalter des Konfessionalismus*, ed. Herbert Jaumann (Wiesbaden: Harrassowitz, 2001), 21–38.

9. Horst Bredekamp, *Antikensehnsucht und Maschinenglauben: Die Geschichte der Kunstkammer und die Zukunft der Kunstgeschichte* (Berline: Wagenbach, 1993), trans. Alison Brown as *The Lure of Antiquity and the Cult of the Machine* (Princeton, NJ: Wiener, 1995); Thomas Kaufmann, *The Mastery of Nature* (Princeton, NJ: Princeton University Press, 1993); Daston and Park, *Wonders.*

10. Jean Bodin, *Colloquium heptaplomeres de rerum sublimium arcanis abditis*, ed. Ludwig Noack (Schwerin: Bärensprung; Paris: Klincksieck; London: Nutt, 1857; rpt. Hildesheim and New York: Olms, 1970), 2; see Ann Blair, *The Theater of Nature: Jean Bodin and Renaissance Science* (Princeton, NJ: Princeton University Press, 1997).

11. Campanella, *City of the Sun*, ed. and trans. Donno, 54–63: "E non accoppiano se non le femine grandi e belle alli grandi e virtuosi, e le grasse a' macri, e le macre alli grassi, per far temperie ... e hanno belle statue di uomini illustri, dove le donne mirano.... E dicono che questo abuso in noi viene dell'ozio delle donne, che le fa scolorite e fiacche e piccole: e però han bisogno di colori e alte pianelle, e di farsi belle per tenerezza, e così guastano la propria complessione e della prole." These were the means traditionally recommended in treatises on the breeding of horses.

12. Bacon, *Works*, ed. Spedding and Ellis, vol. 3, 156–157.

13. Bacon, *Works*, ed. Spedding and Ellis, vol. 3, 159.

14. For the intellectual resources Renaissance philosophers could readily muster and apply to this topic, see the classic surveys of Anthony Close, "Commonplace Theories of Art and Nature in Classical Antiquity and in the Renaissance," *Journal of the History of Ideas* 30 (1969): 467–486; "Philosophical Theories of Art and Nature in Classical Antiquity," *Journal of the History of Ideas* 32 (1971): 163–184. Though Galileo portrayed himself as insisting that one could not trick nature, and suggested that earlier and contemporary engineers believed that one could do so, most of them in fact agreed with him, as had the author of the pseudo-Aristotelian *Mechanica*. See Markus Popplow, *Neu, nützlich und erfindungsreich: Die Idealisierung von Technik in der frühen Neuzeit* (Münster: Waxmann, 1998), 143–176.

15. Popplow, *Neu, nützlich und erfindungsreich*; also see Alex Keller, "Renaissance Theaters of Machines," *Technology and Culture* 19 (1978): 495–508, and Kenneth Knoespel, "Gazing on Machinery: *Theatrum Mechanorum* and the Assimilation of Renaissance Machinery," in Mark Greenberg and Lance Schacterle, eds., *Literature and Technology*, 99–124 (Bethlehem: Lehigh University Press, 1992).

16. Quiccheberg, fol. B ij recto (ed. Roth, 54–57): "Animalia miraculosa & rariora: ut rarae aves, insecta, pisces, conchae ect.... Animalia fusa: ex metallo, gypso, luto, facticiaque materiali: qua arte apparent omnia viva."

17. Quiccheberg, fol. A iij vo (46–47): "Machinarum exempla minuta: ut ad aquas hauriendas, ligna in asseres dissecanda, grana comminuenda, palos impellendos, naves ciendas, fluctibus resistendum: ect. Pro quarum machinularum aut structurarum exemplis, alia maiora rite extrui & subinde meliora inveniri possint."

18. Dioscorides, *De materia medica libri sex*, ed. M. V. Adriani (Florence: Giunti, 1518), sig. AA iii ro.

19. Dioscorides, *De materia medica libri sex*, fol. 209 vo.

20. Dioscorides, *De materia medica libri sex*, sig. AA iii vo; see Paula Findlen, "The Formation of a Scientific Community: Natural History in Sixteenth-Century Italy," in Anthony Grafton and Nancy Siraisi, eds., *Natural Particulars*, 369–400 (Cambridge, MA: MIT Press, 2000), Brian Ogilvie, *The Science of Describing: Natural History in Renaissance Europe* (Chicago: University of Chicago Press, 2006).

21. Bacon, *Works*, ed. Spedding and Ellis, vol. 4, 66; cf. 25.

22. Bacon, *Works*, ed. Spedding and Ellis, vol. 4, 65–66.

23. Bacon, *Works*, ed. Spedding and Ellis, vol. 4, 65.

24. Guido Panciroli, *Rerum memorabilium pars prior . . . pars posterior*, ed. and trans. Heinrich Salmuth, 2 vols. (Frankfurt: Tampach, 1631), vol. 2, 125: "Hodie ad eam subtilitatem deducta est ars ista, ut Rhabarbarum, Nuces pineae, pistaceae, Cinnamonum et aliae species Saccharo condiantur, atque ita quasi recentes adserventur. Efformantur quoque ex Saccharo figurae et imagunculae pulcherrimae: necnon omnis generis fructus repraesentantur, ita ut naturales et feri videantur."

25. Panciroli, *Rerum memorabilium*, vol. 2, 129, quoting Joseph Quercetanus.

26. Elizabeth Gilmore Holt, *A Documentary History of Art* (Princeton, NJ: Princeton University Press, 1981), vol. 1, 164.

27. Holt, *Documentary History*, 161; also see E. H. Gombrich, *Norm and Form* (London: Phaidon, 1966).

28. H. W. Janson, "The Image Made by Chance in Renaissance Thought," *Essays in Honor of Erwin Panofsky* (New York: New York University Press, 1961), 254–266.

29. Katharine Park, "Impressed Images: Reproducing Wonders," in C. A. Jones and P. Galison, eds., *Picturing Science, Producing Art* (New York: Routledge, 1998), 254–271.

30. Leon Battista Alberti, *On Painting and on Sculpture*, ed. Cecil Grayson (London: Phaidon, 972), 120–121: "Artes eorum, qui ex corporibus a natura procreatis effigies et simulacra suum in opus promere aggrediuntur, ortas hinc fuisse arbitror. Nam ex trunco glaebave et huiusmodi mutis corporibus fortassis aliquando intuebantur lineamenta nonnulla, quibus paululum immutatis persimile quidpiam veris naturae vultibus redderetur. Coepere id igitur animo advertentes atque adnotantes adhibita diligentia tentare conarique possentne illic adiungere adimereve atque perfinire quod ad veram simulacri speciem comprehendendam absolvendamque deesse videretur. Ergo quantum res ipsa admonebat lineas superficiesque istic emendando expoliendoque institutum adsecuti sunt, non id quidem sine voluptate."

31. Alberti, *On Painting and on Sculpture*, 121–123: "Quemadmodum enim praestitit natura ex trunco, uti diximus, glebave, ut fieri aliquid posse a te suis operibus simile sentires, ita ab eadem ipsa natura existit promptum habileque aliquid, quo tu quidem modum mediaque habeas certa et rata, quibus ubi intenderis facile possis aptissime atque accommodatissime summum istius artificii decus attingere."

32. Alberti, *On Painting and on Sculpture*, 122–123: "Qualia autem statuariis a natura praestentur media commoda et pernecessaria ad opus bellissime perficiendum, exponendum est."

33. Alberti, *On Painting and on Sculpture*, 98–99: "Fugit enim imperitos ea pulchritudinis idea quam peritissimi vix discernunt. Zeuxis, praestantissimus et omnium doctissimus et peritissimus pictor, facturus tabulam quam in templo Lucinae apud Crotoniates publice dicaret, non suo confisus ingenio temere, ut fere omnes hac aetate pictores, ad pingendum accessit, sed quod putabat omnia quae ad venustatem quaereret, ea non modo proprio ingenio non posse, sed ne a natura quidem petita uno posse in corpore reperiri, idcirco ex omni eius urbis iuventute delegit virgines quinque forma praestantiores, ut quod in quaque esset formae muliebris laudatissimum, id in pictura referret."

34. Alberti, *On Painting and on Sculpture*, 98–101: "Prudenter is quidem, nam pictoribus nullo proposito exemplari quod imitentur, ubi ingenio tantum pulchritudinis laudes captare enituntur, facile evenit ut eo labore non quam debent aut quaerunt pulchritudinem assequantur, sed plane in malos, quos vel volentes vix possunt dimittere, pingendi usus dilabantur."

35. Alberti, *On Painting and on Sculpture*, 133–135: "Ergo non unius istius aut illius corporis tantum, sed quoad licuit, eximiam a natura pluribus corporibus, quasi ratis portionibus dono distributam, pulchritudinem adnotare et mandare litteris prosecuti sumus, illum imitati qui apud Crotoniates, facturus simulacrum Deae, pluribus a virginibus praestantioribus insignes elegantesque omnes formae pulchritudines delegit, suumque in opus transtulit."

36. Giorgio Vasari, *Lives of the Artists*, trans. J. C. Bondanella and P. Bondanella (Oxford: Oxford University Press, 1991), 277; *Vite degli artefici, Parte terza, proemio, Opere* (Milan: Ubicini, 1840), 249: "Il disegno fu lo imitare il più bello della natura in tutte le figure così scolpite come dipinte."

37. Vasari, *Lives of the Artists*, 282 (250–251).

38. Julius Caesar Scaliger, *Exotericarum exercitationum liber quintus decimus* (Paris: Vascosan, 1557), fols. 395 vo–396 ro: "Falsa, inquis, delectant: quia admirabilia. Delectant igitur pueros et stolidos, non senes et sapientes. Certe verum est. Sed quaeso: cur Homerica phasmata (sic enim libet appellare) delectant sapientes? . . . Illud huiusce rei caput est. Mentem nostram esse natura sua infinitam. Quamobrem et quod ad potentiam attinet, aliena appetere: et quod spectat ad intellectionem, etiam e falsis ac monstrorum picturis capere voluptatem, propterea quod exuperant vulgares limites veritatis. Aspernatur enim certorum finium praescriptionem. . . . Ergo picturae quoque laudat sapiens perfectionem: tametsi fictam esse, haud ignorat. Mavultque pulchram imaginem, quam naturali similem designatae. Naturam enim in eo superat ars. Quia multis eventis a primo homine symmetria illa depravata fuit. At nihil impedit plasten, quominus attollat, deprimat, addat, demat, torqueat, dirigat. Equidem ita censeo: nullum unquam corpus humanum tam affabre fuisse a Natura factum (duo scilicet excipio: unum primi hominis: alterum veri hominis, veri Dei) quam perfecte finguntur hodie doctis artificum manibus."

39. Erwin Panofsky, *Idea*, trans. J. J. S. Peake (Columbia: University of South Carolina Press, 1968), 222–223 n. 20.

40. Carlo Ginzburg, *Jean Fouquet: ritratto del buffone Gonella* (Modena: Panini, 1996).

41. Holt, *A Documentary History of Art*, vol. 1, 273–275.

42. Alberti, *On Painting and on Sculpture*, 32–33: "Struttura sì grande, erta sopra e' cieli, ampla da coprire con sua ombra tutti e popoli toscani"; "Onde stimai fusse, quanto da molti questo così essere udiva, che già la natura, maestra delle cose, fatta antica e stracca, più non producea come né giuganti così né ingegni." See Christine Smith, *Architecture in the Culture of Early Humanism* (New York: Oxford University Press, 1992), for a full discussion of this text.

43. Lorenzo Valla, *Gesta Ferdinandi regis Aragonum*, ed. Ottavio Besomi (Padua: Antenore, 1973), 195–196: "De quibus ego horologiis non loquor que et vetera sunt nec tantopere admiranda, et que ipsum per se experimentum docuit. Loquor de eo quod vere est horologium, in quo non tantum ratio horarum, sed etiam, ut sic dicam, sermo agnoscitur; utrunque enim logos significant, rationem et sermonem; quod quodammodo vitam habet, cum sponte sua cietur, et dies ac noctes pro homine opus facit. Nec solum horam oculis ostendit ac praescribit, sed etiam auribus procul et domi manentium nuntiat, campana, que superimposita est, numerum distinguente: quo nihil neque utilius neque iocundius."

44. Valla, *Gesta Ferdinandi regis Aragonum*, 194: "Et certe necesse est ut docti aliquando constituant quibus vocabulis appellande sint ee res que non ita multo superioribus temporibus sunt excogitate. Non enim exhausta sunt mortalitatis ingenia; quod haud dubie fatendum est, nisi invidemus laudes nostras proxime accedere ad solertiam antiquorum in multis, et si non omnibus, honestis atque utilibus."

45. On the origins and transmission of this text see Alex Keller, "A Renaissance Humanist Looks at 'New' Inventions: The Article 'Horologium' in Giovanni Tortelli's *De Orthographia*," *Technology and Culture* 11 (1970): 345–365; Ottavio Besomi, "Dai 'Gesta Ferdinandi Regis Aragonum' del Valla al 'De orthographia' del Tortelli," *Italia Medioevale e Umanistica* 9 (1966): 75–121; Brian Copenhaver, "The Historiography of Discovery in the Renaissance: The Sources and Composition of Polydore Vergil's *De Inventoribus Rerum* I–III," *Journal of the Warburg and Courtauld Institutes* 41 (1978): 192–214; Polydore Vergil, *On Discovery*, ed. and trans. Brian Copenhaver (Cambridge, MA: Harvard University Press, 2002).

46. Giannozzo Manetti, *De dignitate et excellentia hominis*, ed. Elizabeth Leonard (Padua: Antenore, 1975), 77: "Nostra namque, hoc est humana, sunt quoniam ab hominibus effecta cernuntur: omnes domus, omnia opida, omnes urbes, omnia denique orbis terrarum edificia, que nimirum tanta et talia sunt, ut potius angelorum quam hominum opera ob magnam quandam eorum excellentiam iure censeri debeant. Nostre sunt picture, nostre sculpture; nostre sunt artes, nostre scientie."

47. Marsilio Ficino, *Platonic Theology*, ed. James Hankins and William Bowen, trans. Michael Allen, vol. 1 (Cambridge, MA: Harvard University Press, 2001), 254–255: "Quid artificium? Mens artificis in materia separata. Quid naturae opus? Naturae

mens in coniuncta materia. Tanto igitur huius operis ordo similior est ordini qui in arte est naturali quam ordo artificii hominis arti, quanto et materia propinquior est naturae quam homini, et natura magis quam homo materiae dominatur."

48. Ficino, *Platonic Theology*, vol. 1, 200–201: "Vidimus Florentiae Germani opificis tabernaculum, in quo diversorum animalium statuae ad pilam unam connexae atque libratae, pilae ipsius motu simul diversis motibus agebantur: aliae ad dextram currebant, aliae ad sinistram, sursum atque deorsum, aliae sedentes assurgebant, aliae stantes inclinabantur, hae illas coronabant, illae alias vulnerabant. Tubarum quoque et cornuum sonitus et avium cantus audiebantur, aliaque illic simul fiebant et similia succedebant quam plurima, uno tantum unius pilae momento. Sic deus per ipsum esse suum, quod idem re ipsa est ac intellegere ac velle quodve est simplicissimum quoddam omnium centrum, a quo, ut alias diximus, reliqua tamquam lineae deducantur, facillimo nutu vibrat quicquid inde dependet."

49. André Chastel, *Marsile Ficin et l'art* (Geneva: Droz, 1954), 105. Also see the subtle analysis by Stéphane Toussaint, "Ficino, Archimedes and the Celestial Arts," in Michael Allen and Valerie Rees, with Martin Davies, eds., *Marsilio Ficino: His Theology, His Philosophy, His Legacy*, 307–326 (Leiden: Brill, 2002).

50. Joseph Rykwert, *On Adam's House in Paradise*, 2nd ed. (Cambridge, MA: MIT Press, 1981); Stephanie Moser, *Ancestral Images* (Ithaca, NY: Cornell University Press, 1998).

51. Holt, *A Documentary History of Art*, vol. 1, 339.

52. T. D. Kendrick, *British Antiquity* (London: Methuen, 1950); Florike Egmond and Peter Mason, *The Mammoth and the Mouse* (Baltimore: Johns Hopkins University Press, 1997); Moser, *Ancestral Images*.

53. Pamela Long, *Openness, Secrecy, Authorship* (Baltimore: Johns Hopkins University Press, 2001).

54. Pamela Long, "Power, Patronage and the Authorship of *Ars*: From Mechanical Know-How to Mechanical Knowledge in the Last Scribal Age," *Isis* 88 (1997): 1–41.

55. Paolo Rossi, *Philosophy, Technology and the Arts in the Early Modern Era*, trans. S. Attanasio, ed. Benjamin Nelson (New York: Harper & Row, 1970); Alex Keller, "Mathematical Technologies and the Growth of the Idea of Technical Progress in the Sixteenth Century," in Allen Debus, ed., *Science, Medicine and Society in the Renaissance: Essays to Honor Walter Pagel*, 2 vols. (London: Science History, 1972), vol. 1, 11–27. Popplow emphasizes that sixteenth-century writers rarely made contemporary technologies as such the basis for their claims of technical progress, and insists on the role of the theaters of machines in inspiring a different and more precise view in the seventeenth century.

56. Werner Gundersheimer, *Louis Le Roy* (Geneva: Droz, 1966).

57. Louis Le Roy, *De la vicissitude ou variété des choses en l'univers* (Paris: Fayard, 1988), 378: "N'ayant toute l'antiquité rien qu'elle puisse comparer à ces trois."

58. Jean Bodin, *Methodus ad facilem historiarum cognitionem*, in Joannes Wolf, ed., *Artis historicae penus*, 2 vols. (Basel: Perna, 1579), vol. 1, 309–310: "Ac nemini dubium esse potest in eam rem penitus intuenti, quin inventa nostrorum cum maiorum inventis conferri pleraque debeant anteferri ... omitto catapulta veterum et antiqua belli tormenta, quae si cum nostris conferantur, sane puerilia quaedam ludicra videri possint ... una typographia cum omnibus omnium veterum inventis certare facile potest ... habet natura scientiarum thesaurus innumerabiles, qui nullis aetatibus exhauriri possunt."

59. Henry Cornelius Agrippa, *Of the Vanitie and Vncertaintie of Arts and Sciences*, ed. Catherine Dunn (Northridge: California State University Press, 1974), 88; also see Agrippa, *De occulta philosophia libri tres*, ed. V. Perrone Compagni (Leiden: Brill, 1992), 2.1, 250–251: "Et Iulii Caesaris Romae iuxta Vaticanum erecta pyramis et in medio mari extructi arte montes et arces saxorumque moles, cuiusmodi ego in Britannia vidi vix credibili arte congestas. Et legimus apud fidos historicos similibus artibus olim abscissas rupes, completas valles et actos in planum montes, perfossa saxa, adaperta mari promontoria, excavata terrae viscera, diducta flumina, iuncta maribus maria, coercita aequora scrutataque maris profunda, exhaustos lacus, exsiccatas paludes, factas novas insulas rursusque alias restitutas continenti. Quae omnia, etsi cum natura ipsa pugnare videantur, tamen legimus facta et in hunc idem cernimus illorum vestigia, cuiusmodi vulgus daemonum opera fuisse fabulatur, cum eorum artes atque artifices a memoria perierint nec sint qui curent ea intelligere atque scrutari."

60. See William Eamon, *Science and the Secrets of Nature* (Princeton, NJ: Princeton University Press, 1994).

61. Leon Battista Alberti, *L'architettura*, ed. Giovanni Orlandi, trans. Paolo Portoghesi, 2 vols. (Milan: Il Polifilo, 1966), vol. 1, 9–11: "Quid demum, quod abscissis rupibus, perfossis montibus, completis convallibus, coercitis lacu marique, expurgata palude, coaedificatis navibus, directis fluminibus, expeditis hostiis, constitutis pontibus portuque non solum temporariis hominum commodis providit, verum et aditus ad omnes orbis provincias patefecit?"

62. See the great article by Otto Mayr, "Automatenlegenden in der Spätrenaissance," *Technikgeschichte* 41 (1974): 20–32.

63. Gaspar Schott, *Magia universalis naturae et artis*, 2nd ed., 4 vols. (Bamberg: Schönwetter, 1677), vol. 3, 211: "Haec [sc. Magia thaumaturga] enim Naturae non juvat modo ac perficit, ut aliae [sc. Mathematicae disciplinae]; sed evidentissime etiam vincit, dum per machinas, quas ingeniosissime excogitat, nullam non debiliorem virtutem confirmat ac promovet, nullam non debiliorem sistit ac superat, nullum non corporibus motum, progressionem, gyrationem inducit; audax nimirum ac potentissima virium Naturae in corporibus non tam contemplatrix, quam arbitra." On the magical practices and machinery of Schott and Athanasius Kircher, see *La "Technica curiosa" di Kaspar Schott*, ed. and trans. Maurizio Sonnino (Rome: Edizioni dell'Elefante, 2000).

64. Campanella, *City of the Sun*, ed. and trans. Donno, 126–127: "Onde la costellazione che da Lutero cadavero cavò vapori infetti, da Gesuini nostri che fûro al suo

tempo cavò odorose esalazioni di virtù, e da Fernando Cortese che promulgò il cristianesimo in Messico nel medesimo tempo."

65. Girolamo Cardano, *The Book of My Life*, trans. Jean Stoner (New York: Dutton, 1931; rpt. New York: New York Review of Books, 2002), 189–190; *De propria vita liber*, *Opera*, ed. Charles Spon, 10 vols. (Lyons: Huguetan and Ravaud, 1663; rpt. Stuttgart-Bad Cannstatt: Frommann-Holzboog, 1966), vol. 1, 35: "Crevit opinio, minuentur et contemnentur bonae artes, et certa pro incertis commutabuntur. Sed haec alibi, interim nos florente prato gaudebimus. Nam quid mirabilius Pyrotechnia et fulgure mortalium, quod pernitiosius multo est quam Caelestium."

66. Campanella, *City of the Sun*, ed. and trans. Donno, 122–123: "E dicono che a' Cristiani questo apporterà grand'utile; ma prima si svelle e monda, poi s'edifica e piñata."

67. Campanella, *City of the Sun*, 122–125.

68. Giovanni Botero, *The Reason of State*, trans. P. J. and D. P. Waley (New Haven, CT: Yale University Press, 1956), 8.3, 151–153.

# 9

## Leibniz's Theater of Nature and Art and the Idea of a Universal Picture Atlas

Horst Bredekamp

### The Return of the *Kunstkammer*

Twleve years ago in *The Lure of Antiquity and the Cult of the Machine* I ended my interpretation of sixteenth- and seventeenth-century *Kunstkammern* by predicting that although the modes of perception and thought associated with this type of museum had made a comeback, this kind of collection would not serve as a model for the future: "Nobody wants to return to the deliberate chaos of the Kunstkammer as a museum."[1] I could not have been more wrong, for since then the Kunstkammer type of exhibition, which unlike specialized modern museums encompasses exhibits from the two poles of nature and art, has experienced an unexpected renaissance.

As early as 1988 the Villa Hügel in Essen reconstructed Rudolf II's unsurpassed ensemble in *Prag um 1600*,[2] in 1991 the ideal Kunstkammer simulated in New York was said to be a sign of the *Age of the Marvellous*,[3] in the following year the Kunstkammern of the Netherlands were represented as highly ambitious microcosms,[4] and in 1994 a sophisticated *Wunderkammer des Abendlandes* was created in Bonn.[5] These extravagant and somewhat excessively aesthetic experiments, seeking to bring a sense of wonderment back to a soulless world, contrast with reconstructions of certain Kunstkammern that emphasized historical accuracy. Examples include the 1991 reconstruction of the Amerbach cabinet,[6] the *Praunsche Kabinett* in Nuremberg in 1994,[7] or once again, Rudolf II's Kunstkammer as recreated in 1997.[8] The historical reconstruction of the exhibits that made up Braunschweig's Kunstkammer was another event,[9] as was the recovery of the Franke Kunstkammer in Halle[10] and the *Kircherianum* in Rome.[11]

The principle underpinning the Kunstkammer involved bringing together objects from the world of nature, research instruments, and works of art, but not to simply deposit them in a static collection. Instead the exhibits were to be actively utilized and the collection linked with laboratories and libraries. This notion quite explicitly fed into a promenade along the narrow border between art and science. The unparalleled 1993 *L'âme au corps* exhibition in Paris made its way along this thin line with all the assurance of a sleepwalker,[12] and was followed by events such as *Deep Storage*, which reached a number of towns in recent years.[13]

With the gigantomanic Kunstkammer-millenium exhibition of Berlin[14] this kind of event began to be seen as heralding a new religion of science, gradually replacing ruined social utopias.[15] It at the very least raised the question of what a revitalized science might take as its goals, a science that would be radically modern and specialized and yet at the same time would interact with other fields in order to be scrutinized and integrated, inspired and criticized. To draw on a somewhat clichéd term, the Kunstkammer did not call for an "interdisciplinarity" in linking nature and art; instead this approach was a prerequisite for their very existence.

Gottfried Wilhelm Leibniz, the great philosopher, historian, and natural scientist, embodied both poles, and thus he could be taken as an inspiration for the recent museological experiment in Berlin, the *Theatrum Naturae et Artis*.[16] Because he tried to establish a combination of objects of nature and the arts throughout his intellectual life, he is to be discovered not only as one of the most ambitious theorists of art and nature, but also as a highly inspired museologist.

## Leibniz and the Kunstkammer

Both theory and practice motivated Leibniz to establish a forum both for exponents of nature and the arts. First, he was clearly enthusiastic about visiting museums.[17] During his youth he certainly got to know the collections of Leipzig, and during his studies he became familiar with Erhard Weigel's extensive Kunstkammer in Jena, held to be one of the city's seven wonders, and Mauritius Hofmann's *Theatrum* in Altdorf. This was followed by Kunstkammern in Strasbourg and subsequently, of course, in Paris and London, Braunschweig, Kassel, Frankfurt am Main, and Nuremberg, where he had the opportunity to become acquainted with one of the largest contemporary natural history cabinets with exhibits dating back to the sixteenth century. In Italy he visited the museum assembled by Antonio Magliabecchi, librarian to the court of the Medici in Florence, followed in 1689 by Ferdinando Cospi's Kunstkammer, which had absorbed Ulisse Aldrovandi's splendid collection, then Luigi Ferdinando Conte de Marsigli's Kunstkammer in Bologna and Ferrante Imperato's collection in Naples, as well as the vast Kunstkammer in Rome belonging to Athanius Kircher. After trips to numerous collections in Germany, the crowning experience was his visit to Gottorp's Kunstkammer, for this collection, assembled by Olearius Worm, was one of the marvels of the museum world of the day. It included the giant globe, which could be spun with the full force of the hand and served as a model for Louis XIV's globe of the heavens. Finally, in the last year of his life, Leibniz visited August Hermann Franke's cabinet of wonders of the natural world in Halle.

His second main interest was in the theory of collecting. In 1680 he called for natural objects from the Harz mountains to be collected on a systematic basis, justifying his proposal by asserting that nature had been the artist shaping this central German mountain range. He felt that the Harz mountains should be seen as a kind of natural outdoor extension of the Kunstkammer, and that hence objects from the area were destined to be collected and made available to scientific research: "The Harz mountains are simply a wonderful stage on which nature struggles with art to gain the upper hand."[18]

This was an image Leibniz had inherited from mannerist theories of art and the philosophy of nature. The aesthetic appearance of the works of nature in their totality, but above all those figurative forms produced by the earth as so-called chance images, testified to the struggle between nature and humankind, and permeated the whole mannerist era, also shaping sixteenth-century garden design, as Torquato Tasso expressed in a unique word-game describing a garden in *Gerusalemme liberata*:

> Nature would craft in counterfeiting pass,
> And imitate her imitator art.[19]

In referring to this kind of *paragone*, Leibniz sets objects from the natural world in the arena of nature's struggle with human art. This productive conflict was the idée fixe running through all his thoughts on reforming research.

## Theaters of Nature and Art

Leibniz was well aware of the long-term strategies of collecting. He made use of them in two ways, both of which focus on the bipolarity and intertwining of nature and art. The first develops the idea of a *Theater of Nature and Art*, and the second tries to draw out the epistemological consequences of this institution by creating a pictorial *Atlas Universalis*.

As early as 1671, in his first plan to establish a Society of Scholars modeled after the Academies in London and Paris, Leibniz, inspired by Johann Joachim Becher's *Methodus Didactica* and *Catalogus omnium corporum, quae in theatro naturae et artis Becheriano reperiuntur*,[20] but surely also reflecting the anatomical theaters of his time as well as Samuel Quiccheberg's museum theory *Theatrum amplissimum* published in 1565,[21] called for a complex of museums to be set up alongside the rather moribund library: "a Theatrum naturae et artis or a cabinet of the arts, curiosities and anatomy."[22] In the second version he similarly referred to a "chamber of the arts, curiosities, paintings and anatomy, . . . in other words a Theatrum naturae et artis, to receive living impressions and knowledge of all things."[23] A few years later, writing from Paris, Leibniz imagined a "Theatre de la nature et de l'art" in his *Drôle de Pensée*.[24]

The same notion informed his petitions to the king for the establishment of the Berlin Academy of Science. In March 1700, Leibniz explained: "All of these sciences are served by libraries, Iconothecae (collections of engravings, plans, models, and paintings), cabinets of art and curiosities, armories, all kinds of gardens, also zoological gardens, and the great works of nature and art, all of which your Most Serene Highness possesses in sufficient quantities for the Theatro Naturae et Artis"[25]—a clear reference to the royal Kunstkammer, situated in the north wing of the Royal Palace of Berlin.[26] In the July 1700 general decree there is also mention of a "Theater of nature and art,"[27] and two years later Leibniz admonished King Frederick I to create the "Theater of nature and art."[28]

This concept is also a recurrent element in the later plans. In 1704–1705 Leibniz recommended that the Elector of Saxony employ a "theater of nature and art,"[29] and, as he wrote on several occasions between 1697 and 1716, it should also comprise a cabinet of nature and the arts.[30]

In 1713 Leibniz stated in Vienna that theaters of nature and art should be made available to the Academy there,[31] and here he was certainly thinking of the imperial library and Kunstkammer. In 1713 he called once again for the Academy to be developed and supplied with "Theatra naturae et Artis,"[32] and in 1714 he called again for "theatres de la Nature et de l'Art."[33]

## The Idea of a Picture Atlas

Leibniz's reverence for this kind of collection as a theater led again and again to the key epistemological notion that humans require a pictorial image to form something new and to classify. In relation to a theater of nature and art Leibniz thus thought of a second project, the construction of an Atlas Universalis, consisting of masses of systematically collected images.

Leibniz considered it necessary to concern oneself with pictures because the mind works through the senses: "For if our mind employs the images of sensual things, this implies that the thinker cannot stray if he directs his attention to them, because the images are implanted like a chain."[34] For this reason, Leibniz accorded eyesight a special effectiveness in the teaching and grasping of knowledge. He considered the issue of pictures especially intensively when, working from 1678 on as a librarian for Duke Johann Friedrich of Braunschweig-Lüneberg in Hanover, he assayed his experience in the Paris of Louis XIV. In fall 1679, in six writings to establish the literary magazine *Semestria Literaria*, Leibniz developed the plan of an encyclopedia, which would include an Atlas Universalis:

> The Atlas Universalis will be added to this encyclopedia, a work of magnificent practicality, to submit to the human brain easily and with pleasure a great magnitude of tabulae, figures, and well-executed and illuminated drawings and outlines, so that everything can be captured by the eye and designed on paper, thus to be pre-shaped to the spirit and imprinted more forcefully on the mind as if at one gaze all the more rapidly and pleasantly and almost playfully and without the use of words.[35]

Seldom did Leibniz so clearly define his high estimation of pictures and connect it with the beginnings of a theory of perception as in this project paper for the founding of a book magazine. He regards pictures as suitable for conveying information at a higher speed and with greater beauty than text, and above all permitting the highest efficiency of learning, because they speak to our desire to play.

Here he draws support from the ideas of the great art-technological utopias of about 1600—for example, the painted and repainted rings of walls in Tommaso Campanella's *Città del Sole*,[36] as prefigured in Heemskerck's depiction of the walls of Babel[37] or in Johann Valentin

Andreae's *Christianopolis*, in which the entire Creation with all its species and phenomena are painted on the walls of an "Exhibition House of Nature," where "boys are in a sense playfully learning."[38] For Leibniz, pictures are reliable, in contrast to words, which frequently stray. In their concreteness, they firmly plant themselves in the mind.

When Leibniz says that pictures are "imprinted more forcefully on the mind," he apparently relies on the Hobbesian theory of sight, which regards the light-filled appearances as an atomistic bombardment coming from objects themselves. Since seeing is merely a special case of grasping, pictures can implant themselves in the mind much more strongly and directly than letters. This theory is in sharp conflict with Leibniz's thesis of monads having no windows and thus not being able to be the objects of imprints. Leibniz's *theatrum naturae et artis* and its will to offer haptic and visual stimuli to the creation of ideas thus changes the whole frame of his philosophy.[39]

The Consilium, the last of the texts Leibniz devoted to setting up a universal encyclopedia of books, ends with a brief characterization of the Atlas Universalis. It follows his praise for collecting, described as a haptic correspondence to eyesight: "I see nothing that human effort could present more comprehensively."[40] As the last expression of considerations aimed at the idea of a universal encyclopedia of books, this methodological moral contains irony. Book knowledge is demoted to an auxiliary resource to be used only because the tactile and visual vehicles of knowledge are not available everywhere.

The same is true of the proposal for a library, which Leibniz probably composed at the end of January 1680 for the new Prime Minister Franz Ernst von Platen. The paper, which ostensibly served the cause of book knowledge, already favors the picture in its second paragraph. Leibniz's decisiveness in emphasizing pictures surprises once more: "There is nothing in the world that could instruct young men better than figures."[41]

In a manner quite damaging to his professional interests as a librarian, the beginning and end of his praise for an Atlas Universalis suggest the higher value of the iconothèque over the library. A picture catalog would overshadow the books, because the Atlas Universalis

> can surpass a complete library: one finds the portraits of almost all famous men in the world; the representations of countless entries and public events; an entire Theatrum naturae et artis; hunts, sails, and tempests, battles and castles, palaces, gardens, landscapes, infinite

> hieroglyphics, capriccios, ornaments, devices, symbols, and, in sum, whatever of truths and tales can come into human thoughts. If one could have this opus, one would surely have a treasure and an infinite source of information, which one could use not only for courtly pleasures, parades, masquerades, and tournaments, but also for buildings, gardens, machines, and many occasions. In sum, one could call such a collection truly a vibrant library.[42]

Once again, Leibniz here plays the aspect of liveliness off against books, and thus again uses the concept of an Atlas Universalis that outdoes an entire library.

In a pedagogical instructional text of 1685–1686, Leibniz was able to think through his theory of the function of the picture atlas. It reminds one immediately of Comenius's *Orbis pictus sensualium*.[43] As soon as the child reaches an age governed by the imagination, Leibniz argues, pictures would have a "wonderful" effect: "I would favour pictures as soon as he [the prince] would have them, and I often wished that one should draw and print huge prints, like those included in the atlases, which are able to represent at one glance an entire science, art, or profession."[44] These picture plates seemed to him his age's contribution to the theory of learning and knowledge: "Nothing is more important than to grasp things with the mind, that is the central point that must not be forgotten. And because these arguments are graspable and sensual, the satisfaction is doubled. Precisely here, reason has made use of the escort of imagination."[45] Leibniz underscores that the media of reason's imagination include a visual Atlas Universalis as well as a haptically experienceable "théâtre de la nature et de l'art." Leibniz sent copies of this design as a proposal to various courts several times as late as 1714; thus he adhered until the end of his life to the idea of the Atlas Universalis as a still viable impulse of the 1660s and 1670s.

## The Atlas Universalis

Leibniz focused his general calls for the creation and use of an Atlas Universalis in a text probably written in 1678, in which he aimed no less than to develop a first general iconography. In the introduction, Leibniz already proclaims the special effectiveness of eyesight in teaching and grasping knowledge: "But it comes to my mind that the entire encyclopedia could be excellently comprised by a kind of universal atlas. Namely, almost everything that must be taught and instructed can first be presented to the eyes."[46] The proposed arrangement of such an atlas is based on illustrated books and existing collections of pictures and plates. The key words provide an unsurpassably thoughtful iconographic kaleidoscope.[47]

In the sense of a first global compilation of astronomy, geography, government, and culture, Leibniz lists the "topography of the heavens," the "topography of the earth," the heraldic and genealogical tables, every language and script, as well as the folk costumes, deportment, and cult of the various nations and professions. In the tradition of the "uomini illustri," the ensuing gallery of famous men includes the instrumental figures of history up to the then-recent past. Attention to objects then focuses on collections of images of coins and of architectural ornament. A section with coins, inscriptions, pictures on rings, ceremonies, vestments, and hieroglyphics, followed by Cesare Ripas's "Iconologia," develops the complete program of a nontopographic reception of antiquity.

Proceeding from the fine to the applied arts, Leibniz lists mathematics, with perspective and music, as well as architecture, passing through military architecture to reach the means of the art of war. Machine construction then leads to national economy and agriculture, which offer the transition to materials and their refinement. Cloth, stone, wood, glass, and other materials from nature lead to the painters, sculptors, and writers. Elaborations on fluid substances like castable metals lead to the vintner's art, so that the list includes the whole range of chemistry, pharmacy, herbal medicine, botany, anatomy, surgery, and comparative animal anatomy. In their apparently comprehensive pattern, exotic forms, as "Rariora naturae et artis in *Exoticophylaciis*," are terminologically close to the *Theatrum naturae et artis*.

Leibniz places the depictions of what is optically furthest away next to illustrations of what is seen under the microscope. He thus completes the circle to astronomy, mentioned at the beginning of the list, which had produced images no less striking with the aid of the telescope. The Atlas Universalis is framed by a world of pictures that the assisted eye had opened up to very distant and invisibly small realms. In this way, Leibniz's Atlas Universalis reflects not only an *Orbis Sensualium pictus*, like the one Comenius had designed as a picture encyclopedia of the visible world, but also the macro- and microcosm discoverable with instruments.

The idea of an Atlas Universalis as a collection of images opposing the abstraction of signs remains one of the major discoveries Leibniz has to offer. One of the greatest experts, André Robinet, concluded on Leibniz's travels through Italy: "Leibniz did not see anything."[48] If Leibniz really had not seen anything, the amount of energy he dedicated to this nullum would be all the more remarkable. The statement may be taken as a sign of how much work remains to be done in the history of philosophy to escape the luxurious intellectual prison of Western neo-Neoplatonism and to reapproach the controversial bipolarity that Leibniz made use of in his two concepts of a theater of nature and art and an Atlas Universalis.

The statement also offers a key to the understanding of the Kunstkammer in our day. It is astonishing that the so-called virtual worlds of digital simulation have also produced an amazing desire for haptic and visual objects and installations. For Leibniz it would not be at all surprising. Leibniz, the eminent mathematician, who is recognized as one of the fathers of digital number theory, viewed the graspable and the sensual not as an enemy of abstraction, but as its fundamental counterpart.

## Notes

1. Horst Bredekamp, *The Lure of Antiquity and the Cult of the Machine: The Kunstkammer and the Evolution of Nature, Art, and Technology* (Princeton, NJ: Markus Wiener Publishers, 1995), 113. (German version, 1993.)

2. *Prag um 1600: Kunst und Kultur am Hofe Rudolfs II*, exhibition catalog (Essen: Villa Hugel, 1988).

3. Joy Kenseth, ed., *The Age of the Marvelous* (Hanover, NH: Hood Museum of Art, Dartmouth College, 1991).

4. *De wereld binnen handbereik: Nederlandse kunst- en rariteitenverzamelingen, 1578–1735*, exhibition catalog (Amsterdam: 1992).

5. *Wunderkammer des Abendlandes: Museum und Sammlung im Spiegel der Zeit*, exhibition catalog (Bonn: Kunst- und Ausstellungshalle der Bundesrepublik Deutschland, 1994).

6. *Das Amerbach-Kabinett*, 5 vols., exhibition catalog (Basel: 1991).

7. *Das Praunsche Kabinett*, exhibition catalog (Nuremberg: 1994).

8. Eliska Fucikova, ed., *Rudolf II and Prague: The Court and the City*, exhibition catalog (Prague: 1997).

9. *Die Barocke Sammellust: Die Bibliotek und Kunstkammer des Herzogs Ferdinand Albrecht zu Braunschweig Lüneburg (1636–1687)* (Wolfenbüttel: 1988); Rudolf-Alexander Schütte, *Die Kostbarkeiten der Renaissance und des Barock* (Brunswick: Herzog Anton-Ulrich-Museum Braunschweig, 1997); *Weltenharmonie: Die Kunstkammer*

*und die Ordnung des Wissens*, exhibition catalog (Brunswick: Herzog Anton-Ulrich-Museum Braunschweig, 1999).

10. Thomas J. Müller-Bahlke and Klaus E. Göltz, *Die Wunderkammer: Die Kunst- und Naturalienkammer der Frankeschen Stiftungen zu Halle (Saale)* (Halle/Saale: Verlag der Franckeschen Stiftungen, 1998).

11. Eugenio Lo Sardo, ed., *Athanasius Kircher: Il museo del mondo*, exhibition catalog (Rome: De Luca, 2001).

12. Jean Clair, ed., *L'âme au corps. arts et sciences 1793–1993*, exhibition catalog (Paris: Gallimard/Electa, 1993).

13. *Deep Storage: Arsenale der Erinnerung: Sammeln, Speichern, Archivieren in der Kunst*, exhibition catalog (Munich and New York: Prestel, 1997).

14. *Sieben Hügel—Bilder und Zeichen des 21. Jahrhunderts*, exhibition catalog, 7 vols. (Berlin: Henschel, 2000).

15. Ulrich Raulff, "Den Göttern eines neuen Sonnenkults," *Frankfurter Allgemeine Zeitung*, no. 112, 2000, 49.

16. Horst Bredekamp, Jochen Brüning, and Cornelia Weber, eds., *Theater der Natur und Kunst: Theatrum Naturae et Artis; Wunderkammern des Wissens*, Catalog and Essays (Berlin: Henschel, 2000).

17. The reconstruction of these experiences I hope to be able to publish in the near future. For the moment the main material is collected in Wilhelm Ennenbach, "Gottfried Wilhelm Leibniz' Beziehungen zu Museen und Sammlungen," in *Beiträge zu: Leibniz geowissenschaftliche Sammlungen* (Berlin: Institut für Museumswesen, 1978), pp. 1–63.

18. "Maßen der Harz an sich selbst nichts anderes als ein wunderbarer Schauplatz, alda die Natur mit der Kunst gleichsam streitet" (Gottfried Wilhelm Leibniz, *Sämtliche Schriften und Briefe*, ed. Preußische, then Deutsche Akademie der Wissenschaften zu Berlin [Berlin: Akademie-Verlag, 1923ff.], I, III, no. 17, p. 17, l.33–135).

19. Torquato Tasso, *Gerusalemme liberata*, 16, 11f.

20. A number of examples for the use of "Theater of Nature" are to be found in the fundamental study by Ann Blair, *The Theater of Nature: Jean Bodin and Renaissance Science* (Princeton, NJ: Princeton University Press, 1997). But no one combines it with "Theater of Art," as Becher does for the first time in 1668: "Ich habe einmal die Spekulation gehabt, wenn ein Herr wäre, der als Muster an Wohlgefälligkeit die Mittel gäbe, ein *Theatrum Naturae et Artis* solchergestalt aufzurichten, daß alle *Naturalia* und *Artefacta, Instrumente* und *Manufacta* darin wären, so viele man immer bekommen könnte" (Johann Joachim Becher, *Methodvs didactica* (Munich, 1668), 51). See Herbert Hassinger, *Johann Joachim Becher: Ein Beitrag zur Geschichte des Merkantilismus* (= Veröffentlichungen der Kommission für Neuere Geschichte Österreichs, vol. 58) (Vienna: Holzhausen, 1951), 264. See also Leibniz, *Sämtliche Schriften*

*und Briefe*, VI, 2, p. 392, l.10–16, and p. 394, l.5f. On Leibniz's reflection of Becher, see Herbert Breger, "Becher, Leibniz und die Rationalität," in *Johann Joachim Becher*, ed. Gotthardt Frühsorge und Gerhard F. Strasser (Wiesbaden: Harrasowitz 1993), 69–84.

21. Harriet Roth, *Samuel Quicchebergs "Inscriptiones vel Tituli Theatri Amplissimi" von 1565 oder der Anfang der Museumslehre in Deutschland* (Berlin: 2000).

22. "Ein Theatrum naturae et artis oder Kunst-, Raritäten- und Anatomie-Kammer" (Leibniz, *Sämtliche Schriften und Briefe*, IV, 1, no. 43, p. 537, l.5–15).

23. "Kunst- und raritäten-, Schilderey- auch Anatomiae-Cammern, anders als ietzt geschicht bestellte Apothecen, Hortos medicos completos, Thiergärten und also Theatrum Naturae et artis, umb vor allen dingen lebendige impressiones und connoissance zu bekommen, anzurichten" (Leibniz, *Sämtliche Schriften und Briefe*, IV, 1, no. 43, p. 540, l.16–18).

24. Leibniz, *Sämtliche Schriften und Briefe*, IV, 1, no. 49; p. 563, l.26.

25. "Zu allen diesen Wißenschafften dienen Bibliothecken, Iconothecae (oder Collectanea von Kupferstücken, Rissen, Bildungen und Gemählden), Kunst- und Raritätenkammern, Zeug- und Rüst-Häuser, Gärten vieler Art, auch Thier-Behältnisse, und die großen Wercke der Natur und Kunst selbsten, von welchen allen, zum Theatro Naturae et Artis, bey Churfürstl. Durchlaucht kein Mangel" (Hans-Stephan Brather, *Leibniz und seine Akademie: Ausgewählte Quellen zur Geschichte der Berliner Sozietät der Wissenschaften 1697–1716* (Berlin: Akademie Verlag, 1993), 77).

26. Horst Bredekamp, "Leibniz' Theater der Natur und Kunst," in Horst Bredekamp, Jochen Brüning, and Cornelia Weber, eds., *Theater der Natur und Kunst*, Essays, 17.

27. "Observatorio, Laboratorio, Bibliothec, Instrumenten, Musaeo und Rariteten-Cammer, oder Theatro der Natur und Kunst" (Brather, *Leibniz und seine Akademie*, 97).

28. "Anschaffung des Theatri der Natur und Kunst" (Brather, *Leibniz und seine Akademie*, 152).

29. "Observatorio, laboratorio, Bibliothec, Theatro naturae et artis" (Louis A. Foucher de Careill, *Oeuvres de Leibniz*, vol. 7 (Paris, 1875), 219f.).

30. Woldemar Guerrier, *Leibniz in seinen Beziehungen zu Rußland und Peter dem Großen* (St. Petersburg and Leipzig, 1873), no. 13, p. 18; no. 73, p. 97; no. 125, p. 176; no. 127, p. 182; no. 143, p. 208; no. 148, p. 218; no. 240, p. 351.

31. Onno Klopp, "Leibniz' Plan der Gründung einer Sozietät der Wissenschaften in Wien," *Archiv für österreichische Geschichte*, vol. 40, 1868, app. XIII, 239.

32. Klopp, "Leibniz' Plan," app. XVI, 244.

33. Klopp, "Leibniz' Plan," app. XVII, 248.

34. "Cum enim Animus noster utatur imaginibus rerum sensibilium, consequens est, si imagines velut catena quadam implicentur, cogitantem exerrare, dummodo attendat, non posse" (Leibniz, *Sämtliche Schriften und Briefe*, VI, 4, A, no. 78, p. 324, l.2–4).

35. "Zu dieser Encyclopaedie wird kommen der Atlas Universalis kommen, ein werck von vortreflichen Nuzen dem Menschlichen gemüth alles leicht und mit Lust beyzubringen vermittelst einer großen menge Tafeln, figuren und wohlgemachter auch da nöthig und nüzlich illuminirter Zeichnungen oder Abriße, damit alles so einigermaßen mit den augen gefaßet, und auf dem papier entworffen werden kan, desto geschwinder und anmuthiger und gleichsam spielend und wie in einem blick, ohne umbschweiff der worthe, durch das gesicht dem gemüth vorgebildet und kräfftiger eingedrücket werden könne" (Leibniz, *Sämtliche Schriften und Briefe*, IV, 3, no. 116, p. 785, l.1–8).

36. Tommaso Campanella, "La Città del Sole," in Luigi Firpo, ed., *Scritti scelti di Giordano Bruno e di Tommaso Campanella* (Turin: Unione tipografioo-editrice torinese, 1968), 412ff.

37. According to Diodoros, II, 8, 4f.; see Ulrike Wegener, *Die Faszination des Maßlosen: Der Turmbau zu Babel von Pieter Bruegel bis Athanasius Kircher* (Hildesheim, Zurich, and New York: Georg Olms Verlag, 1995), 107ff.

38. Johann Valentin Andreae, *Christianopolis*, ed. and trans. Wolfgang Biesterfeld (Stuttgart: Reclam, 1975), §47, 73.

39. Horst Bredekamp, *Thomas Hobbes visuelle Strategien. Der Leviathan: Das Urbild des modernen Staates; Werkillustrationen und Portraits* (Berlin: Akademie, 1999), 170f.; Horst Bredekamp, *Die Fenster der Monaden. Gottfried Wilhelm Leibniz' Theater der Natur und Kunst* (Berlin: Akademie, 2004), 17–22 and passim.

40. "non video quid amplius humana industria praestare possit" (Leibniz, IV, 3, no. 116, p. 795, l.20).

41. "Es ist nichts in der Welt so beßer sonderlich junge Herrn instruire, als Figuren" (Leibniz, *Sämtliche Schriften und Briefe*, I, 3, no. 17, p. 17, l.7).

42. "Es kan vor eine ganze Bibliothec passiren, und findet man darinn die Bildnüßen fast aller berühmten Personen in der Welt, die repraesentationen unzehlicher einzüge und publiqver solennitäten; Ein ganzes Theatrum naturae et artis: Jagten, schiffarten und Tempesten, schlachten und festungen, Palläste, Garten, Landschafften, unzehliche hieroglyphica, capricen, ornamenten, devisen, symbola, und summa was von Wahrheiten und fabeln zierliches in Menschliche Gedancken kommen kan. Köndte man dieses werck haben, so hätte man gewislich einen schaz, und unerschöpfliche Quelle unzehlicher Nachrichtungen; deren man sich nicht nur bey Fürstlichen Lustbarkeiten, aufzügen, Mascaraden, Tourniren, sondern auch vielmehr bey Gebaüden, Gartenwerck, Machinen, und vielen begebenheiten bedienen köndte. Summa man köndte eine solche collection wohl eine lebendige Bibliothec nennen" (Leibniz, *Sämtliche Schriften und Briefe*, I, 3, no. 17, p. 17, l.12–21).

43. Johann Amos Comenius, *Orbis sensualium pictus. Hoc est, Omnium fundamentalium in Mundo Rerum & in Vitâ Actionum Pictura & Nomenclatura* (Nuremberg: 1658).

44. "J'approuverois merveilleusement les Tableaux des arts, s'il y en avoit, et j'ay souvent souhaitté qu'on fist designer et graver des grandes tailles douces, comme celles qui entrent dans les Atlas, qui representassent d'une seule veue toute une science, art ou profession" (Leibniz, *Sämtliche Schriften und Briefe*, IV, 3, no. 68, p. 551, l.19–22).

45. "Rien n'est plus important que d'apprendre les choses par raison, c'est le moyen de ne les point oublier. Et lorsque ces raisons sont palpables et sensibles, la satisfaction en est redoublée. Jusqu'icy la raison s'est servie de l'escorte de l'imagination" (Leibniz, *Sämtliche Schriften und Briefe*, IV, 3, no. 68, p. 553, l.10–13).

46. "Mihi autem in mentem venit Encyclopaediam totam Atlante quodam Universali egregie comprehendi posse. Primum enim pleraque quae doceri discique oportet oculis subjici possunt" (Leibniz, *Sämtliche Schriften und Briefe*, VI, 4, A, no. 31, p. 86, l.16–19).

47. Leibniz, *Sämtliche Schriften und Briefe*, VI, 4, A, no. 31, pp. 86–90.

48. "Leibniz n'a rien *vu*" (André Robinet, *G. W. Leibniz Iter Italicum [Mars 1689–Mars 1690]: Le dynamique de la République des Lettres.* Nombreux textes inédits (Florence: 1986), 2).

# 10

## Spinoza on the Natural and the Artificial

Alan Gabbey

Metaphysically speaking, there cannot be a real distinction for Spinoza between the artificial and the natural, between Art and Nature. There are two fundamental reasons. First, God and Nature are the same thing under different names, the "same thing" being infinite Substance with an infinity of attributes, of which we know only two: Thought and Extension. This God and this Nature are not the God and Nature of traditional ways of thinking, not even the God and Nature of the Stoics, a source of inspiration for Spinoza. The Stoics' *Natura sive Deus* connotes the presence of God *in* all parts of Nature, and is the inverse of Spinoza's *Deus seu Natura*, which connotes the total identification of God *with* Nature, the power of Nature being identical to the divine power, God's decrees being identical to the laws of nature.[1] There is no real distinction between "natural" processes (as traditionally understood) and the results of human interventions in natural processes or of human activities that use natural processes or materials. For Spinoza, all processes, all states of affairs, both mental and physical, belong to Spinozan Nature: "From the necessity of the divine nature there must follow an infinity of things in an infinity of ways (that is, everything that can fall under infinite intellect)."[2] Nor should it be thought that Nature is God's Artifice, because there is nothing with respect to which Nature could be said to *be* artificial. In any case, there is no such thing as divine *creation*.

Second, one of the attributes of infinite substance is Thought, and the will is a mode of thinking, just as motion is a mode of extension, so "will cannot be called a free cause, but only a necessary one," because "no volition can exist, or be determined to produce an effect, unless determined by another cause, and this cause determined in turn by another, and so on to infinity." This is true whether the will is finite or infinite. Spinoza's God does not have absolute will or *liberum arbitrium*, and, though it sounds like a category mistake, neither does Spinozan Nature. Neither do human beings, because each human being is a part of Nature or, if you like, a part of God. Each human mind consists in modifications of the attribute of Thought, just as each human body consists in modifications of the attribute of Extension, the mind being the (composite) idea of the body. Human will is the open-ended *conatus* of the mind to persevere in its being, so "in the [human] mind there is no absolute will, or if you like no free will, but the mind is determined to will this or that by a cause which is also determined by another cause, and this cause yet again is determined by another, and so on to infinity."[3] Spinoza demonstrates in *Ethics* I, Prop. 29, that "by the nature of things there is nothing contingent, but all things have been determined from the necessity of the divine nature to exist and produce effects in a certain way." Or in the words of *Ethics* I, Prop. 33, "Things could not have been produced by God in any way or in any order other than those in which they were produced."[4] Artifacts are no more contingent than natural objects: all objects, whether artificial or natural, are the necessary effects of causal chains. Hence another reason why Nature cannot be God's Artifice: there is no divine will that could have decreed its creation.

These doctrines undercut a characteristic traditionally attributed to Art and deriving from Aristotle, that its objects are contingent because they are the product of human *liberum arbitrium*, and so are not necessary objects, in contrast to those that are natural.[5] Before Spinoza, debates on the Art-Nature relation took for granted the dependence of Art on *free decisions*, decisions to make this or that machine, to exercise this or that craft, to develop this or that process, to intervene in this or that way in the processes of nature. Spinoza's philosophy presented a serious challenge to this assumption, and brought into critical focus the dependence of views on the Art-Nature relation on doctrines of the human and divine will. In *Ethics* II he refers to, and goes on to reject, the relation between free will—*libera voluntas*—and the contingent: "By God's power ordinary people understand God's free will and dominion over all the things that exist, things which consequently are taken ordinarily to be contingent. For they say that God has the power to destroy all things and return them to nothing."[6]

Now, Spinoza earned his living as a lens grinder, and his tools of the trade were artifacts, made by himself or by others; so he talked about them and other artifacts in the everyday language of the ordinary citizen. His writings in political philosophy treat of states and kingdoms, artifacts of the civic life; so he talked about these too in the everyday language of the ordinary citizen. As every citizen knows, lenses and lathes do not grow on trees, nor do states and kingdoms come about throu[illegible] *aller* among humans in a state of nature. Is this therefore the same Spinoza who wrote the *Ethics*? No doubt about it. Spinoza the lens grinder, Spinoza the political theorist and Biblical exegete, and Spinoza the metaphysician, are one and the same person under different descriptions.

I broached the question by consulting Emilia Giancotti Boscherini's *Lexicon Spinozanum* (1970), where every Latin and Dutch term used by Spinoza is listed in quoted phrases or as paraphrases in context. As a bonus, this presentation conveniently sidelines translations, which can never be trusted where the issues are terminological. Of course, concepts and terms do not always share an isomorphic relation, but it made sense to check at least *ars*, *artificium*, *artificialis*, and *naturalis*. Some of Spinoza's uses of these terms are what you would expect. Half a dozen times he uses *ars* in the sense of "the arts and sciences," or "the military arts," or "the arts of war and peace." In the chapter on aristocracy in the *Tractatus Theologico-Politicus* (1670), he argues that everyone in a free commonwealth may think as they please and say what they think. In particular, freedom of judgment is "of the first importance in fostering the arts and sciences, for only those whose judgment is free and unbiased can attain success in these fields."[7]

The *Lexicon Spinozanum* lists fifteen or more occurrences of *naturalis*, and no occurrences of *artificialis*, though it misses *artificium*, which appears in the *Ethics*.[8] The nonoccurrence of *artificialis* is unexpected. Metaphysically speaking, for Spinoza there cannot be artificial objects or states of affairs, in the sense of things or states of affairs produced voluntarily by humans in an effort to act "against" nature in some way. It would therefore have been difficult for him not to use the term *artificialis* had he wanted to make this point in explicit terms. In keeping with this absence in his published philosophical vocabulary, Spinoza uses *naturalis* not in opposition to "artificial," but to "divine," or to "civil" in the legal sense, or he uses *naturalis* to emphasize that the substantive thus qualified is a part of Nature. Writing on aristocracy in the *Tractatus Politicus*, he argues that if the sword of the dictator be confined to the wicked, evil will be kept under control. To ensure the maintenance of these conditions, there should be a council of syndics subordinate to the supreme council, "to the end that the sword of the dictator should be permanent in the hands not of any natural person, but of a civic person. . . ."[9] Again in the *Tractatus Politicus*, Spinoza characterizes natural right as "the very laws or rules of nature, in accordance with which everything takes place; in other words, the power of nature itself." Had human nature been such that men live according to the dictates of reason, their natural right would have been determined by the power of reason alone. But it is desire, more than reason, that drives us, and natural right is limited by appetites more than by reason. Also, "those desires, which arise not from reason, are not so much actions as passive affections of man." Then Spinoza continues:

> But as we are treating here of the universal power or right of nature, we cannot here recognize any distinction between desires engendered in us by reason, and those engendered by other causes, since the latter, as much as the former, are effects of nature, and display the natural endeavor by which man strives to continue in existence. For a man, be he learned or ignorant, is part of nature, and everything by which any man is determined to action, must be referred to the power of nature, that is, to that power as it is limited by the nature of this or that man. For man, whether guided by reason or mere desire, does nothing save in accordance with the laws and rules of nature, that is, by natural right.

However, echoing the preface to *Ethics* III, Spinoza notes that "most people believe that the ignorant disturb rather than follow the course of nature, and conceive of mankind in nature as one dominion within another. For they maintain that the human mind is not produced by natural causes, but is created directly by God, and is so independent of other things, that it has an absolute power to determine itself, and make right use of reason."[10]

The natural right of the individual, as explained in chapter 16 of the *Tractatus Theologico-Politicus*, is coextensive with the individual's power. The sovereign right of Nature, the set of all individuals, is therefore also coextensive with its power, which is the sovereign power of God. But this holds *within* a given subset of individuals. Each of those individuals has natural right, and collectively they have sovereign right if they contract to set up a democracy through the cession of individual right to the sovereign power. So in that context we can distinguish between individual natural right and the natural right of the democracy, construing the opposition only within the context of a political situation in which the options are democracy and a Spinozan state of nature.

Given that there is no metaphysical distinction between the *natural* and the *artificial*, we must see how Spinoza uses the terms *ars* and *naturalis* in contexts that indicate his view of the art-nature relation as it was debated in conventional terms by contemporaries and predecessors. In the appendix to *Ethics* I, Spinoza argues that final causes are human fictions, and that those who resort to the will of God as the cause of inexplicable events are taking refuge in a sanctuary of ignorance.

> So also are they bewildered when they behold the fabric of the human body. They know nothing about the causes of such a great work, so they conclude that it was constructed not by mechanical but by divine or supernatural art, and was set up in such a way that one part does not damage another. And so it comes to pass that someone who tries to discover the true causes of miracles, who strives to understand natural things, like the man of learning, who is not lost in bedazzlement at them, like the fool, is treated here, there and everywhere, and in noisy declamation, as a heretic and as impious by those whom the people hold in adoration as interpreters of Nature and of the Gods.[11]

It is clear from the context that "divine or supernatural art" is to be understood ironically. In Spinoza's eyes, divine or supernatural art cannot be anything other than an absurdity.

Proposition 2 of *Ethics* III states a central component of Spinoza's metaphysics: "The body cannot determine the mind to thinking, nor can the mind determine the body to motion, or to rest, or to some other [mode] (if there is any)." This is because "all modes of thinking have as their cause God, insofar as he is a thinking thing, not insofar as he is explicated by another attribute," and so "what determines the mind to thinking is a mode of thinking, not of extension." The Scholium then explains that this is better understood from an earlier Scholium, that of Prop. 7 of Part II, according to which "the mind and the body are one and the same thing, conceived now under the attribute of thought, now under the attribute of extension," with the result that "the order or connection of things is one, whether Nature is conceived under this attribute or that." All of this is as clear as daylight, yet people are persuaded "that the body first moves then is at rest on a sole command from the mind, and that it does lots of things that depend solely on the mind's will and inventiveness [*excogitandi ars*]." But no one knows what the body can or cannot do, because no one knows its structure in sufficient detail to be able to explain its wondrous behavior from the laws of nature, even if we knew them in their totality. Nor does anyone know "the cause or the means by which the mind might move the body, nor how many degrees of motion it might give the body, nor with how much speed it might be able to move it."

> From which it follows that when men say this or that action of the body originates from the mind, which has command over the body, they don't know what they are talking about, and are doing nothing but betray in gilded words their blithe ignorance of the true cause of the action.

Such men will object, saying

> that buildings, paintings, and things of that sort, which are the result of only human invention [*humana ars*], cannot turn out to be causally derivable just from the laws of nature, taken as it is to be merely corporeal. Nor is building some temple within the power of the human body, unless determined and directed by the mind.

But experience shows us many things that spring from the laws of nature alone, yet which the same men would never believe could happen without the control of a mind. Spinoza then adds a decisive example:

> the fabric of the human body, which in its contrivance [*artificium*] surpasses to the highest degree all things that have been constructed by human art [*humana ars*]; and here I pass over what I would have shown earlier [*Ethics* I, Prop. 16], that an infinity of things follow from Nature, under whatever attribute it be considered.[12]

Here is a Spinozan counterpoise between "excogitandi ars" and "humana ars." Like the "divine or supernatural art" of the appendix to *Ethics* I, "excogitandi ars" is intended ironically. Unrepentent Cartesians, who have a method for guiding the mind in its quest for truth, might believe that this methodological advantage gives the mind extra purchase over the body. But all of this is in vain if we are ignorant of the structure of the body—and of the true nature of mind and the will—and of the successive correlations of mind and body as each maps out the causal path prescribed for it by the laws of nature. On the other hand, Spinoza's contrast between the incomparable intricacy of the human body and the constructions of *ars humana* should be read at face value. It is a fact of the matter, known to all, that there is such a contrast, so there is nothing wrong in using ordinary language to refer to it. All the better if the term *ars* can bridge etymologically such widely differing degrees of mechanical complexity within the infinite domain of the infinitely many things that follow from Nature according to its eternal laws.

Spinoza was undoubtedly aware of the distinctive contrast between his position on "the Art-Nature relation" and those of his contemporaries and predecessors. Yet he did not address the question as an itemized article in his philosophical program. He probably thought his metaphysics had made the whole debate redundant as a philosophical issue, and had transformed it into the more significant question of whether artifacts were any sort of evidence that free choice and the commands of the imperious mind are necessary conditions for technological engagements with the external world. From a Spinozan philosophical standpoint, the questions traditionally debated on the art-nature relation since antiquity look effete and seem no longer of philosophical import.[13] In imitating Nature, can Art produce things that are truly natural? Can Art assist Nature in bringing to fruition natural processes? Can Art act against Nature, or divert it to different ends? Can Art do something that Nature cannot do? For Spinoza, questions of this sort would have made no sense, or would have received responses to the effect that the implied difficulties were practical, not philosophical.

Take the discipline of mechanics as traditionally understood in Spinoza's day. I do not know how much Spinoza knew of Renaissance or seventeenth-century writings on mechanics, but had he perused them closely, he would have found curious anomalies. Guido Ubaldo del Monte, writing in 1577, praised mechanics because "it operates against nature or rather in rivalry with the laws of nature."[14] Confronted with this text, the author of the *Ethics* would have denied that it was possible for mechanics to *compete against* nature and its laws, especially if the laws of mechanics *are* the fundamental laws of nature, or are at least dependent on them. Did Spinoza believe this was the position of Descartes, his more immediate point of reference, or was he aware that the status of mechanics in Descartes' thinking, and its role in his natural philosophy, are complicated issues? There isn't enough evidence to warrant instructive conclusions about Spinoza's own view of the nature of mechanics and its relation to physics. According to the *Lexicon Spinozanum*, there is only one instance of *mechanica* outside the correspondence, in the *Treatise on the Emendation of the Intellect* (before 1662). To reach the state of happiness that comes from the knowledge of the union of mind and Nature, we must understand enough of Nature to establish the social order that will maximize the number of those who reach this state, paying close attention to moral philosophy, the education of children, the development of medicine, and lastly, "because through art many difficult things are made easy, and we can save a great deal of time and bring great convenience into our lives, Mechanics is in no way to be disparaged."[15] This is the ordinary view of mechanics as the practical discipline of public usefulness, but we cannot infer from it that Spinoza believed the principles of mechanics ground the laws of nature in some way, or (which is more likely) that he believed the laws of nature are sets of necessary regularities that underpin mechanical law.[16]

Still, it is likely that Spinoza knew Descartes' letter to Constantin Huygens of March 1638,[17] from which he would have concluded that sixty years after Ubaldo the problem really had not gone away or been fully resolved. Descartes contrasts Nature acting autonomously in the ordinary way with Nature being assigned supernumerary duties to meet the requirements of human artifice:

> I still cannot see any likelihood of releasing my *Monde* on the world, at least not for a long time; and without that, neither would I be able to finish the Mechanics you wrote to me about, for it depends entirely on my *Monde*, mainly in what concerns the speed of motions. And you have to explain what the laws of nature are, and how she acts in the ordinary way, before you can show properly how she can be applied to effects to which she is not accustomed.[18]

There is something odd about the idea of Nature being unaccustomed (even in a manner of speaking) to effects that by the canons of Descartes' own natural philosophy must instantiate the laws of nature that he had presented in *Le Monde* and was to present in revised form in *Principia Philosophiae* (1644), where he declared starkly that the only difference between artifacts and natural bodies is the difference of scale in their constituent components. Whatever their dimensional differences, "it is as natural for a clock, composed of wheels of a certain kind, to indicate the hours, as for a tree, grown from a certain kind of seed, to produce the corresponding fruit."[19]

Spinoza shared that view, yet nothing in his writings betrays the tension in Descartes between the idea of art (including mechanics) acting against nature and natural things and processes. Spinoza had resolved these tensions, though he wrote nothing on mechanics and very little on natural philosophy. He does distinguish between the natural and the artificial, between natural philosophy and the arts, but the distinction is between two perspectives within which can be viewed the same thing, not between one state of affairs that expresses natural process, and another that originates from the volitions of humans intervening in the natural order of things. Viewing the same thing from different perspectives, or understanding it under different descriptions, is a key trope in

Spinoza's philosophical style, the key word being the adverb *quatenus* ("insofar as," "to the extent that"), which does an incalculable amount of expository work throughout the *Ethics*.[20] Insofar as we practice the arts as citizens, individually or within a community, or exercise the mechanical arts for human benefit, there is a difference between the artificial and the natural, though demarcations may not always be precise. Insofar as we contemplate these things from a Spinozan metaphysical standpoint, there cannot be a distinction between the artificial and the natural, between Art and Nature, because all that pertains to artifacts pertains also to Nature and its immutable laws. In particular, the human volitions or desires that are the partial cause of artifacts pertain to Nature and its laws, in which there is no room for fictions such as *liberum arbitrium* or *libera voluntas*.

## Notes

1. Lagrée, "Juste Lipse," 52.

2. "Ex necessitate divinae naturae, infinita infinitis modis (hoc est, omnia, quae sub intellectum infinitum cadere possunt) sequi debent" (*Ethics* I, Prop. 16. *Opera* II, 60). Except where otherwise indicated, all translations are my own.

3. "Voluntas non potest vocari causa libera, sed tantùm necessaria ... unaquaeque volitio non potest existere, neque ad operandum determinari, nisi ab aliâ causâ determinetur, & haec rursùs ab aliâ, & sic porrò in infinitum." "In Mente nulla est absoluta, sive libera voluntas; sed Mens ad hoc, vel illud volendum determinatur à causâ, quae etiam ab aliâ determinata est, & haec iterùm ab aliâ, & sic in infinitum" (*Ethics* I, Prop. 32 and Dem.; II, Prop. 48. *Opera* II, 72, 129). The human will as the *conatus* of the mind to persevere in its being is explained in *Ethics* III, Prop. 9, Scholium (*Opera* II, 147–148).

4. "In rerum naturâ nullum datur contingens, sed omnia ex necessitate divinae naturae determinata sunt ad certo modo existendum, & operandum." "Res nullo alio modo, neque alio ordine à Deo produci potuerunt, quàm productae sunt" (*Ethics* I, Props. 29, 33. *Opera* II, 70, 73).

5. *Nichomachean Ethics* 1139a20–1140b30. In his article on *ars*, Micraelius notes: "Artis objecta sunt ea, quae fieri & non fieri possunt, i.e. contingentia, quorum fabricatio pendet ab artificis arbitrio" (Micraelius, *Lexicon*, col. 155). "Objects of art are those that can come into being or not, that is, they are contingent, whose construction depends on a decision of the artificer." See further Mikkeli on Zabarella's account of Aristotle's necessary-contingent distinction (*Aristotelian Response*, 25–29). I thank Constance Blackwell (personal communication) for underlining the importance of understanding the artifical-natural relation in terms of the contrast beween the contingent and the necessary.

6. "Vulgus per Dei potentiam intelligit Dei liberam voluntatem, et jus in omnia, quae sunt, quaeque propterea communiter ut contingentia considerantur. Deum enim potestatem omnia destruendi habere dicunt, et in nihilum redigendi" (*Ethics* II, Prop. 3, Scholium. *Opera* II, 87).

7. "Quod haec libertas apprime necessaria est ad scientias, & artes promovendum; nam hae ab iis tantum foelici cum successu coluntur, qui judicium liberum, & minime praeoccupatum habent" (*Tractatus Theologico-Politicus*, chap. 20. *Opera* III, 243. *Theological-Political Treatise*, trans. Shirley, 234).

8. *Ethics* III, Prop. 2, Scholium; IV, Prop. 57, Scholium. *Opera* II, 143, 253.

9. "Ut scilicet dictatorius ille gladius perpetuus esset non penes personam aliquam naturalem, sed civilem, cujus membra plura sint, quàm ut imperium inter se possint dividere (per Art. 1. & 2. Cap. 8), vel in scelere aliquo convenire" (*Tractatus Politicus*, chap. X, sec. II. *Opera* III, 354. *A Political Treatise*, trans. Elwes, 380 (translation slightly modified)). In the *Tractatus Politicus*, aristocracy is treated in chapters VIII–X.

10. "IV. Per Jus itaque naturae intelligo ipsas naturae leges, seu regulas, secundum quas omnia fiunt, hoc est, ipsam naturae potentiam." "V. . . . Equidem fateor, cupiditates illas, quae ex ratione non oriuntur, non tam actiones, quàm passiones esse humanas. Verum quia hîc de naturae universali potentiâ, seu Jure agimus, nullam hîc agnoscere possumus differentiam inter cupiditates, quae ex ratione, & inter illas, quae ex aliis causis in nobis ingenerantur: quandoquidem tam hae, quàm illae effectûs naturae sunt, vimque naturalem explicant, quâ homo in suo esse perseverare conatur. Est enim homo, sive sapiens, sive ignarus sit, naturae pars, & id omne, ex quo unusquisque ad agendum determinatur, ad naturae potentiam referri debet, nempe quatenus haec per naturam hujus, aut illius hominis definiri potest. Nihil namque homo, seu ratione, seu solâ cupiditate ductus, agit, nisi secundum leges, & regulas naturae, hoc est (*per Art. 4. hujus Cap.*), ex naturae jure." "VI. At plerique, ignaros naturae ordinem magis perturbare, quàm sequi, credunt, & homines in naturâ veluti imperium in imperio concipiunt. Nam Mentem humanam a nullis causis naturalibus statuunt produci, sed a Deo immediatè creari, a reliquis rebus adeò independentem, ut absolutam habeat potestatem sese determinandi, & ratione rectè utendi" (*Tractatus Politicus*, chap. II, secs. IV, V, VI. *Opera* III, 277, 277–278. *A Political Treatise*, trans. Elwes, 292, 292–293 (translation modified)).

11. "Sic etiam, ubi corporis humani fabricam vident, stupescunt, & ex eo, quòd tantae artis causas ignorant, concludunt, eandem non mechanicâ, sed divinâ, vel supernaturali arte fabricari, talique modo constitui, ut una pars alteram non laedat. Atque hinc fit, ut qui miraculorum causas veras quaerit, quique res naturales, ut doctus, intelligere, non autem, ut stultus, admirari student, passim pro haeretico, & impio habeatur, & proclametur ab iis, quos vulgus, tanquam naturae, Deorumque interpretes, adorat" (*Ethics* I, appendix. *Opera* II, 81).

12. "Proposition II. Nec Corpus Mentem ad cogitandum, nec Mens Corpus ad motum, neque ad quietem, nec ad aliquid (si quid est) aliud determinare potest. Demonstratio. Omnes cogitandi modi Deum, quatenus res est cogitans, & non quatenus alio attributo explicatur, pro causâ habent (per Prop. 6. p. 2); id ergo,

quod Mentem ad cogitandum determinat, modus cogitandi est, & non Extensionis." "Scholium . . . Mens, & Corpus una, eademque res sit, quae jam sub Cogitationis, jam sub Extensionis attributo concipitur. Unde fit, ut ordo, sive rerum concatenatio una sit, sive natura sub hoc, sive sub illo attributo concipiatur . . ." ". . . persuasi sunt, Corpus ex solo Mentis nutu jam moveri, jam quiescere, plurimaque agere, quae à solâ Mentis voluntate, & excogitandi arte pendent." "Deinde nemo scit, quâ ratione, quibusve mediis Mens moveat corpus, neque quot motûs gradûs possit corpori tribuere, quantâque cum celeritate idem movere queat. Unde sequitur, cùm homines dicunt, hanc, vel illam actionem Corporis oriri à mente, quae imperium in Corpus habet, eos nescire, quid dicant, nec aliud agere, quàm speciosis verbis fateri, se veram illius actionis causam absque admiratione ignorare." "At dicent ex solis legibus naturae, quatenus corporea tantùm consideratur, fieri non posse, ut causae aedificiorum, picturarum, rerumque hujusmodi, quae solâ humanâ arte fiunt, possint deduci, nec Corpus humanum, nisi à Mente determinaretur, ducereturque, pote esset ad templum aliquod aedificandum." "Addo hîc ipsam Corporis humani fabricam, quae artificio longissimè superat omnes, quae humanâ arte fabricatae sunt, ut jam taceam, quòd suprà ostenderim, ex naturâ, sub quovis attributo consideratâ, infinita sequi" (*Ethics* III, Prop. 2, Dem. and Schol. *Opera* II, 141–143).

13. From the vast literature on the subject, of particular relevance here for background are Close, "Art and Nature"; Schmitt, *John Case*, chap. V ("John Case on Art and Nature"), 191–216; Mikkeli, *Aristotelian Reponse*, 111–118; Des Chene, *Physiologia*, chap. 7 ("Nature and Counternature"), 212–251, especially 239–251, and his contribution to this volume (chapter 6). See also my "Beween *ars* and *philosophia naturalis*," especially 142–145.

14. *Mechanicorum liber* (Pesaro, 1577), preface. Quoted from Drake and Drabkin, *Mechanics in Sixteenth-Century Italy*, 241.

15. ". . . & quia arte multa, quae difficilia sunt, facilia redduntur, multumque temporis, & commoditatis in vitâ eâ lucrari possumus, ideò Mechanica nullo modo est contemnenda" (*Opera* II, 9).

16. On the difficulties on this issue that arise in the case of Descartes, see my "Descartes' Physics and Descartes' Mechanics: Chicken and Egg?"

17. The letter was published in 1659 by Clerselier, in his *Lettres de M. Descartes* (t. II, 1659, 377–380).

18. ". . . je ne voy encore aucune apparence que je puisse donner au moins de longtemps mon Monde au monde; et sans cela, je ne sçaurois aussi achever les Mechaniques dont vous m'écrivez, car elles en dépendent entierement, principalement en ce qui concerne la vitesse des mouvemens. Et il faut avoir expliqué quelles sont les loix de la nature, & comment elle agit à son ordinaire, avant qu'on puisse bien enseigner comment elle peut estre appliquée à des effets ausquels elle n'est pas accoustumée" (*Oeuvres de Descartes*, II, 50).

19. *Principia Philosophiae* IV, art. 203. *Principles of Philosophy*, trans. Miller and Miller, 285–286.

20. A striking example is the Corollary to Prop. 11 of *Ethics* II: "The human mind is a part of the infinite intellect of God. So when we say that the human mind perceives this or that, all we are saying is that God has this or that idea, not insofar as he is infinite, but insofar as he is explicated through the nature of the human mind, or insofar as he constitutes the essence of the human mind. And when we say that God has this or that idea, not only insofar as he constitues the nature of the human mind, but insofar as he has at the same time the idea of another thing together with the human mind, then we say that the human mind perceives the thing only partially, or inadequately." "Hinc sequitur Mentem humanam partem esse infiniti intellectûs Dei; ac proinde cùm dicimus, Mentem humanam hoc, vel illud percipere, nihil aliud dicimus, quàm quòd Deus, non quatenus infinitus est, sed quatenus per naturam humanae Mentis explicatur, sive quatenus humanae Mentis essentiam constituit, hanc, vel illam habet ideam; & cùm dicimus Deum hanc, vel illam ideam habere, non tantùm, quatenus naturam humanae Mentis constituit, sed quatenus simul cum Mente humanâ alterius rei etiam habet ideam, tum dicimus Mentem humanam rem ex parte, sive inadaequatè percipere" (*Ethics* II, Prop. 11, Coroll. *Opera* II, 94–95).

## References

Boscherini, Emilia Giancotti. 1970. *Lexicon Spinozanum*. 2 vols. The Hague: Nijhoff.

Close, A. J. 1969. Art and Nature in Antiquity and the Renaissance. *Journal of the History of Ideas* 30: 467–486.

Descartes, René. 1657, 1659, 1667. *Lettres de M. Descartes, où sont traittées les plus belles questions de la morale, de la physique, de la médecine et des mathématiques*. Ed. Claude Clerselier. 3 vols. Paris.

Descartes, René. 1964–1974. *Oeuvres de Descartes, publiées par Charles Adam & Paul Tannery*. Rev. ed., with the collaboration of the Centre National de la Recherche Scientifique. Ed. P. Costabel, J. Beaude, and B. Rochot. 11 vols. Paris: Vrin. Original ed., Paris: Cerf, 13 vols., 1897–1913.

Descartes, René. 1983. *Principles of Philosophy*. Trans. and notes Valentine Rodger Miller and Reese P. Miller. Synthese Historical Library, vol. 24; Collection des Travaux de l'Académie Internationale d'Histoire des Sciences, no. 30. Dordrecht: Reidel.

Des Chene, Dennis. 1996. *Physiologia: Natural Philosophy in Late Aristotelian and Cartesian Thought*. Ithaca: Cornell University Press.

Drake, Stillman, and I. E. Drabkin, eds. 1969. *Mechanics in Sixteenth-Century Italy: Selections from Tartaglia, Benedetti, Guido Ubaldo, & Galileo*. University of Wisconsin Publications in Medieval Science. Madison: University of Wisconsin Press.

Gabbey, Alan. 1993a. Between *ars* and *philosophia naturalis*: Reflections on the historiography of early modern mechanics. In *Renaissance and Revolution: Humanists, Scholars, Craftsmen and Natural Philosophers in Early Modern Europe*, 133–145. Ed. J. V. Field and Frank A. J. L. James. Cambridge: Cambridge University Press.

Gabbey, Alan. 1993b. Descartes' Physics and Descartes' Mechanics: Chicken and Egg? In Stephen Voss, ed., *Essays on the Philosophy and Science of René Descartes*, 310–323. Oxford: Oxford University Press.

Lagrée, Jacqueline. 1994. *Juste Lipse et la restauration du stoicisme*. Paris: Vrin.

Micraelius, Johann. 1653. *Lexicon philosophicum terminorum philosophis usitatorum ordine alphabetico sic digestorum, ut inde facile liceat cognosse, praesertim si tam Latinus, quam Graecus Index praemissus non negligatur, quid in singulis disciplinis quomodo sit distinguendum et definiendum*. Jena.

Mikkeli, Heikki. 1992. *An Aristotelian Response to Renaissance Humanism: Jacopo Zabarella on the Nature of Arts and Sciences*. Societas Historica Finlandiae, Suomen Historiallinen Seura, Studia Historica 41. Helsinki: Suomen Historiallinen Seura.

Schmitt, Charles B. 1983. *John Case and Aristotelianism in Renaissance England*. McGill-Queen's Studies in the History of Ideas, vol. 5. Kingston and Montreal: McGill–Queen's University Press.

Spinoza, Baruch. [1925] 1972. *Spinoza Opera*. Ed. Carl Gebhardt. 4 vols. Heidelberg: Carl Winter. Rpt. 1972.

Spinoza, Baruch. 1951. *A Theologico-Political Treatise and A Political Treatise*. Trans. R. H. M. Elwes. New York: Dover.

Spinoza, Baruch. 1989. *Tractatus theologico-politicus* (Gebhardt edition, 1925). Trans. Samuel Shirley, introduction Brad S. Gregory. Leiden: E. J. Brill.

# 11

## Eighteenth-Century Wetware

Jessica Riskin

> To be a machine, to feel, think, know good from evil like blue from yellow . . .
>
> —Julien Offray de la Mettrie, *Man a Machine* (1747)[1]

The word *machine*, especially in reference to machinery of the pre- and early industrial age, calls to mind clockwork: rigid, precise, passive. The contrast between machines, thus construed, and living organisms seems easy: creatures are soft, fluid, and active. Yet this way of distinguishing between artificial and organic machinery was, by the early eighteenth century, already obsolete. In keeping with one of this book's organizing arguments—that the blurring of the line between nature and artifice has long been in progress—this chapter examines the unclocklike (soft, fluid, internally active) machines designed by Enlightenment philosophers and engineers to test the boundaries between life and mechanism. These machines constitute what is referred to in the title of this piece as "eighteenth-century wetware."

*Wetware* is the name that computer scientists and engineers give to the human brain and nervous system, to contrast them with computer hardware and software. Rudy Rucker, a popular science writer, novelist, and mathematician in the Department of Mathematics and Computer Science at San Jose State University, coined the term to serve as the title of a 1988 novel, in which he defined *wetware* as referring to "all [the brain's] sparks and tastes and tangles, all its stimulus/response patterns—the whole biocybernetic software of [the] mind."[2] Rucker's definition makes manifest the dual action of his new word: even as it distinguishes animal from artificial machinery, *wetware* also unites the two, and has in fact come to be used in ways that undermine the contrast between animals and machines: in reference to artificially intelligent systems that are modeled closely on human neurology, or ones that incorporate biological components, or ones that resemble biological systems in texture and substance, or any combination of these, for example "biomimetic," "chemomechanical" systems made of polymer gels.[3]

*Wetware*, then, with its Silicon Valley derivation and its cutting-edge applications, is the expression of a particular moment, the turn of the twentieth to the twenty-first century. The neologism voices one of the current moment's organizing ambivalences: we believe that the processes of life and consciousness are essentially mechanistic, and can therefore be simulated, yet we are equally firmly persuaded that the essences of life and consciousness will ultimately be beyond the reach of mechanical reproduction.

Although the conflicting assumptions expressed in the word *wetware* and the machinery to which it refers seem utterly specific to the present, they were in fact also characteristic of the second half of the eighteenth century. That period saw the emergence of artificial life in a flurry of attempts to simulate with machinery the physiological processes and cognitive behaviors of living creatures.

Here and throughout, I use the word *simulation* and all its forms in their modern sense, which originated around the middle of the twentieth century, to refer to an experimental model from which one can discover properties of the natural subject. *Simulation* in its eighteenth-century usage meant artifice, and had a negative connotation, implying fakery. I employ it here despite the anachronism in order to suggest that eighteenth-century projects in artificial life had a pivotal place in the history of attempts to simulate (in its modern sense) life processes.[4] The first designers of artificial life intended their projects to resemble natural life in texture and substance, sometimes even making use of biological components. The resulting simulations, like present-day "wetware," made manifest both their makers' assumptions about the differences between animals and machines, and their impulse to undermine these differences.

Part of the surprise and the interest in this similarity between late eighteenth and late twentieth-century approaches to artificial life is that there was a long, intervening period, during the nineteenth and early twentieth centuries, when people thought very differently about the possibility of simulating life. This conspicuous changeability, over the past two and a half centuries, in how designers of artificial life have conceived of their project has been strangely absent from recent discussions of their early productions. For example, Gaby Wood's *Edison's Eve* treats Descartes' animal-machine, the clockwork androids of the mid- to late eighteenth century, the mechanical tricks of the magician Jean-

Eugène Robert-Houdin during the mid-nineteenth century, and the robots currently inhabiting MIT's Artificial Intelligence Laboratory all as expressions of the same impulse, a sort of rationalism gone mad. Wood describes this mad-scientist impulse as also always provoking the same response, from the journalists who gave effusive accounts of a set of Swiss mechanical musicians during the 1770s, to the lady-spectators who crossed themselves and fainted during exhibitions of the notorious Chess-playing Turk of the late eighteenth and early nineteenth centuries, to E. T. A. Hoffmann's mystical treatment of artificial life in "Die Automaten" (1821): in each of these instances, Wood discerns the Freudian "Uncanny."[5]

Here I will suggest, in contrast, that the project of artificial life, and also the surrounding cultural representations and assessments of that project, transformed foundationally from each generation to the next. The story of the origins of modern artificial life lies, not in a changeless quest emerging from timeless human impulses, but rather in the experimenters', philosophers', and critics' continually shifting understandings of the boundary between intelligent and rote, animate and mechanical, human and nonhuman. In what follows, I describe the eighteenth-century intense interest in producing artificial life as a discrete and sui generis moment, examine what set it apart from previous and subsequent ways of conceiving the relations between animal and artificial machinery, and close with some speculation about the similarity between the two moments in the history of artificial life, the second half of the eighteenth century and the second half of the twentieth.

The emergence of artificial life in the mid-eighteenth century was crucially informed by a particular philosophical development, namely a materialist, mechanist understanding of life and thought. Materialists repudiated Descartes' separation between mind and body, and insisted that all the functions that might be ascribed to mind and soul actually resided in the stuff of which living creatures were made. Mechanists argued that interaction among the body's parts, animal machinery, was directly responsible for all vital and mental processes. These materialist and mechanist accounts of life worked in both directions. Not only did they shape how people thought about living creatures, but reciprocally they also changed how people thought about matter and mechanism. If life was material, then matter was alive, and to see living creatures as machines was also to vivify machinery.

Thus materialism and mechanism were themselves transformed during the second half of the eighteenth century by their application to the explanation of life. Materialists began to invoke a vital property of matter called "sensibility" that, many physiologists believed, was inherent in organic substance. Mechanists began to draw on such qualities in their explanations, and therefore to throw off the restrictions of seventeenth-century mechanism, no longer confining themselves to the primary qualities of size, shape, state of motion, number, and solidity. The altered, eighteenth-century meanings of materialism and mechanism obtain in works such as Julien Offray de La Mettrie's *L'Homme-machine*, the title of which could be misleading, since the book does as much to animate machinery as it does to mechanize life. La Mettrie's human machinery senses and feels, and is in fact eroticized. La Mettrie writes: "If what thinks in my brain is not a part of that vital organ, and consequently of the whole body, why does my blood heat up when I am lying tranquilly in bed thinking ... Why does the fever of my mind pass into my veins?"[6] He knows that thinking is carried out by his machinery because of the sensual agitations produced by thought.

The mechanists and mechanicians of the eighteenth century described animal machinery that was sensitive and passionate. Seeing animals as machinery, they began also to see machinery as animal, and to design machines accordingly. The results, like modern "wetware," called attention to certain differences of texture, substance, and mode of action between animal and artificial machinery, and simultaneously worked to undermine these differences. It is because of their dual function that I call these machines "eighteenth-century wetware."

One species of such machines was comprised of automata: mechanical figures of people and animals. During the mid- to late eighteenth century, a particular style dominated their design. Eighteenth-century designers tried to simulate life's textures and substances, and even physiology, making automata from this period look very different from those of the previous or subsequent periods. Mechanical animals of the previous period, the seventeenth and early eighteenth centuries, present artistic renditions of animal movements but no attempt to imitate physiological processes. Consider, for example, an artificial swan (figure 11.1), presented to the Paris Academy of Sciences in 1733 by a mechanician named Maillard. The swan paddled through the water on a paddle wheel while a set of gears swept its head slowly from side to side.[7] It was intended to represent, rather than to simulate, a natural swan.

Strikingly, the very pinnacle of mechanist physiology in the mid-seventeenth century did not correspond with attempts to simulate

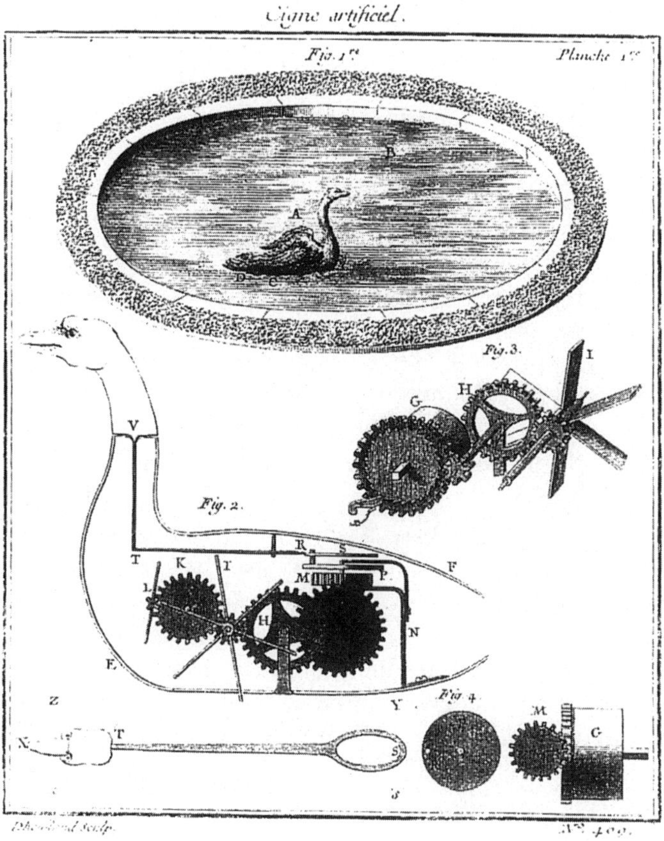


**Figure 11.1**

Maillard's artificial swan. Reprinted from Gallon, ed., *Machines et Inventions approuvées par l'Académie Royale des Sciences depuis son établissement jusqu'á present; avec leur description*, (Paris, 1735–1777), 1:133–135. Courtesy of Department of Special Collections, Stanford University Libraries.

animals using machinery. Seventeenth-century mechanist physiologists drew analogies between animals and machines, but they did not use machines to simulate life. Descartes compared animals to automata and even built automata himself, but he did not design these as physiological simulations.[8] His philosophical heirs, though, did try to simulate life. This disparity shows up the crucial divergence between analogies and simulations, two conceptual devices that function quite differently: analogies work by preserving a certain distance between the two things being likened, whereas simulations operate by collapsing that distance.

The period between the 1730s and the 1790s was one of simulation, in which mechanicians tried earnestly to collapse the gap between animate and artificial machinery.

The period of simulation was surrounded on both sides by contrasting moments in which analogies between life and machinery were rife, but simulations rare. Nineteenth-century automata, like their seventeenth-century ancestors, were not simulations, but instead were artistic renditions of animal and human activities. Indeed, in the nineteenth century there was a great proliferation of such renditions. Particularly after midcentury, when automata began to be mass-produced, they were suddenly everywhere. To get a sense of the transformation, consider that during the 1840s, automata sold in Paris for prices in the thousands of francs, whereas in 1868, one could buy an automaton for 8 francs, 50 centimes.[9] Mechanicians began to design very complicated displays, ultimately filled with the preoccupations of the Belle Epoque: dandies, circus and street performers, magicians, the exotic, workers at work and schoolchildren at their lessons and shopkeepers in their shops. But despite their elaborateness, these nineteenth-century automata were markedly less ambitious than automata of the preceding century. Or anyway, their ambition was not to simulate. Once again, philosophers, physicists, physiologists, and engineers drew analogies, particularly during the second half of the nineteenth century, between human and animal bodies, on the one hand, and machinery, on the other, often resting on the new concepts of energy and work.[10] But, by and large, those who drew such analogies did not use machinery to simulate living beings; indeed they tended to reject such simulation on principle. The second half of the eighteenth century was an exceptional moment, then, for the very literal way in which it construed the similarity between animal and artificial machinery.

The difference between eighteenth- and nineteenth-century approaches to artificial life is encapsulated in the contrast between "Pierrot écrivain" (figure 11.2), a late nineteenth century automaton by Gustave Vichy, one of the most successful designers of the period, and the Writer (figures 11.3a, 11.3b, and 11.3c), built during the early 1770s by a Swiss clock-making family named Jaquet-Droz.[11] Whereas Pierrot did not actually write, but merely waved his pen over his paper in a rough imitation of writing, the Jaquet-Droz Writer not only wrote, but could (and can) be programmed to write any message of up to forty characters. He remains in working condition at the *Musée d'art et d'histoire* in Neuchatel, Switzerland, where he is accompanied by a Draughtsman (figures

**Figure 11.2**
*Pierrot écrivain*, Gustave Vichy. Musée National de Monaco.

11.4a, 11.4b, and 11.4c) who uses a bit of charcoal to draw four pictures, and by a Lady-musician (figures 11.5a and 11.5b) who plays a harpsichord, following her hands with her eyes as she plays.

The Jaquet-Droz automata do not just carry out the processes of writing, drawing, and playing music, they are also anatomical and physiological simulations. Their skeletal structures were likely designed with the help of the village surgeon.[12] Both the Lady-musician and the Draughtsman also breathe. The Draughtsman periodically blows the charcoal dust from his paper and surveys his work, and the Lady-musician sighs in time to the music. Her breathing was what spectators

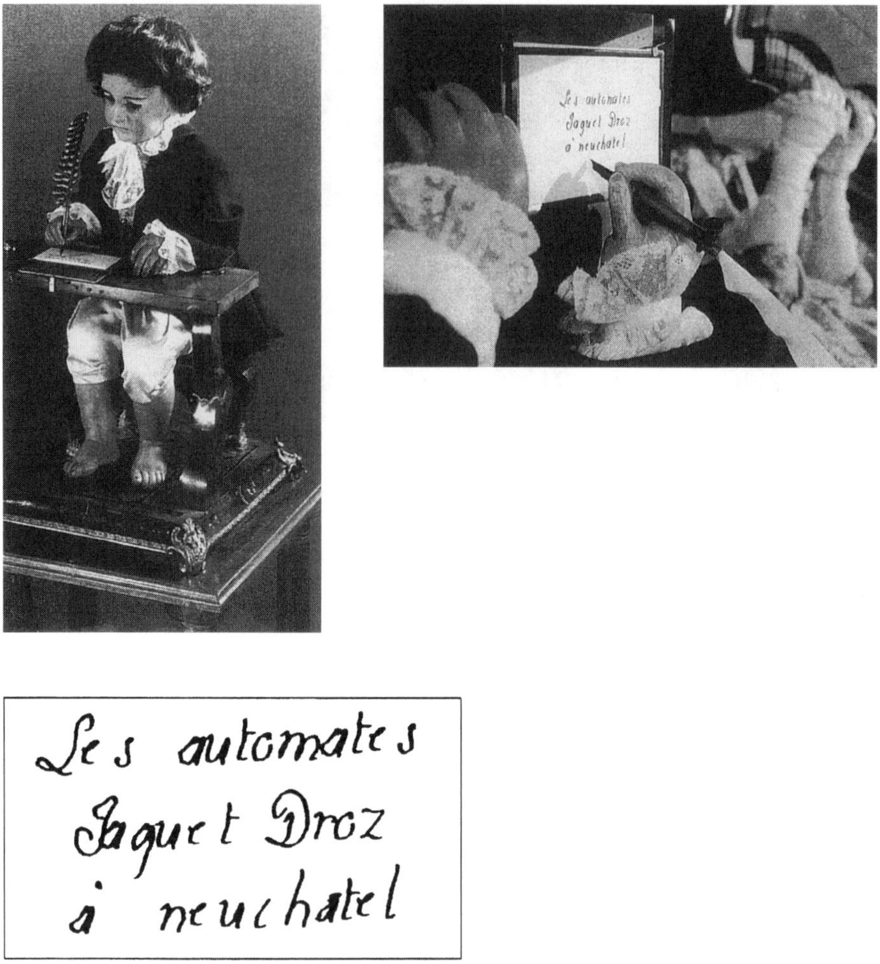


**Figure 11.3**
*The Writer*, Jaquet-Droz Family. Musée d'art et d'histoire, Neuchâtel, Switzerland.

most often commented on. It made her seem not only alive, but emotional. She appeared moved by the music she played.[13]

Breathing automata were quite popular in the late eighteenth century. They originated with Jacques Vaucanson's android Flute-player of 1738, who needed to breathe in order to play his flute. He was the first automaton musician actually to play his instrument, rather than being a music box with a decorative figure on top, and he had three sets of bellows, giving him three different blowing pressures. In addition to breathing, the Flute-player also had lips that he could flex in four directions, a supple tongue, and fingers with a skin of soft leather.[14]

Physiological correctness, then, was a new and pervasive interest on the part of automaton designers of the mid- to late eighteenth century. Automata of this period were physiologically correct sometimes to the point of being scatological. The leading example is Vaucanson's defecating Duck of 1738. In addition to cavorting with its bill and wings, bending its neck and flexing its feet, the mechanical Duck digested its food—or so Vaucanson claimed—by means of a "Chymical Elaboratory" in its stomach. It swallowed bits of grain, and after a moment, it excreted them at the other end in an altered state.[15] This was the main attraction that drew people from all over Europe. The digestion was later demonstrated to be fraudulent (instead of being digested, the grain was caught in a reservoir at the base of the throat, while the rear end was loaded before the demonstration with fake excrement),[16] but that does not detract from the interest of Vaucanson's choice of subject. Why a defecating duck? Because, I think, the subject epitomized messy, organic, animal nature. The snooping protagonist of Jonathan Swift's 1730 poem, "The Lady's Dressing Room," gradually discovering that his true love's beauty is a triumph of art over nature, has a final epiphany when he discovers her chamber pot: "Thus finishing his grand survey, disgusted Strephon stole away, repeating in his amorous fits, O Celia, Celia, Celia shits."[17] Here was the most natural of products, the antithesis of art.

Eighteenth-century projects in artificial life produced machines with soft skin, flexible lips, and delicate, jointed fingers. These machines not only wrote, drew, and played musical instruments, but also breathed, ate, and defecated. They performed functions, in other words, that their designers took to epitomize the animate and the organic.

One function that many took to epitomize the organic was spoken language. The materialist-mechanist understanding of intelligence operated at its most literal in the widespread consideration of speech, the defining function of human intelligence, as an essentially physiological process. Eighteenth-century designers of artificial life assumed that the sounds of spoken language depended on an organic structure in the throat and mouth, and it was this dependence that provided the interest in designing speaking machines. The assumption that a talking machine required simulated speaking organs had not always dominated thinking

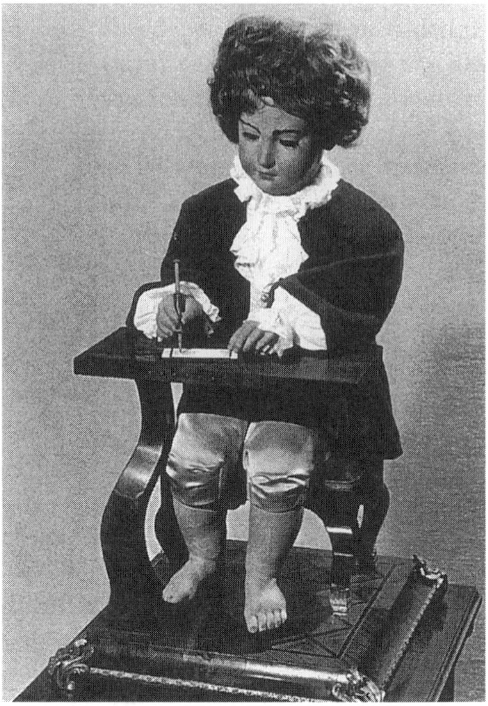

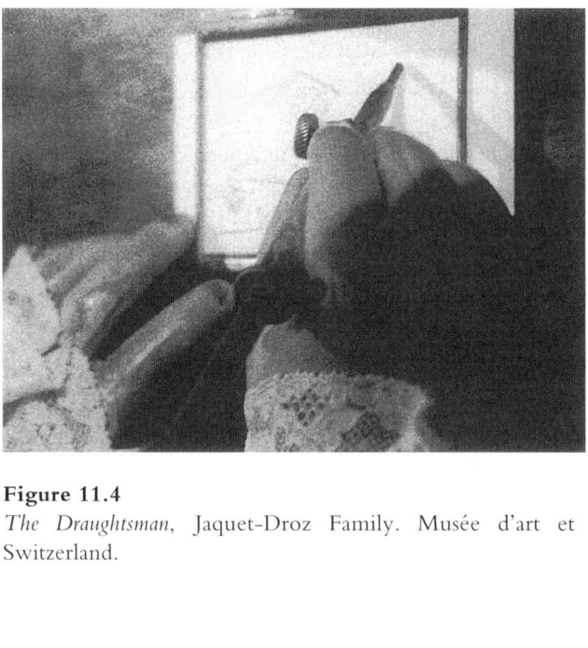

**Figure 11.4**
*The Draughtsman*, Jaquet-Droz Family. Musée d'art et d'histoire, Neuchâtel, Switzerland.

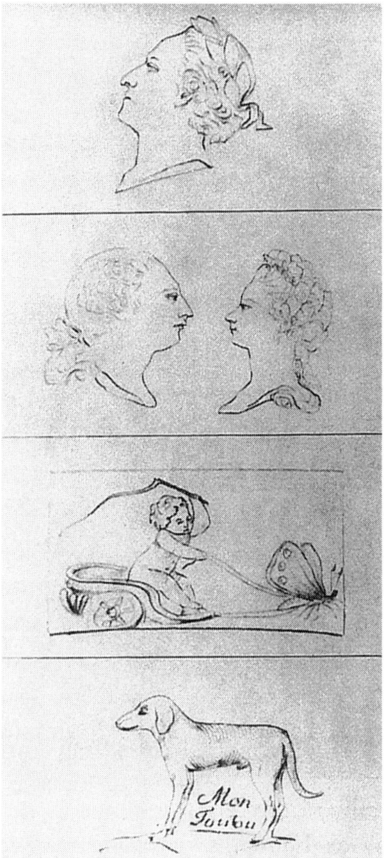


**Figure 11.4**
(continued)

about artificial speech. In 1648, John Wilkins, the first secretary of the Royal Society of London, described plans for a speaking statue that would synthesize, rather than simulate, speech by making use of "inarticulate sounds." He wrote, "we may note the trembling of water to be like the letter L, the quenching of hot things to the letter Z, the sound of strings, to the letter Ng, the jirking of a switch to the letter Q, etc."[18] But in the eighteenth century, builders of speaking machines mostly assumed that it would be impossible to create artificial speech without building a talking head: reproducing the speech organs and simulating the process of speaking.

**Figure 11.5**
*The Musician*, Jaquet-Droz Family. Musée d'art et d'histoire, Neuchâtel, Switzerland.

Throughout much of the century, there was a great deal of skepticism about artificial speech on the grounds that the human larynx, vocal tract, and mouth were too soft, supple, and malleable to be simulated mechanically. Around 1700, Denys Dodart, personal physician to Louis XIV, presented several memoirs to the Paris Academy of Sciences on the subject of the human voice, in which he argued that the voice and its modulations were caused by constrictions of the glottis, and that these were "inimitable by art."[19] Bernard le Bouyer de Fontenelle, who was then Perpetual Secretary of the Academy, commented that no wind instrument produced its sound by such a mechanism (the variation of a single opening) and that it seemed "that Nature had the design of placing [the instruments of the voice] altogether outside the realm of imitation.... Nature can use materials that are not at our disposal, and she knows how to use them in ways that we are not at all permitted to know."[20]

In 1738, following the public presentation of Vaucanson's automata, the abbé Desfontaines predicted that, despite these triumphs in artificial life, the mechanical imitation of speech would be impossible because of the inimitability of the "larynx and glottis ... the action of the tongue, its folds, its movements, its varied and imperceptible rubbings, all the modifications of the jaw and the lips."[21] And in 1775, Court de Gébelin maintained, "the trembling that spreads to all the parts of the glottis, the jigging of its muscles, their shock against the hyoid bone that raises and lowers itself, the repercussions that the air undergoes against the sides of the mouth ... these phenomena" could only take place in living bodies.[22]

However, during the last three decades of the century, several people took on the project of simulating the organs and process of speech. The first was Erasmus Darwin, who in 1771 reported that he had "contrived a wooden mouth with lips of soft leather, and with a valve over the back part of it for nostrils." Darwin's talking head had a larynx made of "a silk ribbon ... stretched between two bits of smooth wood a little hollowed." It said "mama, papa, map, and pam" in a "most plaintive tone."[23]

The next to simulate speech was a Frenchman, the abbé Mical, who presented a pair of talking heads to the Paris Academy of Sciences in 1778 (figure 11.6). The heads contained several "artificial glottises of different forms [arranged] over taut membranes." By means of these glottises, the heads performed a fairly insipid dialogue in praise of Louis

**Figure 11.6**
"The abbé Michal's Talking Heads," reprinted from Alfred Chapius and Édouard Gélis, *Le Monde des Automates Étude Historique et Technique*, (*Paris 1928*), 2:205.

XVI: "The King gives Peace to Europe," intoned the first head; "Peace crowns the King with Glory," replied the second; "and Peace makes the Happiness of the People," added the first; "O King Adorable Father of your People their Happiness shows Europe the Glory of your Throne," concluded the second head. The Academicians appointed to examine Mical's talking heads emphasized that their enunciation was "very imperfect," but granted their approval to the work anyhow because it was done in imitation of nature and contained "the same results that we admire in dissecting . . . the organ of the voice."[24] Several more people built talking heads before the turn of the century, among them a Hungarian engineer named Wolfgang von Kempelen, who claimed to follow nature "absolutely" in designing his speaking machine. The resulting apparatus had bellows for lungs, a glottis of ivory, a leather vocal tract with a hinged tongue, a rubber oral cavity and mouth whose resonance could be altered by opening and closing valves, and a nose with two little pipes as nostrils.[25]

After this flurry of activity in the 1770s, 1780s, and 1790s, there was a marked decline in interest in speech simulation. A few people over the course of the nineteenth century, including Charles Wheatstone and Alexander Graham Bell, built their own versions of Kempelen's and Mical's speaking machines and of other talking heads from the earlier period.[26] But for the most part, designers of artificial speech turned their attention once again to speech synthesis.[27] In 1828, Robert Willis, a professor of applied mechanics at Cambridge, wrote disparagingly that most "writers who have treated on the vowel sounds appear never to have looked beyond the vocal organs for their origin. Apparently assuming the actual forms of these organs to be essential to their production ... [they have considered] vowels in fact more in the light of physiological functions of the human body than as a branch of acoustics." In fact, Willis argued, vowels "are not at all beyond the reach of human imitation in many ways, and are not inseparably connected with the human organs."[28] In addition to promoting synthesis over simulation, many also returned to the conviction that simulation of the vocal organs was impossible. Around 1850, Claude Bernard wrote in his notebook, "The larynx is a larynx ... that is to say ... [its] mechanical or physical conditions are realized nowhere but in the living organism."[29]

Disenchantment with speech simulation was so pronounced that when a German immigrant to America named Joseph Faber designed quite an impressive talking head in the late 1840s (figures 11.7 and 11.8), he could not get anyone to take notice of it. Faber's talking head was modeled on Kempelen's and Mical's, but was far more elaborate. It had the head and torso of a man dressed like a Turk, and inside were bellows, an ivory glottis and tongue, a variable resonance chamber, and a mouth cavity with a rubber palate, lower jaw, and cheeks. The machine could pronounce all the vowels and consonants, and was connected by way of levers to a keyboard of seventeen keys, so that Faber could play it like a piano. He first exhibited the machine in New York City in 1844, where it aroused very little interest. He then took it to Philadelphia, where he had no better luck. P. T. Barnum found Faber

**Figure 11.7**
"Euphonia," Joseph Faber. Circus World Museum, Baraboo, Wisconsin.

and his talking head there, renamed the machine the "Euphonia," and took them on tour to London, but even Barnum could not make a success of it. Finally the Euphonia was exhibited in Paris in the late 1870s, where it was mostly ignored, and soon thereafter all traces of it disappear.[30] The moment for talking heads had passed.[31] In the early part of the twentieth century, designers of artificial speech moved on from mechanical to electrical speech synthesis.[32] The simulation of the organs and process of speaking—of the trembling glottis, the malleable vocal tract, the supple tongue and mouth—was specific to the last decades of the eighteenth century.

At that time, speaking, like defecation, seemed a quintessentially natural act, and this was what provided the interest of trying to simulate it. The next step was to reproduce the body itself, and in fact, prosthetic devices underwent a major transformation in this period. The change was largely in materials. Mechanical prostheses had originated in the sixteenth century as heavy, cumbersome, iron things with very limited movements. The hands designed by the French surgeon Ambroise Paré worked by springs and catches, and he also built a leg with a knee lock

**Figure 11.8**
"Euphonia," Joseph Faber. Circus World Museum, Baraboo, Wisconsin.

that could be fixed in either the standing or sitting, equine position (figures 11.9a, 11.9b, and 11.9c).[33]

During the first decade of the eighteenth century, a mechanician to the French court named Sébastien designed two artificial hands for a Swedish military officer named Gunterfield who had lost both arms above the elbow. These hands had flexible fingers that Gunterfield could control using his stumps by means of a network of threads. The finished product enabled him to don and doff his hat. But the things were uncomfortable and awkward, and he decided he would rather do without.[34] In 1732 an inventor named Kriegseissen applied for and received the approval of the Paris Academy of Sciences for an arm and hand made of copper leaves (figure 11.10). The contraption was designed for a below-the-elbow amputee. The amputee's upper arm fitted into the hollow upper portion of the artificial arm, and the movements of the wrist and hand were controlled by means of pulleys fastened on either side of the elbow and cords of catgut passing through them to the

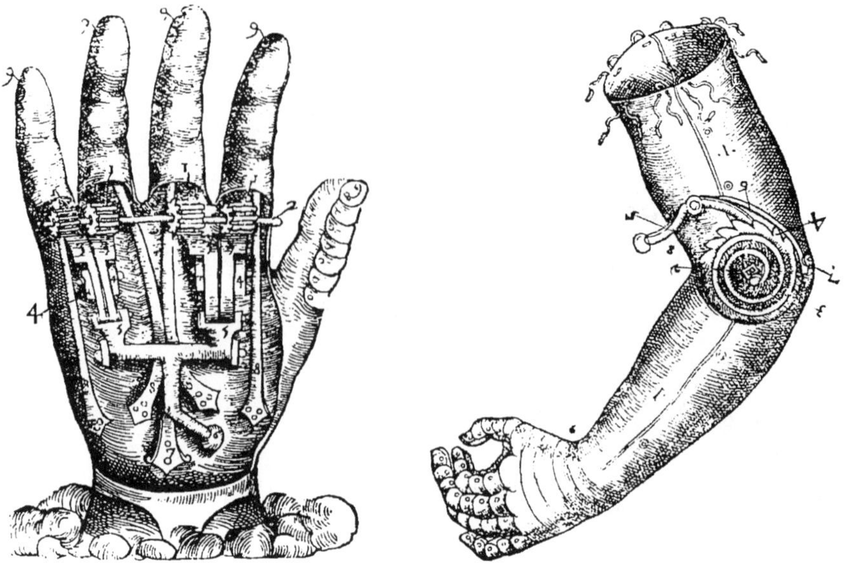

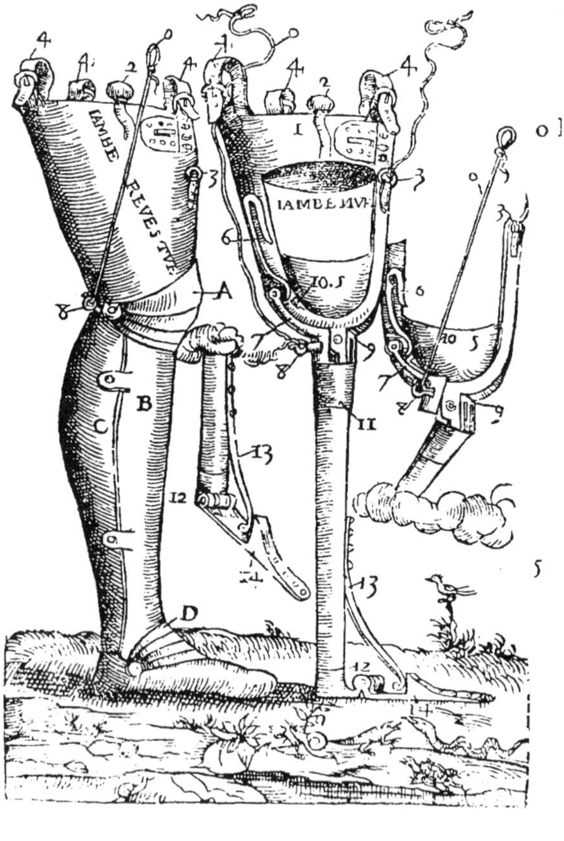

**Figure 11.9**
Ambroise Paré's prosthetic hand, leg, and arm. Reprinted from Ambroise Paré, *The Collected Works of Ambrose Paré* (Pound Ridge, NY: 1968), 881–882.

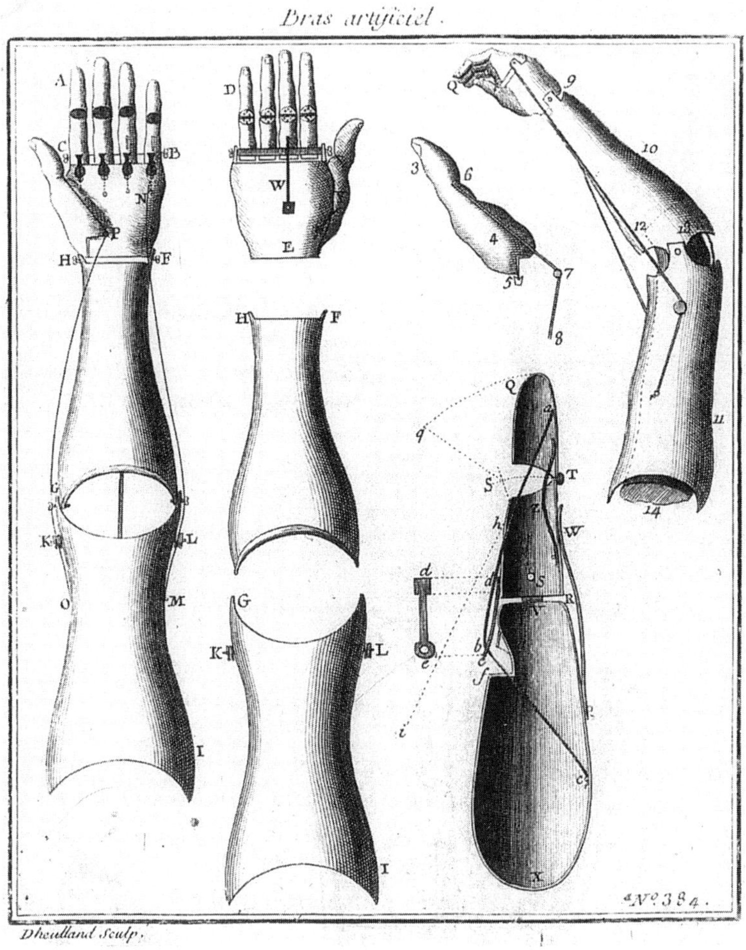


**Figure 11.10**
Kriegseissen's copper arm. Reprinted from Gallon, ed., *Machines et Inventions approuvées par l'Académie Royal des Sciences depuis son établissement jusqu'á present; avec leur déscription* (Paris: 1735–1777), 6:71–73. Courtesy of Department of Special Collections, Stanford University Libraries.

thumb and fingers. By using his own elbow to bend the lower arm in toward the upper arm, the wearer tightened the two cords on either side, which curled the whole hand inward into a fist. Springs on each joint of each finger and the back of the thumb restored them to their straight positions.[35] Next, around 1760 a mechanician named Laurent received a knightship for improving on Kriegseissen's design so that it could be used by an above-the-elbow amputee. The beneficiary of this improved copper arm was reportedly able to write "very legibly" with it.[36]

The major innovations in the field of articulated prosthetic limbs came in the 1780s and 1790s, at the hands of the Jaquet-Droz family, the automaton designers. After the success of their automata, they were asked by a tax collector named La Reynière to design two artificial hands for his son, who had lost his own in a hunting accident. The result was a pair of prostheses made from the same materials that the Jaquet-Droz family had used in their automata: leather, cork, parchment, and papier-mâché on a steel frame. They were very light, about 480 grams, and reportedly very versatile. The Jaquet-Droz operation continued to design prosthetic hands and arms of this sort through the 1790s.[37]

In the same period, anatomical models for teaching underwent the same transformation as prosthetic devices. They became less like models and more like simulations, reproducing their organic subjects in texture and substance. For example, the King's midwife, Mme. du Coudray, whose life story is told in Nina Gelbart's biography,[38] designed a "birthing machine" to use in teaching midwifery (figure 11.11a and 11.11b). This machine, of which du Coudray produced many copies to send to midwives and surgeons all over France, had skin and soft organs made from flesh-colored linen and leather, some dyed redder and some paler, and stuffed with padding. The earlier models were built on pelvic bones taken from real skeletons; many of the later ones used wood and wicker. As a "supplement" to the machine, one could buy "liquids," an opaque red fluid and a clear one, along with a set of sponges. The sponges, saturated with the fluids, were to be planted inside the birthing machine by the demonstrator and made to release their fluids at the appropriate moments.[39]

The birthing machine, like the Jaquet-Droz artificial limbs with leather skin, the talking heads with tongues and glottises, and the automata that breathed and defecated, all reflected the assumption that an artificial model of a living creature should be soft, flexible, sometimes also wet and messy, and in these ways resemble its organic subject. This was

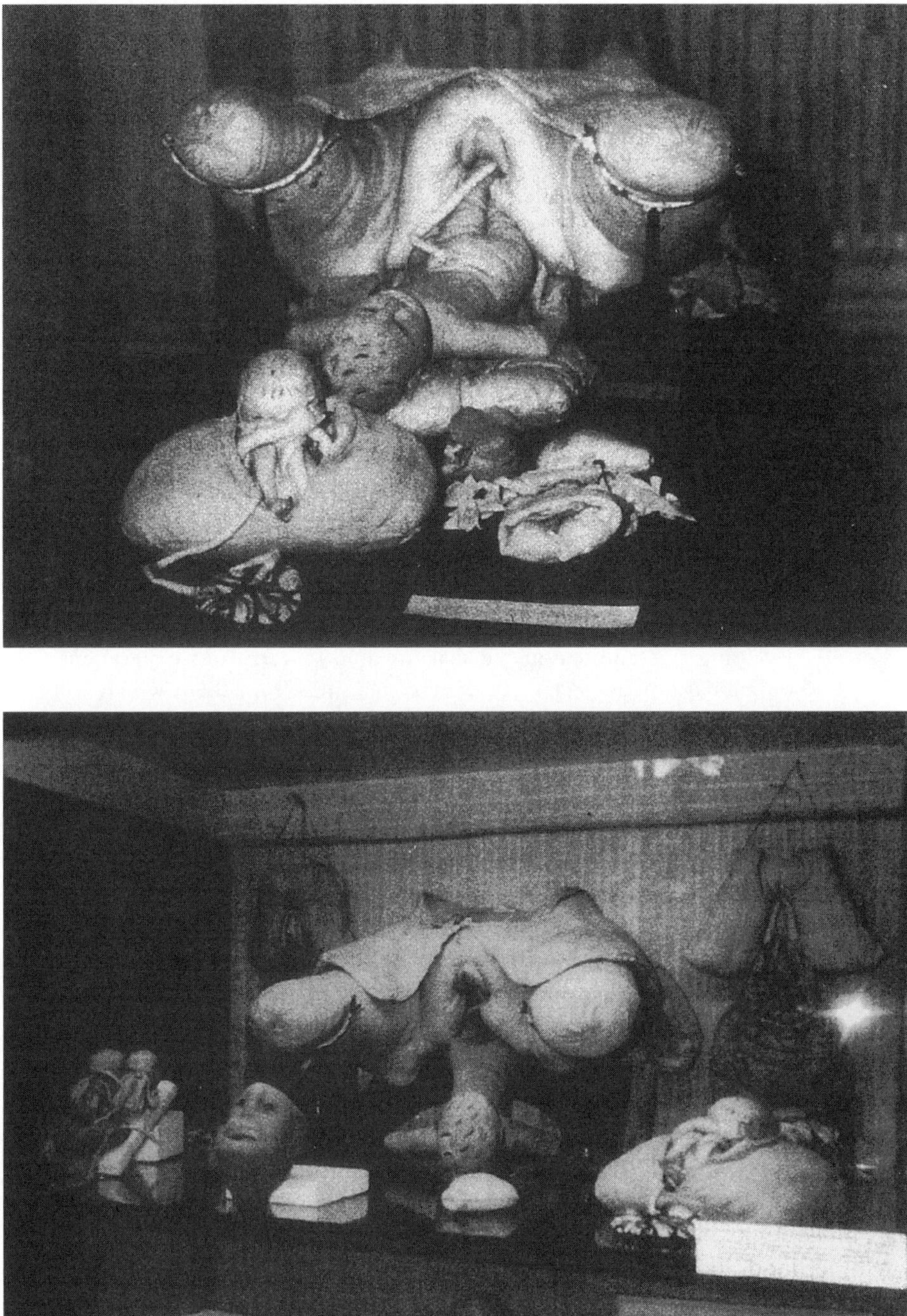

**Figure 11.11**
Mme. du Coudray's birthing machine. Reprinted from Nina Rattner Gelbart, *The King's Midwife: A History and Mystery of Madame Du Coudray*. (Berkeley: 1998), 62.

the flip side of a materialist-mechanist understanding of life. If living creatures were simply the matter and moving parts they were made of, then artificial creatures could potentially be very much like them.[40]

The "wetware" approach to artificial life was exemplified, finally, in the work of designers of so-called moving anatomies, mechanical models of physiological processes. The phrase "moving anatomy" was Vaucanson's. He used it to refer to his initial project (before the Duck and the Flute-player), which he described as a machine containing "several automata, and in which the natural functions of several animals are imitated by the movement of fire, air and water." Very little is known about this first machine except that Vaucanson took it on a successful tour of France.[41] Later he returned to his moving anatomy project, and in 1741 he presented to the Académie de Lyon his plan[42]

> to create an automatic figure whose motions will be an imitation of all animal operations, such as the circulation of the blood, respiration, digestion, the movement of muscles, tendons, nerves and so forth.... By using this automaton we shall be able to carry out experiments on animal functions, and ... draw conclusions from them which will allow us to recognize the different states of human health.

This machine seems never to have been finished. But more than twenty years later, still in pursuit of his "moving anatomy," now in the more modest form of a hydraulic model of the circulatory system alone, Vaucanson applied to Louis XV for support. The King approved Vaucanson's request to have the machine built in Guyana, where he proposed to use "elastic gum" to make the veins. These veins would have been the first flexible rubber tubes.[43] Again the project lapsed, but its conception is another example of the new interest in using lifelike materials to imitate animal and human parts. The difference between a model of the circulatory system with metal tubes and one with rubber veins, like the difference between an artificial arm made of iron and one made of cork and leather, was not just the materials but the concept, the idea that a machine could—and should—have a lifelike texture. Vaucanson's anatomical and physiological projects were also all conceived as wet machines: they centrally involved fluids acting either chemically or hydraulically.

Along with being soft, moist, and malleable, the substances from which moving anatomies were made were also lifelike in that they seemed to act purposefully. Fluids and airs tended to maintain a balance, to fill a vacuum, to equalize pressures. These active and purposeful tendencies seemed to designers of moving anatomies to be crucial to the operation of animal machinery, and they were crucial also to the way they understood their artificial machines. An example is the moving anatomy designed by the French surgeon François Quesnay. Like many of his contemporaries, he believed that the source of motion in animal bodies was a fluid principle so volatile that "a mere nothing excites and puts [it] into action." This fluid, sometimes called "animal spirits" or the "vital principle," was "distributed by means of the threads of the nerves" all throughout the animal. As to its mode of action, it "extraordinarily confused Physicists," because it could only be comprehended by "recourse to active Elements"—that is, by its own property of activeness.[44] Quesnay also appealed to the active properties of ordinary fluids, in particular their striking tendency to seek an equilibrium. He emphasized that bleeding did not diminish the amount of blood in a given vessel, because when a surgeon depleted the blood in one of a patient's blood vessels, an equal amount of blood came from branching vessels to replace it, and he persuaded himself of the truth of this purposeful activity of blood and other fluids by building a mechanical model of the circulatory system.[45]

A competitor of Quesnay's, a fellow surgeon named Claude-Nicolas Le Cat, also designed moving anatomies. In 1739, he published a description (now lost) of an "automaton man in which one sees executed the principal functions of the animal economy," circulation, respiration, and "the secretions."[46] His idea, like Vaucanson's and Quesnay's, was that one could experiment on this automaton to test the effects of therapies. It is not clear what became of this early project, but Le Cat returned to the idea in 1744, when, according to the proceedings of the Academy of Rouen, he read a memoir there making the same proposal, to build "an artificial man or automaton, in which he hopes to show all the operations of a living man, the circulation of blood, the movement of the heart, the play of the lungs, the swallowing of food, its digestion, the evacuations, the filling of the blood vessels and their depletion by bleeding, even"—and here Le Cat exhibited that peculiarity of contemporary materialist-mechanist philosophy, the treatment of language as a bodily function—"speech and the articulation of words."[47]

A great crowd was assembled to hear the memoir, and one witness described the scene as follows: "Monsieur Le Cat told us of a plan for an artificial man. . . . His automaton will have respiration, circulation, quasi-digestion, secretion and chyle, heart, lungs, liver and bladder, and God forgive us, all that follows from it. Let him have a fever, we will bleed him, we will purge him and he will but too much resemble a man."[48] Now, to say that one could build a machine man capable not only of respiration and digestion but of speaking, of having a fever, of being treated and cured of illness held implications not only for what human beings were like, but also for what machinery was like. What sort of machinery did Le Cat have in mind? Like Quesnay, Le Cat gave the primary role in animal machinery to its liquid components, also dividing the liquids into two categories, the "liqueurs"—which were the tangible liquids such as blood, chyle, lymph, and bile—and the "fluids"—which were rarefied, intangible media that flowed through the nerves: the motor fluid, the sensitive fluid, and above all the animal or vital fluid. These motive fluids were the primary force in the "animal machine." They acted on the solids, which acted, by means of their property of "organic spring," on the liqueurs, which reacted in their turn to maintain a continual oscillation.[49] Central to Le Cat's model of animal machinery, as to Quesnay's, was his assumption of the active nature of organic matter and vital fluids.

Eighteenth-century wetware, then, made manifest, not a reduction of animals to machinery, but a convergence in people's understanding of animals and of machines. Not only did they begin to understand animals as machine-like, but they also, at the same time, began to understand machines as animal-like: soft, malleable, sometimes warm, with fluid parts that acted, not only by constraint, but by inner purpose.

These projects in artificial life represented one moment in an ongoing dialectical engagement between our understandings of life and of machinery, in which living creatures and machines have continually redefined each other, both by being identified with one another, and by being opposed to one another. Eighteenth-century wetware, like its present-day analogue, arose from an initial assumption of an unbreachable rift dividing cold, hard, dry, machinery, its inanimate parts moving only by constraint, from warm, soft, wet, living creatures, their organic parts driven by vital purpose. It was the articulation of certain differences between natural and artificial life that triggered the invention of machines that undermined those differences. But these machines in turn led people to rethink what constituted life, and to define natural life by contrast with artificial life. And so people's assumptions about what is essential to life, and what is within the purview of machinery, have continually transformed one another.

Present-day builders of automata, when they encounter the work of their eighteenth-century predecessors, invariably ask why no one in that earlier period tried to simulate the action of the five senses, sensation being currently the single most obvious function to give an artificial creature. The question is all the more intriguing when one considers that eighteenth-century materialist-mechanists such as La Mettrie subscribed to the sensationist doctrine that ideas were not innately implanted in the mind, but were created by the action of the senses and the nervous system, and therefore could not be abstracted from body. The eighteenth-century conviction that life, consciousness, and thought were essentially embodied in animal and human machinery has striking parallels in current artificial intelligence. A prominent school of AI, called artificial life, is founded in the principle that intelligence must be "physically grounded" and "embodied."[50] Rodney Brooks, director of the AI Lab at MIT, has left behind the purely software model of AI, and instead builds robots with sensors and feedback loops, giving them vision, hearing, and touch. The eighteenth-century materialist-mechanist insistence that the functions of mind were all carried out by the brain, as distinct from the soul, and Brooks's claim that the software of the mind cannot be abstracted from its hardware, come down to strikingly similar conceptions of the nature of thought. We have returned to a rigorously literal view of the sameness of living and artificial machinery. Brooks's writings about his robots, in their insistence that intelligence cannot be disembodied, have a distinctly eighteenth-century sound, and indeed, a recent book on artificial life identifies Vaucanson's Duck as the progenitor of the field.[51]

Why, then, did Brooks's eighteenth-century equivalents not try to simulate sensation? Perhaps the answer is that what we find blindingly obvious, endowing an automaton with the ability to sense, is not at all obvious but the product of the centuries-long interaction between our understandings of life and of machinery. This chapter has examined a single phase of the engagement. In the subsequent phase, during the early nineteenth century, critics of eighteenth-century "wetware" drew new lines between living creatures and machines. They observed, for example, that no automaton had been truly self-moving, and reaffirmed the Aristotelian principle that animal life was defined by the ability of living creatures to produce their own source of motion. Animals were self-moving and machines were not. Next, around the middle of the nineteenth century, Hermann von Helmholtz and others undermined the notion that living creatures produced their own power by establishing the concepts of energy and its conservation. Animals, like machines, simply converted energy into work. Life and machinery had been similar in their ability to move autonomously, then antithetical because of the reliance of machinery on external sources of motion, and were now similar again, because animals too consumed energy.

Nineteenth-century critics of eighteenth-century projects in artificial life also decided that animal life was defined by its ability to maintain a stable internal environment in response to the external conditions in which it found itself. It was chiefly Claude Bernard, the same who insisted that a "larynx was a larynx," who redefined living creatures in these terms.[52] Animals were responsive to their environments and machines were not. Or were they? The mid-twentieth-century mechanical tortoises designed by a Cambridge University neurologist and engineer named Grey Walter had two sensors, one for light and the other for touch. They were designed to explore and respond to a simple environment consisting of lightbulbs and obstacles.[53] Moreover, when Walter and Norbert Wiener and others became interested in the possibility of designing artificial creatures that would be responsive to their environments, they traced responsive machinery—machines that employed what they now called feedback—back to the eighteenth century, in particular to James Watt's steam engine governor.[54] That is, they retrospectively designated as responsive machines whose designers had by no means understood them as such. It was only after machines were compared and then contrasted and then compared again with living beings that they came to seem capable of responding to their environments. By the same token, it was by being contrasted and then compared and then contrasted again with inanimate machinery that animal machinery came to seem defined by its responsiveness to the world around it.

In the long history of this dialectical engagement between our understandings of animals and of machines, I have suggested that the second half of the eighteenth century and the second half of the twentieth century represented similar moments, in which people have been preoccupied by the possibility of simulating life, whereas during the intervening period they mostly renounced this project. After the first decades of the nineteenth century, and until about the middle of the twentieth, people became largely disenchanted with the close simulation of life.[55] Even when they drew analogies between animals and machines, such as the analogy between labor and mechanical work—and despite the growing presence of automation all around them—nineteenth-century scientists and engineers mostly rejected the possibility of mechanically simulating life processes. Helmholtz, for example, accused eighteenth-century mechanicians of hubris, alleging that they had considered "no problem beyond . . . [their] power."[56]

Such moral indictments of artificial life—depictions of the quest to synthesize a living creature as hubristic, and of the results as monstrous—were absent from eighteenth-century commentaries. These indictments were rather a development of the early nineteenth century, coeval with Mary Shelley's *Frankenstein* (1818). People began around the same time to debunk the frauds of eighteenth-century artificial life. In 1821, Robert Willis, the same who disparaged the practice of creating artificial speech by simulating the vocal organs, published a pamphlet denouncing the more famous, but fraudulent, automaton that Kempelen had designed, in addition to his talking head, the chess-playing Turk.[57] Similarly in 1858, the French magician and automaton maker Jean Eugène Robert-Houdin exposed the charlatanism at the heart—or really the stomach—of Vaucanson's Duck. He also commented sarcastically on a notice in the *Journal des savants* from a century and a half earlier that had described an automaton in which "with the exception of the operation of the soul, everything that takes place in the body may be witnessed." Robert-Houdin wrote, "What a pity the mechanician stopped so soon! For it would have cost him so little, while making so exquisite a resemblance to the fairest work of the Creator, to add to his automaton a soul moving by clockwork."[58] Robert-Houdin's own automata were not true automata, in that they all involved hidden levers or pedals attended by human operators (often his son). Not until Walter's mid-twentieth-century mechanical tortoises did the next period of intensive interest in physiological simulation begin.

Why did people mostly turn away from the close simulation of living processes in the early part of the nineteenth century? This turning away does, it is true, seem to be in keeping with certain contemporaneous developments, in particular a Romantic distaste for rational and mechanical systems and the development of vitalism in biology. On the other hand, this chapter has shown, I hope, that the simulations of the latter eighteenth century transformed the meaning of mechanist philosophy to accommodate such previously nonmechanical phenomena as emotions, desires, and vital fluids. Moreover, physiologists of the nineteenth century—like those of the seventeenth century—as well as physicists establishing the concepts of energy and its conservation drew frequent analogies between animals and machines, making their disenchantment with simulation seem all the more curious.

To understand this curious disenchantment, we might return to the distinction I have been suggesting between drawing such analogies and using machinery to simulate animal life. These are two quite different endeavors, since an analogy rests on an assumed distinction between its two terms, while a simulation works to collapse the distinction. Thus we can make sense of the fact that, in the very same text in which he described his mechanical simulation of the circulatory system, Quesnay urged, "Let us stop representing the human body as a hydraulic machine." The objectionable analogy belied the "active" nature of the animal machine, and the "organic action" of its "flexible" parts.[59] Seventeenth- and nineteenth-century physiologists drew analogies, but did not simulate; Quesnay simulated, but disapproved of mechanistic analogies.

Why were the seventeenth and nineteenth centuries periods of analogy, and the late eighteenth and late twentieth centuries periods of simulation? I have returned repeatedly to the double-edged nature of simulations: they represent transformations, not only of people's understanding of animals, but of their understanding of machines as well. Analogies in contrast tend to hold one side of the equation fixed, and use it to say something about the other side: we know what machines are, and animals turn out to be a lot like them. Thus when he objected to the analogy between the body and a hydraulic machine, Quesnay had in mind an older, more static notion of a machine as something rigid that moved purely by constraint. But simulations transform both sides: we are not entirely sure what animals are, or what machines can be, and we can find out about both by trying to build an animal-machine. With his moving anatomy, Quesnay did not merely mechanize the circulatory system. He also transformed machinery into something active, flexible, and organic.

It makes sense, in light of this tendency of simulations to transform both sides of the equation, that the second half of the eighteenth century and the second half of the twentieth century have both been periods of simulation. The beginnings of the Industrial Revolution and the beginnings of the Information Revolution were both periods of extreme fluidity in people's understandings of what machines were—and indeed in the nature of machines. Once the industrial period was fully underway, artificial machinery and its relations to living creatures stabilized, replacing the fluidity required by a simulation with two terms that were, for the moment, fixed: life and mechanism. Only when the Information Revolution introduced a new kind of machinery did this fixity give way to a new fluidity, and the possibility of using machinery to simulate life again seemed intriguing.

In other words, the modern makers of automata that see, hear, and feel in fact have a great deal in common with the eighteenth-century makers of automata that breathed and spoke and defecated. They too use machines to simulate life precisely because their understanding of what machines are is no better established than their understanding of what life is.

## Notes

A version of this chapter appeared in *Representations* © 2003 by The Regents of the University of California. Reprinted from *Representations* No. 83 by permission of the University of California Press.

1. Julien Offray de La Mettrie, *Man a Machine and Man a Plant*, trans. Richard A. Watson and Maya Rybalka. (Indianapolis: Hackett, 1994), 71–72. (First published as *L'Homme-machine* in 1747.)

2. Rudy Rucker, *Wetware* (New York: EOS, 1988), 66.

3. *Biomimetic* means imitation of biological systems, and *chemomechanical* means that chemical energy is converted directly into mechanical work, as is the case in living creatures. The terms, along with *wetware* and *soft machines*, are found in Yoshihito Osada and Simon B. Ross-Murphy, "Intelligent Gels," *Scientific American*, May 1993, 82–87.

4. I am grateful to Evelyn Fox Keller for pointing out the difference between eighteenth- and twentieth-century meanings of *simulation* and pressing me to clarify my use of the term. For arguments that eighteenth-century machines were simulative in the modern sense, see André Doyon and Lucien Liaigre, "Méthodologie comparée du biomécanisme et de la mécanique comparée," *Dialectica* 10 (1956): 292–335; Georges Canguilhem, "The Role of Analogies and Models in Biological Discovery," in A. C. Crombie, ed., *Scientific Change: Historical Studies in the Intellectual, Social, and Technical Conditions for Scientific Discovery and Technical Invention, from Antiquity to the Present* (New York: Heinemann, 1961), 507–520, especially 510–512; Derek J. De Solla Price, "Automata and the Origins of the Mechanistic Philosophy" in *Technology and Culture* 5, no. 1 (winter 1964): 9–23. David M. Fryer and John C. Marshall, "The Motives of Jacques Vaucanson," Technology and Culture 20, no. 1 (January 1979): 257–269.

5. Gaby Wood, *Living Dolls: A Magical History of the Quest for Mechanical Life* (London: Faber and Faber, 2002), xvi, 59, xiv. (The American edition was issued under the title *Edison's Eve*.) Similarly Tom Standage, in his recent account of the Turk, emphasizes how "little has changed" since its performances; see his *The Turk: The Life and Times of the Famous Eighteenth-Century Chess-Playing Machine* (New York: Walker and Co., 2002), 246.

6. La Mettrie, *Man a Machine* (1747), 63–64.

7. "Diverses machines inventées par M. Maillard. Cygne artificiel," in M. Gallon, ed., *Machines et inventions approuvées par l'Académie royale des sciences depuis son établissement jusqu'à present; avec leur Déscription* (Paris, 1735–1777), vol. 1, 133–135.

8. Georges Canguilhem has observed that the "Cartesian animal-machine remained as a manifesto, a philosophical war-machine, so to speak" in contrast with eighteenth-century physiologists' and mechanicians' "elaboration of detailed plans with a view to the construction of simulators" (Canguilhem, "Analogies and Models in Biological Discovery," in Crombie, *Scientific Change* (1963), 510–511. On the importance of automata as "models of intelligibility" in Descartes' philosophy, see Peter Dear, "A Mechanical Microcosm: Bodily Passions, Good Manners, and Cartesian Mechanism," in Christopher Lawrence and Stephen Shapin, eds., *Science Incarnate: Historical Embodiments of Natural Knowledge* (Chicago: University of Chicago Press, 1998), 51–82 (quote on 59). I have found one possible exception to the general rule that seventeenth-century automata were not simulative, a "statue" designed by a Württemburg physician named Reyselius. According to reports, this artificial man demonstrated circulation, digestion, and respiration with great "resemblance to man in all the internal parts" (*Journal des savants* 1677: 352). On the artificial man of Reyselius, see also Thomas L. Hankins and Robert J. Silverman, *Instruments and the Imagination* (Princeton: Princeton University Press, 1995), 182; André Doyon and Lucien Liaigre, *Vaucanson, mécanicien de genie* (Paris: PUF, 1966), 117–118, 162–163. Going back still earlier, one might well take Leonardo da Vinci's uses of cords and wires to model the muscles as simulations. (See Paolo Galluzzi, "Leonardo da Vinci: From the 'elementi macchinali' to the man-machine," *History and Technology* 4 (1987): 235–265.) I am grateful to Michael John Gorman for pressing me to consider the simulative aspects of sixteenth- and seventeenth-century automata and models, and for pointing me to Dear's and Galluzzi's articles. There does seem to me to be an important difference between a model or illustration, which is meant to depict its natural subject, and a simulation, which is meant to reproduce it. Whereas a model assumes a gap between itself and its subject, a simulation tries to collapse the gap.

9. Christian Bailly, Sharon Bailly and Eric Desmarest, *Automates* (Paris, 1993), 20, 26; Henri Nicolle, *Les jouets—ce qu'il y a dedans* (Paris, 1868).

10. Anson Rabinbach has studied the cultural importance of such analogies in his book *The Human Motor: Energy, Fatigue, and the Origins of Modernity* (Berkeley: University of California Press, 1990).

11. Pierre Jaquet-Droz built clocks, watches, and automata together with his son Henri-Louis and his adopted son Jean-Frédéric Leschot. On the Jaquet-Droz family, see Charles Perregaux and F.-Louis Perrot, *Les Jaquet-Droz et Leschot* (Neuchatel, 1916); Alfred Chapuis and Edmond Droz, *The Jaquet-Droz Mechanical Puppets* (Neuchatel: Simonin, 1956); and F. M. Ricci, *Androïdes, les automates des Jaquet-Droz* (Lausanne: Sciptar, 1979).

12. Perregaux and Perrot, *Les Jaquet-Droz et Leschot* (1916), 31–34.

13. See Richard Altick, *The Shows of London* (Cambridge: Belknap, 1978); Simon Schaffer, "Enlightened Automata," in William Clark, Jan Golinski, and Simon Schaffer, eds., *The Sciences in Enlightened Europe* (Chicago: University of Chicago Press, 1999), 126–165, especially 138; Alfred Chapuis and Edmond Droz, *Automata: A Historical and Technological Study*, translated by Alec Reid (New York: Central Book Company, 1958), 280–282.

14. Jaques Vaucanson, *Le mécanisme du fluteur automate* (Paris, 1738).

15. "Lettre de M. Vaucanson, à M. l'abbé D.F.," in Vaucanson, *Le Mécanisme du fluteur automate* (1738), 19–22.

16. Jean-Eugène Robert-Houdin exposed Vaucanson's fraud (see below); Robert-Houdin, *Memoirs of Robert-Houdin*, trans. Lascelles Wraxall (New York, 1964), 104–107. (First published in French as *Confidences d'un prestidigitateur* (Blois, 1858).) See also Doyon and Liaigre, *Vaucanson* (1966), 124–127.

17. Jonathan Swift, "The Lady's Dressing-Room" (1730), in Robert A. Greenberg and William B. Piper, eds., *The Writings of Jonathan Swift* (New York: Norton, 1973), 537–538.

18. John Wilkins, *Mathematicall Magick. Or, The Wonders that may be performed by Mechanicall Geometry* (London: M. F. for S. Gellibrand, 1648), 177–178. See Hankins and Silverman, *Instruments and the Imagination* (1995), 181.

19. Denys Dodart, "Sur les causes de la voix de l'homme et de ses différents tons," November 13, 1700, in *Histoire de l'Académie royale des sciences*, Année 1700, *Mémoires*, 244–293; Dodart, "Supplément au Mémoire sur la voix et sur les tons," April 14, 1706, and "Suite de la première partie du Supplément," in *Histoire de l'Académie royale des sciences*, Année 1706, *Mémoires*, 136–148, 388–410; Dodart, "Supplément au Mémoire sur la voix et les tons," March 16, 1707, in *Histoire de l'Académie royale des sciences*, Année 1707, *Mémoires*, 66–81. For Fontenelle's commentary on Dodart's memoirs, see Bernard le Bovier de Fontenelle, "Sur la formation de la voix" in *Histoire de l'Académie royale des sciences*, Année 1700, 17–24; Année 1706, 136–148; Année 1707, 18–20. See also Jean-Pierre Séris, *Langages et machines à l'âge classique* (Paris: Hachette, 1995), 231–235.

20. Fontenelle, "Sur la formation de la voix" (1707), 20.

21. Abbé Desfontaines, "Lettre CLXXX sur le Flûteur et l'Aristippe moderne," in *Observations sur les écrits modernes* XII (1738), 341. (Reproduced in Doyon and Liaigre, *Vaucanson* (1966), 162.)

22. Antoine Court de Gébelin, *Le monde primitif, analysé et comparé avec le monde moderne* (Paris, 1775), vol. 2, 83–84. See Séris, *Langages et machines* (1995), 239.

23. Erasmus Darwin, *The Temple of Nature; or, The Origin of Society* (London: J. Johnson, 1803), 119–120. See Hankins and Silverman, *Instruments and the Imagination* (1995), 199.

24. *Têtes parlantes inventées et exécutées par M. l'abbé Mical. (Extrait d'un ouvrage qui a pour titre: Système de prononciation figurée, applicable à toutes les langues et executé sur les langues française et anglaise)*, VZ-1853, Bibliothèque nationale, Paris; *Procès verbaux* September 3, 1783, Archives de l'Académie des Sciences, Paris; Antoine Rivarol, "Lettre à M. le president de ***, sur le globe aérostatique, sur les têtes parlantes et sur l'état present de l'opinion publique à Paris," in *Oeuvres complètes de Rivarol* (Paris, 1808), vol. 2, 207. See Séris, *Langages et machines* (1995), 245; André Chapuis and Edouard Gélis, *Le Monde des automates* (Paris: Haraucourt, 1928), vol. 2, 204–206.

25. Wolfgang von Kempelen, *Le Mécanisme de la parole, suivi de la déscription d'une machine parlante* (Vienna: B. Baver et Se trouve chez J. V. Degen, 1791), 413–414. See Hankins and Silverman, *Instruments and the Imagination* (1995), 190–197; Séris, *Langages et machines* (1995), 245–246. In 1779, probably at the instigation of Leonhard Euler, the St. Petersburg Academy of Sciences sponsored a prize competition to determine the nature of the vowels and to construct an instrument like *vox humana* organ pipes to express them. C. G. Kratzenstein, a member of the Academy, won the prize. He used an artificial glottis (a reed) and organ pipes shaped according to the situation of the tongue, lips, and mouth in the pronunciation of the vowels. Hankins and Silverman, *Instruments and the Imagination* (1995), 188–189; Séris, *Langages et machines* (1995), 247. Hankins and Silverman write that "Wilkins, Mical, Kempelen, Darwin, and several of Vaucanson's contemporaries shared a consistent approach to the imitation of the voice. All of them defined vowels and other speech sounds in terms of the configurations of the human organs of speech" (Hankins and Silverman, *Instruments and the Imagination* (1995), 198). This seems to me to be true of Mical, Kempelen, and Darwin but not Wilkins; see above.

26. On Wheatstone's and Bell's reproductions, see J. L. Flanagan, "Voices of Men and Machines," *Journal of the Acoustical Society of America* 51 (1972): 1375–1387; Flanagan, *Speech Analysis, Synthesis and Perception* (Berlin: Springer Verlag, 1965), 166–171; M. R. Schroeder, "History of Synthetic Speech," *Speech Communication* 13 (1993): 231–237; Hankins and Silverman, *Instruments and the Imagination* (1995), 218–219. Hankins and Silverman write that many "artificial glottises, similar in design to those that seem to have activated Mical's heads, were made in the nineteenth century," but the physiologists who made these (Johannes Müller, Édouard Fournié) "did *not* aim to copy articulate speech, but simply to disclose the operation of the larynx" (Hankins and Silverman, *Instruments and the Imagination* (1995), 199). "Nineteenth-century physicists who studied the voice were generally much less interested in reproducing articulate speech than their eighteenth-century counterparts had been" (p. 209). Following are some exceptions. Richard Potter, in the 1870s, "imitated vowels with an apparatus that would have seemed familiar to the eighteenth-century investigators mentioned above: a free reed connected to a hollow india rubber sphere that could be deformed to copy the shape of the mouth and produce a variety of vowels." The Liverpool phonetician R. J. Lloyd wrote in 1890 that the best apparatus for studying speech would have some resemblance to the vocal organs, but not too close a resemblance, and used glass bottles. "One of Lloyd's methodological descendants, Sir Richard Paget, contrived plasticine models of the vocal tract in his study of imitation vowel sounds during the 1920s" (p. 210). "Copying the manifest appearance of the organs of speech was the ultimate

end of the French physiologist Georges René Marie Marage, who worked in the late nineteenth and early twentieth centuries. . . . Marage's resonant cavities exactly copied the shape of the oral cavity. In fact, they were cast from molds of the mouth, complete with lips and teeth" (pp. 210–211). "In 1905, E.W. Scripture discussed another attempt based on the model of the vocal anatomy—one that did not even bother to make a copy of the human form. His 'vowel organ' involved fitting a human skull with artificial cheeks and lips to re-create a resonance chamber. Rubber glottises imitated the larynx" (pp. 211–212). "The first patent for a talking doll was awarded to J. M. Maelzel, the inventor of the metronome, in 1824. It consisted of a bellows, reed and cup-shaped resonator" (p. 213).

27. Hermann von Helmholtz, for example, built a machine using tuning forks and resonance chambers to produce the vowel sounds, described in Helmholtz, *On the Sensations of Tone as a Physiological Basis for the Theory of Music*, trans. A. J. Ellis (New York: Dover, 1954), 399. On Helmholtz's speech synthesizer, see Flanagan, *Speech Analysis, Synthesis and Perception* (1965), 172–174; Timothy Lenoir, "Helmholtz and the Materialities of Communication," *Osiris*, 2nd series, 9 (1993): 185–207; Schroeder, "History of Synthetic Speech" (1993): 232–233; Hankins and Silverman, *Instruments and the Imagination* (1995), 203–205.

28. Robert Willis, "On the Vowel Sounds, and on Reed Organ-Pipes," read November 24, 1828 and March 16, 1829, published in the *Transactions of the Cambridge Philosophical Society* 3 (1830): 231–268. See Hankins and Silverman, *Instruments and the Imagination* (1995), 201.

29. Claude Bernard, *Cahiers des notes*, M. D. Grmek, ed. (Paris: Gallimard, 1965), 171. See Séris, *Langages et machines* (1995), 248.

30. David Lindsay, "Talking Head," in *Invention and Technology*, summer 1997, 56–63; Hankins and Silverman, *Instruments and the Imagination* (1995), 214–216.

31. See Hankins and Silverman, *Instruments and the Imagination* (1995), 216, where the authors identify a partial return to "more humanoid apparatus" in "the last years of the nineteenth century." Investigators such as Lloyd, Marage, Scripture, and Paget "approached the problem from a physiological and phonetic point of view. If Faber had demonstrated his machine either fifty years earlier (in Kempelen's time) or fifty years later than its introduction in the 1840s, the Euphonia might have been greeted by an enthusiastic audience." It seems to me, however, that despite Lloyd, Marage, Scripture, and Paget, the simulative approach to artificial speech never regained the dominance it had had during the late eighteenth century.

32. On the early history of electrical speech synthesis, see Flanagan, *Speech Analysis, Synthesis and Perception* (1965), 171–172; Flanagan, "Voices of Men and Machines" (1972), 1381–1383; Dennis H. Klatt, "Review of Text-to-Speech Conversion for English," *Journal of the Acoustical Society of America* 82, no. 3 (summer 1987): 741–742; Schroeder, "A Brief History of Synthetic Speech" (1993).

33. Ambroise Paré, "Of the Meanes and Manner to Repaire or Supply the Naturall or accidentall defects or wants in mans body," in *The Collected Works of Ambroise*

*Paré*, trans. Thomas Johnson (New York: Milford House, 1968). Paré also presents designs for prostheses to replace missing eyes, ears, noses, teeth, tongues, and penises. These, like his prosthetic limbs, are as remarkable for their unlikeness as for their likeness to the parts they replace. They fulfill either an aesthetic purpose (in the case of the eyes, ears, and noses) or a functional one (the tongues and penises) but not both (except perhaps in the case of the teeth).

34. Fontenelle, "Éloge du père Sébastien Truchet Carme," in *Éloge des académiciens* (La Haye, 1740), vol. 2, 366–367.

35. Gallon, *Machines et inventions approuvées par l'Académie des sciences*, vol. 6, 71–73.

36. Jacques Delille, *Épitre à M. Laurent, . . . à l'occasion d'un bras artificiel qu'il a fait pour un soldat invalide*, 2nd ed. (London, 1761). See Reed Benhamou, "From *Curiosité* to *Utilité*: The Automaton in Eighteenth-Century France," in John Yolton and Leslie Ellen Brown, eds., *Studies in the Eighteenth Century*, 17 (1987): 100.

37. Meanwhile in 1786, a cousin and perhaps collaborator of the Jaquet Droz family, the Director of the French Mint in Paris, a man named Jean-Pierre Droz, designed an artificial hand to improve the safety of workers at the mint, who had to slide metal strips under the balance arm of the machine that stamped them, and frequently had bad accidents. Droz's artificial hand was intended to take on this dangerous task. See Perregaux and Perrot, *Les Jaquet-Droz et Leschot* (1916): 31–36, 89–91, 100–111, 140; Linda Marlene Strauss, *Automata: A Study in the Interface of Science, Technology and Popular Culture, 1730–1855*, PhD dissertation, University of California, San Diego, 1987, 109.

38. Nina Rattner Gelbart, *The King's Midwife: A History and Mystery of Madame du Coudray* (Berkeley: University of California Press, 1998).

39. Gelbart, *The King's Midwife* (1998), 60, 116, 207.

40. The eerily accurate wax models that anatomists began to use during the late seventeenth and early eighteenth centuries—such as those of the Italian naturalist and physiologist Felice Fontana—provide an interesting comparison. They are uncanny in their visual resemblance to their natural subjects, but make no attempt to simulate texture or substance, and therefore seem to me to belong more in the older tradition of illustration than the newer one of simulation. Nevertheless, they evoke actual flesh to such a degree that their relation to projects in mechanical simulation seems well worth investigating. On eighteenth-century wax anatomical models, see F. Gonzalez-Crussi and Rosamond Wolff Purcell, *Suspended Animation* (San Diego: Harcourt, 1995) and Thomas Schnalke, *Diseases in Wax: The History of the Medical Moulage* (Berlin: Quintessence Books, 1995). I am grateful to Paula Findlen for pointing out the relevance of wax anatomical models to the emergence of artificial life in the eighteenth century.

41. The machine is described as an "anatomie mouvante" in Commission extraordinaire du Conseil Plumitif, no. 10, *Archives nationales* V7 582, cited in Doyon & Liaigre, *Vaucanson* (1966), 110; see also 18, 34. The second description is from Acte

de la Société Colvée-Vaucanson du 26-1-1734, *Archives nationales*, Minutier Central, Notaire CXVIII, cited in Doyon and Liaigre, *Vaucanson* (1966), 18.

42. *Registre contenant le Journal des Conférences de l'Académie de Lyon*, quoted in Doyon and Liaigre, *Vaucanson* (1966), 148, and in translation in Beaune, "Classical Age of Automata" (1989), 457. See also Doyon and Liaigre, "Méthodologie comparée du Biomécanisme" (1956), 298.

43. On the plans for a model of the circulatory system, see Marie Jean Antoine Nicolas de Caritat, marquis de Condorcet, "Éloge de Vaucanson" (1782), in A. Condorcet O'Connor and M. F. Arago, eds., *Oeuvres de Condorcet* (Paris: FirminDidot, 1847), vol. 2, 655; Éliane Maingot, *Les automates* (Paris: Hachette, 1959), 18; Doyon and Liaigre, *Vaucanson* (1966), 152–161; Strauss, *Automata* (1987), 71–72.

44. François Quesnay, *Essai phisique sur l'oeconomie animale* (Paris, 1736), 219–223.

45. Quesnay, *Observations sur les effets de la saignée* (Paris, 1730), iv–vi. See also Quesnay, *L'Art de guerir par la saignée* (Paris, 1736) and *Traité des effets et de l'usage de la saignée* (Paris, 1750). The *Traité* is a later version of the *Observations* and of *L'Art de guérir* (1736). See Doyon and Liaigre, "Méthodologie comparée du biomécanisme et de la mécanique comparée," in *Dialectica* 10, no. 1 (1956): 297.

46. This description appeared in conjunction with Le Cat's *Traité de la saignée* (1739) as its "experimental part." The machine was invented "to confirm by experience [Le Cat's] theory of bleeding." "Précis sur la Vie de Mr. Le Cat," in *Mémoires de Trévoux*, Novembre 1768, 334–335. See also Baillère-Delaisment, *Éloge de Mr. Le Cat* (Rouen, 1769), 53. See Doyon and Liaigre, "Méthodologie comparée du biomécanisme" (1956), 299.

47. *Registre-journal des assemblées et délibérations de l'Académie des sciences* . . . établi en 1744: 3 (manuscrit *non classé* de la Bibliothèque publique de Rouen), cited in Doyon and Liaigre, "Méthodologie comparée du biomécanisme" (1956), 300.

48. Le Cornier de Cideville–Fontenelle, December 15, 1744, in Abbé Tougard, *Documents concernants l'histoire littéraire du XVIIIe siècle* (1912), vol. 1, 52–54, on 53. See Doyon and Liaigre, "Méthodologie comparée du Biomécanisme" (1956), 300.

49. Claude Nicolas Le Cat, *Traité des sensations et des passions en general, et des sens en particulier* (Paris, 1767), vol. 1, xi, xix–xxv, xxix–xxxi, 40–50, 60–61.

50. Rodney A. Brooks, "Elephants Don't Play Chess," in *Robotics and Autonomous Systems* 6 (1990): 3–15; "Intelligence without Reason," MIT AI Lab Memo 1293, April 1991; "Intelligence without Representation," in *Artificial Intelligence Journal* 47 (1991): 139–159; Rodney A. Brooks, C. Breazeal Ferrell, R. Irie, C. Kemp, M. Marjanovic, B. Scassellat, and M. Williamson, "Alternate Essences of Intelligence," American Association for Artificial Intelligence, 1998.

51. Steven Levy, *Artificial Life* (New York: Vintage, 1992), 19–20.

52. Bernard, *Leçons sur les phénomènes de la vie* (Paris, 1885, 2nd ed., first published in 1878), vol. 1, 112–124. T. H. Huxley wrote, similarly, that "the living body is not

only sustained and reproduced: it adjusts itself to external and internal changes." Huxley, "On the Hypothesis that Animals are Automata, and its History" (1874), in *Collected Essays* (New York, 1894–1898), vol. 2, 200.

53. W. Grey Walter, "An Imitation of Life," in *Scientific American*, May 1950, 42–63; Walter, *The Living Brain* (New York: Norton, 1953), chapters 5 and 7. See also Owen Holland, "Grey Walter: The Pioneer of Real Artificial Life," in Christopher G. Langton and Katsunori Shimohara, eds., *Artificial Life V* (Cambridge: MIT Press, 1997), 34–41.

54. Norbert Wiener, *Cybernetics, or Control and Communication in the Animal and the Machine* (Cambridge: MIT Press, 1948), 11–12; Otto Mayr, *The Origins of Feedback Control* (Cambridge: MIT Press, 1970).

55. An exception is the French physiologist Étienne-Jules Marey, who is best known for his photographic study of animal motion during the 1880s and 1890s. A couple of decades earlier, Marey built artificial insects and birds to study the "mechanical conditions" of flight. See his "Mécanisme du vol chez les insectes," *Revue des cours scientifiques de la France et de l'étranger*, no. 16 (March 20, 1869): 253–256; Victor Tatin, "Expériences sur le vol mécanique," in Marey, *Physiologie expérimentale: Travaux du laboratoire de M. Marey* (Paris, 1876–1880), vol. 2, 86–108. Marey also describes using artificial bird and insect wings in *Animal Mechanism* (New York: Appleton, 1873). On Marey's artificial insect, see also Michel Frizot, ed., *E. J. Marey. 1830/1904 La Photographie du mouvement* (Paris: Centre Pompidov, 1977), 96–98. Further, Marey built an artificial heart to study the circulation of the blood, described in Marey, *La Circulation du sang à l'état physiologique et dans les maladies* (Paris: G. Masson, 1881). On Marey's completion of Vaucanson's project, see Hankins and Silverman, *Instruments and the Imagination* (1995), 185–186. Finally, on Marey's study of the mechanics of the body in general, see Rabinbach, *The Human Motor* (1990), chapter 4. Marey's move from mechanical simulation to photography during the 1880s suggests that his central interest was in using machinery to analyze natural life rather than to produce artificial life.

56. Helmholtz, "On the Interaction of Natural Forces" (1854), in *Popular Lectures on Scientific Subjects*, trans. E. Atkinson (New York: Appleton, 1873), 155.

57. Robert Willis, *An Attempt to analyse the automaton chess player, of Mr. de Kempelen* (London: Booth, 1821).

58. Robert-Houdin, *Memoirs* (1858), 103–107. Robert-Houdin referred to the artificial man of Reyselius.

59. Quesnay, *Traité des effets de la saignée* (1750), 17–18.

# 12

## Overtaking Nature? The Changing Scope of Organic Chemistry in the Nineteenth Century

John Hedley Brooke

I take my title from a plate in Michael Maier's *Atalanta Fugiens* (1618), which depicts the chemist following in the footsteps of a female Nature. Reason is his staff, he wears the spectacles of experience, and he carries the lamp of reading. Nature strides ahead as his guide. He conspicuously lags behind. One way of writing a history of chemistry would be to trace a gradual closure of the gap between them. In the age of the genome, cloning, and transgenic animals, have we not finally caught up with and even overtaken Nature? To speak of a gradual closure would, however, be seriously misleading because it would imply a smooth, even triumphal, process of simulating nature's processes and products. In reality the path was never smooth and each step in the imitation of nature was controversial. From William Newman's account of the "alchemical debate" in the late Middle Ages it is clear that a polarity existed between positions that were already extreme. Whereas Avicenna had harbored both ontological and epistemological objections to metallic transmutation, Roger Bacon was prepared to say that alchemical gold was superior to natural gold. It supposedly contained the four elements in a better proportion and offered the prospect of restoring the human body to that condition of elemental equality enjoyed by Adam and Eve and those resurrected at the end of time (Newman 1989, 437). We have come a long way since then, but it would be naive to suppose that in closing the gap between Nature and Art, modern chemists have carried a skeptical public with them. In his reflections on the fate of the older Nature-Art distinctions, the late Reijer Hooykaas noted how "natural" foodstuffs and "natural" medicines were still frequently perceived to be more effective than their artificial counterparts. He could even report the case of the Belgian ecclesiastical authorities who had prescribed "natural" water (from rivers and springs) for baptism, implying that synthetic water would not do. For Hooykaas (1999, 267) it was evident that "the old argument used against the alchemists still lingers on." In his discussion of the old arguments, Newman also noted a modern parallel, already drawn by previous commentators: Avicenna would have been on the side of today's public who imagine that synthetic indigo, for example, was not the real thing, only an impressive imitation. There can be no doubt that the old arguments *have* left a legacy.

In this chapter, I focus on the rise of synthetic organic chemistry in the nineteenth century. Its relevance to our theme is obvious. For the first time naturally occurring organic products were prepared artificially. There was a certain "catching up" with Nature. Famously in 1828 Friedrich Wöhler obtained urea, without, as he put it, the use of kidneys. A massive explosion occurred in the production of carbon-based compounds that were not even known to Nature. By the 1860s Marcellin Berthelot could argue that organic chemistry was founded on synthesis. The philosophical and cultural ramifications of this new phase in the imitation of nature were not, however, as straightforward as protagonists sometimes pretended. I will also argue for the paradoxical thesis that in the very process of increasing its scope, of increasing the reservoir of producible compounds, organic chemistry lost something of its identity through assimilation to what had been known as mineral, or inorganic, chemistry.

To appreciate the complexity of the nineteenth-century debates, it is helpful to recall the range of objections leveled against the aspirations of the alchemists, who, through the imitation of Nature, had sometimes sought to surpass her. The ontological objection voiced by Avicenna had presupposed a dichotomy between artificial and natural products, in which their intrinsic differences were stressed. To change an inferior into a superior metal through artifice was intrinsically impossible, even though superficial imitations might be achieved. Avicenna's epistemological objection highlighted the alchemist's ignorance of something crucial that was hidden to the senses: the species-determining characteristics of a metal. A victim of this nescience, the alchemist could not manipulate what he did not know (Newman 1989, 427–428; 2004, 37–44). Skeptical arguments could also be grounded in Galen's distinction between homogeneous products made only by Nature and the heterogeneous products of art (Hooykaas 1999, 282). A comparable distinction, drawn by Thomas Aquinas, focused on the inability of the alchemist to make real gold. Using a vocabulary of "substantial forms," he insisted that "the substantial form of gold is not [induced] by the heat of fire—which alchemists use—but by the heat of the sun in a determinate place where the mineral power flourishes" (Newman 1989, 438). To add to the theoretical objections were the practical, encapsulated later by the

Cambridge Platonist Ralph Cudworth when he distinguished between two kinds of art. The work of a vital agent (Cudworth's "plastick principle") was the work of a power immanent in nature that had "considerable preeminences" above human art. Cudworth explained that human art had to work, as it were, from the outside of an object. It usually involved "a great deal of tumult and hurlyburly, noise and clatter." It required the use of axes, saws, and hammers and for the chemist, he might have added, elaborate apparatus and high temperatures. Nature, by contrast, was art incorporated in matter. Coming from within, it was achieved silently, "vitally and magically" (Cudworth [1678] 1968). There were therefore practical obstacles to a complete imitation of nature—in the sense that one would have to replicate both the conditions for and the workings of an elusive immanent power.

Much had happened by the early years of the nineteenth century to attenuate the force of such objections. In the mechanical philosophies of the seventeenth century an absolute distinction between nature and art had largely collapsed, the universe itself likened to a piece of clockwork by Descartes and Boyle among others. In his *Principia Philosophiae* (1644), Descartes had declined to recognize any difference between the workings of nature and human-made machines. It was no less natural for a watch to indicate the time than it was for a tree to produce fruit. Speaking of nature as full of small engines, the mechanist Henry Power ([1664] 1970) declared all things to be artificial because nature itself was nothing other than the art of God. We will return to this theological point, but we must not forget that practical advances in chemistry had also begun to change the terms of the debate. In 1617 Angelo Sala had decomposed "copper vitriol" (copper sulfate) into "copper ash" (copper oxide), vitriolic acid, and water, resynthesizing it again from these same components, and showing that the artifact had the same properties as the original. Such practical innovations made it easier for Robert Boyle to reinterpret the once homogeneous "forms" of minerals in heterogeneous, corpuscular terms (Hooykaas 1999, 282; Emerton 1984). As Thomas Kuhn (1952) pointed out in one of his early papers, a corpuscular chemistry had the peculiar attraction, voiced by Boyle, that out of anything it should be possible to make almost anything.

With these and later developments were associated cultural and philosophical meanings of remarkable subtlety. In the triumphalist histories of chemical materialism, it is often supposed that the matching of Nature through chemical manipulation must have challenged theological as well as vitalist sensibilities. Sometimes it did. But, as I have argued elsewhere, there was no simple correlation between vitalism and religious sympathies (Brooke 1995, chapter 4). We have just seen in Henry Power a correlation between a mechanical philosophy and the affirmation of nature as divine art. One finds the same in Boyle, whose voluntarist theology of Creation, as with that of Francis Bacon and Isaac Newton, also helped to legitimate experimental methods. If God had been free to make any world God wished, the only way to discover which world *had* been made was through empirical inquiry, not rationalist reflection.

Boyle's Christian theology had also been instrumental in eliminating from scientific explanation the action of any subordinate intelligent agents that would detract from divine sovereignty. In other words, theological reflection could contribute to a critique of any vitalist philosophy in which intelligence was ascribed to the vital agent. If there were theological objections to the presumption of the chemist in seeking to match, or even improve on, nature it was always possible to counter by reminding one's critic that a doctrine of imago dei implied a creativity in humans that could find legitimate expression in making new things and old things anew. The image of humans as God's fellow workers has a long history. It has recurred even among conservative religious thinkers and, in principle at least, has sanctified claims for our being cocreators with the deity. For Isaac Newton, such a delegation of divine power was no subtraction from it. Rather, it was an enhancement:

> If any think it possible that God may produce some intellectual creature so perfect that he could, by divine accord, in turn produce creatures of a lower order, this so far from detracting from the divine power enhances it; for that power which can bring forth creatures not only directly but through the mediation of other creatures is exceedingly, not to say infinitely greater. (Newton, quoted in Dobbs 1991, 36)

To add to the complexity, there have sometimes been correlations between vitalism and pantheistic philosophies of nature. In David Hume's celebrated critique of natural theology, an ordering principle inherent in nature was set up in opposition to an external Designer to account for the *appearance* of design in organic structures. These few examples suggest that questions concerning the ability of chemists to imitate and even replicate natural processes cannot be mapped in any simple way onto wider cultural concerns.

This helps to set the scene for the complexity of nineteenth-century debates concerning the conclusions to be drawn from the artificial synthesis of organic compounds. Early in the century one finds not (as the old textbooks used to say) a general skepticism about the possibility of organic synthesis. Depending on local context and predisposition, one finds a diversity of views, ranging from the optimism of French chemists such as Fourcroy and Chevreul to the admonition of the English Romantic poet Samuel Taylor Coleridge. For Coleridge it was presumptuous of the chemists to seek to penetrate what he called the "sacred adyta" of organic life. By contrast, in 1800 Fourcroy looked forward, as did Chevreul later, to the day when vegetable matter would be synthesized (Hooykaas 1999, 295; Brooke 1968, 88). To regard the distinction between organic and inorganic compounds as absolute was particularly objectionable to Chevreul who, in 1824, protested that this would be "to admit the futility of all the attempts that have for their objective to produce compounds identical with or analogous to those that are now considered as peculiar to organized beings." Evidently there were such attempts. However, not all the Paris chemists were as sanguine about their outcome. Gay-Lussac's collaborator Louis Jacques Thenard, while seeing no impediment in principle, worried about the high temperatures that would be necessary to coax carbon, hydrogen, and oxygen into union, temperatures that would surely destroy the desired organic product.

Back in England, William Prout shows us that there could be divisions among the vitalists. Uninhibited by the reservations of Coleridge, Prout contributed to a chemistry of digestion when, in 1824, he identified hydrochloric acid in the gastric juices. His was a vitalism that was not inimical to chemical research. He could study the dynamics of digestion without doubting that metabolic processes were under the control of an agent that was not itself the product of organization (Brock 1985, 71–72). Yet for Prout there were also cultural, even theological, reasons why the scope of the organic chemist was unlikely to be endless. In Prout's excursion into natural theology, the science of organic chemistry helped to underline the ingenuity of a Creator whose powers far transcended those of lesser chemists:

> Amidst the wonders of creation, it is perhaps difficult to say what is most wonderful; but we have often thought, that the Deity has displayed a greater stretch of power, in accommodating to such an extraordinary variety of changes, a material so unpromising and so

> refractory as charcoal, and in finally uniting it with the human mind; than was requisite for the creation of the human mind itself. (Prout 1834, 446)

We may smile at this kind of chemicotheology, but it reminds us that there was a long way to travel before Nature could ever be completely overtaken: the thought of synthesizing organs and organisms, let alone an organism with such a mind, was as daunting as ever.

The travelling began in earnest during the second quarter of the nineteenth century. To place some order on the journey we might recall Roald Hoffmann's charming claim that "chemists do have a special way of playing with the natural" (Hoffmann in Hoffmann and Schmidt 1997, 36). Five ways in fact. First chemists see it as a challenge to make any molecule nature can. They have chalked up innumerable successes but an immediate qualification has to be addressed: "They may manage their synthesis less efficiently than nature, but then nature has had a few million more years to optimize." A second aspiration is to make molecules that are absent from nature, which can be "real fun." A third is to make molecules resembling natural ones but "better" in some respect, as with polymers stronger than steel. The fourth is to produce molecules cleverly differing (but only slightly) from natural ones in order to trick bacteria and viruses. Last, but not least, synthetic molecules play a crucial role in adding to our comprehension of nature: "how it got to be the beautiful way it is."

The first two of these aspirations and certainly the last were clearly met during the course of the nineteenth century. Examples of the third would not be difficult to find, as in the development of the German dye industry. I am less sure about instances of the kind of trickery mentioned under the fourth aspiration. The point I wish to stress, however, is that even with the first two cases of "playing with the natural," a more intricate analytical structure is required if we are to capture the nuances of debate.

In an important essay on the changing role of synthesis in organic chemistry, Colin Russell (1987, 169) has observed that to speak of making a natural organic product can conceal differences between three procedures: the formation of the natural product from any other simpler natural product, from an artificial product, and from the elements directly. In the formation of an artificial organic compound, it would be necessary to distinguish between production from any other artificial product, from the elements directly, and from a natural one of simpler structure. This may look like splitting hairs, but when I first examined Wöhler's urea synthesis forty years ago I was struck by the relevance of such distinctions to a proper understanding of its reception (Brooke 1968).

The further distinction between an accidental and a planned synthesis becomes important when charting the increased scope of organic chemistry during the nineteenth century. Wöhler had certainly not planned to make urea after the manner of later chemists who would introduce systematic methods for achieving intended goals. Nevertheless there are facets of Wöhler's experiment and its impact that are worth recalling precisely because they illustrate the uncertainties—his own and those of his contemporaries—concerning the significance of his result. In chemical textbook traditions Wöhler's synthesis of urea has all the characteristics of a foundation myth: a triumphal step in matching Nature's art, in eliminating vital forces. The reality was rather different from later rhetoric. One complication had already been exposed before I began my own investigation. Douglas McKie (1944) had observed that Wöhler's synthesis was not a true synthesis at all because his cyanates were ultimately obtained from hooves, horns, and desiccated blood—that is, from other natural products. This immediately reminds us of Russell's distinctions and may explain Wöhler's own uncertainty. When Wöhler inquired of Berzelius whether his artificial formation of urea could be regarded as an example of the formation of an organic from an inorganic substance, he volunteered one complication himself: "It is striking that for the production of cyanate (and also of ammonia) an organic substance is still ultimately required, and a 'Naturphilosophe' might say that the organic part had not disappeared from the animal charcoal or from the cyanogen compounds made from it" (Brooke 1968, 92). That was a possible reaction, though there would have been rejoinders available based on Scheele's earlier formation of potassium cyanate from the cyanide he had obtained from potassium carbonate, charcoal, and ammonia. An actual reaction from a skeptical vitalist came from the physiologist Johannes Müller (1838, vol. 1, 3), who removed urea from the ranks of organic matter, "being rather an excretion than a component of the animal body." This looks like the kind of evasion that only confirms the significance of Wöhler's success—an evasion quite unnecessary for those French chemists who had predicted such a step, or indeed for anyone who might sympathize with the far older prediction of Francis Bacon in his *New Atlantis* (1627) that art will make compounds hitherto only made by nature, grow minerals in artificial mines, transform vegetable and animal species, make new metals with desirable properties, make rain, and change the climate.

And yet it can be so much easier to predict than produce. There were two further respects in which Wöhler might be said not to have delivered. He might have shown that it was possible to form an organic compound artificially, but even if he had shown that it might ultimately be obtained from its elements (of which he was uncertain in his letter to Berzelius), he had emphatically not shown that it might be made from its elements directly. Given a prevailing belief that this is what Nature could and did do, the match was far from perfect. As late as 1844, Jean Baptiste Dumas could say that for a plant to form neutral nonnitrogenous matter, "it suffices . . . to unite carbon with water or its elements"; and similarly to produce neutral nitrogenous matter, "it suffices to unite carbon and ammonium with the elements of water" (Dumas and Boussingault 1844, 30). This assumption that plants fabricated their material from the four elements directly would naturally induce comparisons with what the chemist could do with those elements alone—comparisons that could still be unflattering to the chemist.

There was a second respect in which to say that Wöhler succeeded in imitating Nature could be misleading. The product may have been the same, but surely not the process. Recalling Ralph Cudworth's stipulation that for a perfect match it would be necessary to simulate the silent, delicate operations of immanent forces, there was surely a sense in which Wöhler's achievement, however striking, fell short. The high temperatures and concentrated acids employed by the chemist could not be a strict replication of conditions within an organism—in which case, retaining a vocabulary of "vital force" in describing the latter would not seem unreasonable. Neither Berzelius, nor Liebig, nor Wöhler himself ceased to use the term. There were, of course, reasons drawn from the study of physiology, rather than chemistry, for the use of vitalist language. The ability of a body to heal itself or to maintain its temperature would be examples. Indeed, accounting for the gradual elimination of vital forces from physiology requires a far more complex analysis than is possible here. The proven possibility of an organic synthesis did become part of the rhetoric of those midcentury German mechanists who rebelled against the vitalism of Müller. But there was still the gap between Nature's processes and those of the chemist. This has to be one of the reasons Liebig did not regard synthetic organic chemistry as destructive of the vital force. Having noted the chemist's ability to prepare formic acid, oxalic acid, and urea in the laboratory, he merely concluded in his textbook that the vital force "shares" many properties with chemical forces (Liebig 1840, vol. 1, xciii). It is difficult not to be attracted by the idea that, even while speaking of a vital force, Liebig himself did much to dispel it through the chemical conjectures about bodily processes that he published in his *Animal Chemistry* (Brock 1997, 214). Not everyone, however, was persuaded by his method, which Claude Bernard compared with an attempt to determine what was going on in a house by comparing what went through the door and what came out of the chimney. Bernard, with his distinctive concept of the "milieu intérieur," had no need of conventional vitalist language, but even he made a point of observing that while the chemist could remake nature's products in the laboratory this was not to be confused with imitating her procedures (Brooke 1968, 102; Merz 1965, vol. 2, 426).

I have been suggesting that, in the years following Wöhler's success, there were good reasons for uncertainty about its wider significance. It provided a vivid example of isomerism but only equivocal evidence that the chemist might overtake Nature. In his fine biography of Wöhler's student Hermann Kolbe, Alan Rocke (1993, 241) has noted the consequent ambivalence: "For the next quarter century, chemists were ambivalent, at times predicting a glorious future of artificial organic compounds creating better living through chemistry, at other times racked with doubt about the possibility of extending Wöhler's work to ever more complex substances." During that quarter century, the scope of organic chemistry was extended on many fronts and at such a pace that Liebig was soon complaining to Pelouze that "one becomes dizzy from so many discoveries" (Brock 1997, 92), while Wöhler, in a famous image, compared the growth of the science to that of a jungle. The sheer proliferation of products cried out for taxonomic assistance and the well-known schemes of Dumas, Liebig, Laurent, and Gerhardt were the result (Kapoor 1969, 1970; Fisher 1973; Brooke 1976). On matters synthetic, Kolbe extended the current scope in 1845 when he published what has been described as the first total synthesis of an organic compound—that of acetic acid by a route from the elements involving the chlorination of carbon disulfide, the transformation of tetrachloroethylene into trichloroacetic acid, and the reduction of the latter by electrolytic hydrogen. This was still not from the elements directly, but there was no mediation of natural products. Kolbe could look to a future in which sugar and starch would be made artificially. But as Rocke has pointed out, an increased scope could mean more than the multiplication of examples. It could also mean the articulation of general synthetic techniques. With his English collaborator Edward Frankland, Kolbe was associated with the first two methods that allowed systematic multiplication: carboxylation through nitrile formation and the formation of hydrocarbons through electrolysis of the potassium salts of organic acids. In the fifteen years preceding Berthelot's *Chimie Organique fondée sur la Synthèse* (1860), there was an impressive expansion of innovative synthetic methods. These included Frankland's use of organometallics, Gerhardt's preparation of organic acid anhydrides, Hofmann's methods for the amines, Alexander Williamson's for the ethers, and Wurtz's complexification of hydrocarbons by reacting a mixture of alkyl halides with sodium. And this is not to mention Berthelot's total synthesis of acetylene from which benzene could be derived, or the first commercial synthetic dyestuffs mauve and fuchsine. Hardly surprising, then, that there would be complaints, if only in private, when Berthelot commandeered the whole field of organic synthesis for himself (Rocke 1993, 240). It is, however, in Berthelot where we catch sight of a deeper cultural meaning given to organic synthesis, again with theological ramifications.

In responding to the question what the difference is spiritually between the natural and the synthetic, Roald Hoffmann has urged the view that "the difference is *entirely* spiritual and aesthetic," adding "and no less important for being so" (Hoffmann and Schmidt 1997, 25). There is a sense in which Berthelot sought to purge the natural of residual spiritual connotations, finding aesthetic satisfaction in the practice of creation—of creating, through synthesis, products that had not been there before. The polemical zeal with which he attacked older spiritualities has often been noted. His thesis was that the synthesis of organic compounds demonstrated that "in reality, and without reservation," the chemical forces governing organic matter were the same as those governing inorganic matter (Berthelot 1864, 17). There was a unity to chemistry demonstrable from the fact that inorganic forces could produce the same effects as the organic, reproducing the same compounds.

By excluding reservations, Berthelot almost certainly overstated his case, leaving no room, for example, for Louis Pasteur's concern that artificially produced hydrocarbons should be subjected to the action of polarized light before asserting their identity with optically asymmetric natural products (Brooke 1971, 371–372). Other reservations were also possible. As I have noted in the case of Wöhler's urea, to reproduce an organic compound artificially is not the same thing as establishing the same effects in both organic and inorganic domains. Unless Berthelot could show that the chemist was starting from the same materials as Nature and following the same process, the analogy could easily break down. Having effected a number of syntheses using carbon monoxide, he tried to circumvent the problem by proposing that this was also Nature's method. In vegetables, he surmised, carbon dioxide must be first reduced to carbon monoxide, water to hydrogen, subsequent synthesis occurring by the action of nascent hydrogen on the monoxide. This was not an unreasonable conjecture, given the legacy from Gerhardt's chemistry that all organic compounds were ultimately reducible to carbon monoxide, ammonia, and water. But the fact is that it was a conjecture, leaving the demonstration incomplete (Brooke 1971, 372–373). A third reservation was surely sensed by Berthelot himself. Were the drastic conditions employed by the chemist really to be found in a vegetable? Here the disanalogy creeps in because, by Berthelot's (1864, 182) own admission, whereas the chemist achieved his syntheses via a series of separate and deliberate steps, "in the vegetable kingdom, on the contrary, all the ingredients are in immediate contact in the nascent state." This was not a throwback to vitalism but it was a difference. In fact for all his proclamation of the end of vitalism, even Berthelot (1864, 183) had to look to the future for what he really needed: "One day without doubt, when we are better acquainted with the laws that preside over the natural synthesis ... we shall be able to produce in our laboratories, in more complete fashion, not only the substances manufactured in vegetables, but also that ensemble of chemical processes actually employed."

A British contemporary of Berthelot, William Odling, reviewed the progress of organic synthesis in his *Lectures on Animal Chemistry* (1866). Stressing the "great progress recently made," Odling contributed to the mythology surrounding the state of play forty or fifty years earlier. Not until the human mind had been emancipated from the tyranny of the vital force was the very possibility of organic synthesis entertained (Odling 1866, 78). We have already seen that such a generalization should not be entertained. In fact Odling himself came up with a better explanation for earlier inhibitions. Although chemists had long been able to combine groups of elements together, it was only recently, he reported, that they had produced stearin (from glycerin and fatty acid) or hippuric acid (from benzene and glycine). Why was this? "The neglect of these syntheses," he speculated, "did not arise so much from want of interest in the production of the bodies, as from want of knowledge of their intimate constitution. No sooner was the constitution of the above . . . compounds satisfactorily made out than they were obtained artificially by Berthelot and Dessaignes respectively" (Odling 1866, 55). In this explanation we recognize an intimacy between planned synthesis and knowledge of chemical constitution. It brings us into contact with the fifth of Hoffmann's ways chemists play with the natural. There was to be a symbiotic relationship between synthesis as an instrument for illuminating structure and structural knowledge as an instrument for suggesting methods of synthesis. Examples of the interdependence began to proliferate once a workable theory of structure was established in the late 1850s and early 1860s. Thus Kekulé's model for benzene yielded "the first dramatic payoff of structure theory for the chemical industry . . . in 1868–1869 with the synthetic production of the important natural dye alizarin" (Rocke 1993, 284). Observing that the use of syntheses to determine structure could be part of a number of different strategies, Russell (1987, 175) has followed Ladenburg in listing the major synthetic achievements of the second half of the century. These include alanine (Strecker, 1850), mustard oil (Zincke, 1855), glycine (Perkin and Duppa, 1858), racemic acid (also Perkin and Duppa, 1860), malic acid (Kekulé, 1860), guanidine (Hofmann, 1861), taurine (Kolbe, 1862), choline (Wurtz, 1867), allantoin (Grimaux, 1877), tyrosine (Erlenmeyer, 1883), and indigo (Baeyer, 1885).

From where, however, did theories of chemical constitution (and later of structure) come? There are many histories of chemical theory that tell us and no meaningful survey can be given in this brief compass. But there is one point worth stressing and it leads to my paradoxical thesis. The major theories of organic constitution during the first sixty years of the nineteenth century were developed in the closest possible association with the concepts of inorganic chemistry. It is true that in the early years of the century organic compounds were sometimes conceived as holistic in the way their elements were integrated, by contrast with the dualistic precepts of Lavoisier's chemistry. Once Berzelius had released himself from that disanalogy by modeling organic compounds on inorganic salts, an influential unification of chemistry had been tentatively achieved (Brooke 1995, chapter 6). The theory of organic radicals was an offspring of this underlying quest for unification and undoubtedly contributed to it. In a joint paper on the state of organic chemistry in 1837, Liebig and Dumas could assert the commonality of the laws governing the organic and inorganic realms. In the case of inorganic compounds the radicals were simple (elements); in the organic case they were complex (as in methyl, ethyl, benzoyl). Otherwise there was no difference. By 1840, Berzelius's electrochemical dualism was coming under attack from the French chemists who had discovered that an electronegative element, chlorine, could replace hydrogen, an electropositive element, in certain organic compounds without introducing any fundamental change in property. This led to well-known and colorful animosity. It also led to exaggerated claims, as when Liebig inferred that the properties of an organic compound might not depend on the nature of its constituent elements at all, only on their disposition in the molecule. The new theories that accompanied the challenge to Berzelius's authority nevertheless still retained the most intimate links with inorganic chemistry. In the type theories, first of Dumas and then of Gerhardt, organic and inorganic compounds were still seen as analogous. In fact, in the mature type theory of Gerhardt's *Traité de Chimie Organique* (1853–1856), the inorganic "types" hydrogen, water, ammonia, and methane served as the classifying variables for all compounds whatsoever.

Here then is the paradox: while organic chemistry was being institutionalized (notably in Liebig's Giessen and through Liebig's progeny) as a distinctive branch of chemical analysis and practice, the theoretical underpinning of the science, on which synthetic methods would often be premised, was being driven by a quest for union with inorganic precepts.

A few examples may help to make the point. One of the weaknesses of the radical theory during the 1830s had been the relatively few radicals actually isolated. Gay-Lussac's cyanogen had been conflated with a cyanide radical but this was something of an exception. For committed proponents of the theory such as Frankland and Kolbe, there was therefore strong motivation to produce further examples. Their attempts to isolate the methyl and ethyl radicals were theoretically driven and led to their expansion of synthetic methods—Frankland through the launch of organometallic chemistry, Kolbe through his electrolytic experiments on the salts of organic acids (Russell 1996, 101–108; Rocke 1993, 138–139). In retrospect we recognize that both had actually obtained the dimers of their cherished hydrocarbon radicals, but this very synthetic potential was released through the drive to assimilate the organic to the inorganic. Frankland's organometallic compounds symbolized a fusion of organic with inorganic chemistry. When commenting on the significance of his acetic acid synthesis, Kolbe explicitly referred to a "continuous chain" that now linked inorganic to organic compounds (Rocke 1993, 60).

It would be difficult to deny that the synthetic potential released by Frankland and Kolbe, Gerhardt, Williamson and Wurtz was itself the product of a serious engagement with the theoretical controversies of their time and that one feature of these controversies was how organic and inorganic chemistry could best be assimilated to each other. Gerhardt had proclaimed a new basis for that unification as early as 1848 in his *Introduction à l'Etude de la Chimie par le Système Unitaire*, when he had attacked what he saw as a spurious unification offered by electrochemical dualism—spurious because the need to find an electropositive and an electronegative grouping in every organic compound had introduced an unwelcome arbitrariness. It was more constructive to view the molecule as a unitary species, subject to substitution or double decomposition reactions. Gerhardt's claims are particularly instructive because he wished to have the true unification of chemistry credited to himself—on which he sought to capitalize when applying for a chair at the Collège de France in 1850. I have suggested elsewhere that claims to have unified the two branches of chemistry have recurred many times since and appear to have played a strategic role in career advancement (Brooke 1987, 150–151). Thus to consolidate the argument, we find Wurtz in 1862 claiming that no definitive unification had been achieved—until then, and by him. Berthelot, as we have seen, was playing the same game at the same time. Wurtz based his own claim on the recent convention that theories based on radicals and theories based on types need not be mutually exclusive. If they were integrated the case for unification could be made on three grounds. Organic radicals resembled inorganic elements in passing unchanged through chemical reactions. Functional groups in organic compounds could be arranged in the same manner as the elements in inorganic archetypes. The third ground was that on which Wurtz claimed proprietary interest: certain organic radicals, as with certain elements, could show a valency other than one. Having made a special study of ethylene oxide, he enthused about the analogies it displayed with the oxide of divalent barium. His very own ethylene oxide was the precious "link" between mineral and organic chemistry. In short there was "but one chemistry" (Wurtz 1862).

The paradoxical features of a historical process in which organic chemistry, through novel synthetic methods, sought to catch up with and even overtake Nature, but which in so doing almost lost its identity, can be seen in a lecture given by A. W. Hofmann at London's Royal Institution in 1853. His subject matter, his title, even his identity as a star student of Liebig underlined his work as an *organic* chemist. In Russell's

(1987, 171) words, he was a "living demonstration of the separation of organic chemistry." And yet when we look at the content of his lecture what do we find? The question he chooses to address is whether there is any well-defined boundary between organic and inorganic chemistry. He surveys the various demarcation criteria that might be proposed (including the inability of chemists to synthesize organic compounds) but is obliged to reject them. His conclusion is that "the separation of chemical science into inorganic and organic is by no means found in nature" (Hofmann 1853, 134). Russell has suggested that external pressures and not merely a bid for originality might have shaped his message. This was just the time he was seeking government support for the Royal College of Chemistry when it would be expedient to present the chemical community as a cohesive group (Russell 1987, 171). Russell may also be right in saying that the peculiar identity of organic chemistry to which its synthetic aspirations contributed may have survived until well after 1860. But the paradox is already visible in Hofmann's lecture because he drew attention to it himself: "I am almost afraid, gentlemen, that you will object to me that in denying the distinction of inorganic and organic compounds, I lose the very ground upon which I stand, and that any other title for the lectures I intend to give you would have been better than the one . . . I have chosen" (Hofmann 1853, 134).

In exposing such paradoxical elements it has been a pleasure to find that they have recently been noted elsewhere. With specific reference to the urea synthesis, Peter Ramberg (2000, 183) observes that an "intriguing aspect of the Wöhler Myth is its paradoxical dual role as both the founding moment of organic chemistry and as the agent that unified organic and inorganic chemistry under the same principles." He adds: "That these effects might in fact be contradictory has gone largely unnoticed."

Questions concerning the imitation and improvement of nature have a long history and require long perspectives for their illumination (Brooke and Cantor 1998, 314–346). In this chapter I have focused on a few of the ways the terms of the debate were changed by the development of synthetic organic chemistry in the nineteenth century. The issues were not as straightforward as they have often been perceived to be. The many gains in matching Nature's performance had to be balanced by the reflection that one was not necessarily matching Nature's methods. For all the advances in the biochemical sciences in the late nineteenth and early twentieth centuries, that balancing act still had to be repeated. There continued to be contexts in which the gap between

Nature and her chemical imitators had to be taken seriously. It would be an interesting challenge to root them out. One concluding example will have to suffice. Emil Fischer's contribution to protein chemistry was subject to a criticism aired by Eduard Pfluger in 1909 and one that still surfaced in the 1930s. The objection, quite simply, was that the synthesis of polypeptides gave no information about the structure of biologically active proteins. Fisher's own summary of his work in 1906 shows how there was still a gap to be closed: he spoke of his synthetic products as "very close relatives" of the natural peptones. The problem was "how close?" By 1907 when he had strung fourteen amino-acid units together and obtained a strong biuret reaction he was predicting that if he could only get up to twenty units he would be into the world of proteins (Fruton 1979, 5–6). Among several complications (including alternative ring structures mooted by Fischer himself) was the phenomenon of the denaturation of proteins. Critics of the peptide model could argue that denatured proteins were linear polypeptides but that protein in vivo had to have a more highly organized structure. Even J. H. Northrop who in 1930 argued so convincingly for the protein nature of his crystallized pepsin could still be skeptical about the assimilation of synthetic peptide substrates to natural proteins. The synthetic substrates, he observed, were hydrolyzed about a hundred times more slowly than a protein substrate. In the race between Nature and the chemist, Nature had the knack of accelerating.

## References

Bacon, F. 1627. *New Atlantis.* London.

Berthelot, M. 1864. *Leçons sur les Méthodes Générales de Synthèse en Chimie Organique.* Paris.

Berthelot, M. 1860. *Chimie Organique fondée sur la Synthèse.* Paris.

Brock, W. 1985. *From Protyle to Proton: William Prout and the Nature of Matter.* Bristol: Adam Hilger.

Brock, W. 1997. *Justus von Liebig: The Chemical Gatekeeper.* Cambridge: Cambridge University Press.

Brooke, J. 1968. Wöhler's urea and its Vital Force? A verdict from the chemists. *Ambix* 15: 84–114.

Brooke, J. 1971. Organic synthesis and the unification of chemistry: A reappraisal. *British Journal for the History of Science* 5: 363–392.

Brooke, J. 1976. Laurent, Gerhardt and the philosophy of chemistry. *Historical Studies in the Physical Sciences* 6: 405–429.

Brooke, J. 1987. Methods and methodology in the development of organic chemistry. *Ambix* 34: 147–155.

Brooke, J. 1995. *Thinking about Matter: Studies in the History of Chemical Philosophy*. Aldershot: Ashgate.

Brooke, J., and G. Cantor. 1998. *Reconstructing Nature: The Engagement of Science and Religion*. Edinburgh: T & T Clark.

Cudworth, R. [1678] 1968. *The True Intellectual System of the Universe*. In G. Cragg, ed., *The Cambridge Platonists*. Oxford: Oxford University Press.

Dobbs, B. J. 1991. *The Janus Faces of Genius*. Cambridge: Cambridge University Press.

Dumas, J. B., and J. B. Boussingault. 1844. *Essai de Statique Chimique des Etres Organisés*. 3rd ed. Paris.

Descartes, R. 1644. *Principia Philosophiae*. Amsterdam.

Emerton, N. 1984. *The Scientific Reinterpretation of Form*. Ithaca, NY: Cornell University Press.

Fisher, N. 1973. Organic classification before Kekulé. Parts 1 and 2. *Ambix*: 20: 106–131, 209–233.

Fruton, J. 1979. Early theories of protein structure. *Annals of the New York Academy of Sciences* 325: 1–15.

Gerhardt, C. 1848. *Introduction à l'Etude de la Chimie par le Système Unitaire*. Paris.

Gerhardt, C. 1853–56. *Traité de Chimie Organique*. Paris.

Hoffmann, R., and S. Schmidt. 1997. *Old Wine, New Flasks: Reflections on Science and Jewish Tradition*. New York: Freeman.

Hofmann, A. W. 1853. "Royal Institution Lecture on Organic Chemistry. *Medical Times and Gazette* 6: 131–134.

Hooykaas, R. 1999. *Fact, Faith and Fiction in the Development of Science: The Gifford Lectures Given in the University of St Andrews 1976*. Dordrecht: Kluwer.

Kapoor, S. 1969. Dumas and classification in organic chemistry. *Ambix* 16: 1–65.

Kapoor, S. 1970. The origins of Laurent's Organic Classification. *Isis* 60: 477–527.

Kuhn, T. 1952. Robert Boyle and structural chemistry in the seventeenth century. *Isis* 43: 12–36.

Liebig, J. von. 1840. *Traité de Chimie Organique*. Trans. C. Gerhardt. 3 vols. Paris.

Maier, M. 1618. *Atalanta Fugiens*. Oppenheim.

McKie, D. 1944. Wöhler's synthetic urea and the rejection of vitalism. *Nature* 153: 608–610.

Merz, J. 1965. *A History of European Thought in the Nineteenth Century*. New York: Dover.

Müller, J. 1838. *Elements of Physiology*. Trans. W. Baly. 2 vols. London.

Newman, W. 1989. Technology and alchemical debate in the late Middle Ages. *Isis* 80: 423–445.

Newman, W. 2004. *Promethean Ambitions: Alchemy and the Quest to Perfect Nature*. Chicago: University of Chicago.

Odling, W. 1866. *Lectures on Animal Chemistry*. London.

Power, H. [1664] 1970. *Experimental Philosophy in Three Books Containing New Experiments Microscopical, Mercurial, Magnetical*. London. In M. B. Hall, *Nature and Nature's Laws*, 122–125, 128–130. New York: Walker.

Prout, W. 1834. *Chemistry, Meteorology and the Function of Digestion Considered with Reference to Natural Theology*. 2nd ed. London.

Ramberg, P. 2000. The death of vitalism and the birth of organic chemistry: Wöhler's urea synthesis and the disciplinary identity of organic chemistry. *Ambix* 47: 170–195.

Rocke, A. 1993. *The Quiet Revolution: Hermann Kolbe and the Science of Organic Chemistry*. Berkeley: University of California Press.

Russell, C. 1987. The changing role of synthesis in organic chemistry. *Ambix* 34: 169–180.

Russell, C. 1996. *Edward Frankland: Chemistry, Controversy and Conspiracy in Victorian England*. Cambridge: Cambridge University Press.

Wurtz, C. A. 1855. Sur une nouvelle classe de radicaux organiques. *Annales de Chimie* [3] 44: 275–313.

Wurtz, C. A. 1862. On ethylene oxide, considered as a link between organic and mineral chemistry. *Journal of the Chemical Society* 15: 387.

# 13

## Reconfiguring Nature through Syntheses: From Plastics to Biomimetics

Bernadette Bensaude-Vincent

As Maurice Merleau Ponty noted in 1956, "We cannot think about nature without realizing that our idea of nature is impregnated with artifacts."[1] Each age tends to interpret nature through models derived from one of its most advanced technologies. Nature was a clock in the context of seventeenth-century mechanical theory, then it was described as a laboratory by eighteenth-century chemists. The breeders' activities were behind Darwin's natural selection, and computers are behind the notions of genetic code and program.

Does this mean that we should reverse Aristotle's view and say that "nature is a copy of art"? Such a statement would immediately raise the question: Where does the concept of "artifact" itself come from? Art is always preceded by nature whether it be considered an imitation, a transformation, or an improvement on nature. So we would be quickly trapped in a circle, if we discussed the question in abstraction.

The present paper is an attempt to disentangle this circle through a review of various strategies of chemical synthesis in the twentieth century. In characterizing the various concepts of nature involved in three differerent practices of synthesis—polymer chemistry, combinatorial chemistry, and biomimetic chemistry—I will argue that the representations of nature and artifacts are mutually constructed. Like prey and predator defining their own identities though their relation, nature and artifact are continuously reconfigured through their changing relations. One and the same process gives rise both to the meaning of *natural* and of *artificial.*

### Plastic Artifacts and Rigid Nature

In contrast to wood or metals, synthetic polymers are molded. They are polymerized and shaped simultaneously. In more philosophical terms, matter and form are generated in one single gesture. This specific process undoubtedly increases the potential uses of such materials. However, the triumph of synthetic polymers originated in commercial strategies as much as in their intrinsic properties. Their history is extremely important for determining how *synthetic* became a synonym of *artificial* and how the plasticity of synthetic polymers deeply transformed the perception of nature.[2]

Celluloid is always referred to as the first artificial plastic, although it was made from cotton treated with nitric acid mixed with camphor, then subjected to heat and pressure. Its artificiality derived from the function assigned to this new material rather than from its composition. Celluloid was initially designed and manufactured by John Wesley Hyatt in 1870 as imitation ivory for billiard balls. As Robert Friedel has pointed out, this was a marketing strategy rather than a representation of the intrinsic value of the material because celluloid could only have the appearance of ivory without offering its density and elasticity, two properties that matter for billiard balls.[3] In fact, celluloid, like the parkesine presented by Alexander Parkes at the London Exhibition in 1862, was an invention with no specific purpose. Unlike natural materials it was not attached to one specific function. Instead it could be used for many things, such as combs, buttons, collars, and cuffs. It was a "chameleon material" that could imitate tortoiseshell, amber, coral, marble, jade, onyx, or other materials, depending on its color. Far from being an advantage, this enormous potential created uncertainty among celluloid manufacturers as to the proper image and function of their product. Although it was a better material than the natural products that it replaced for certain uses, celluloid was viewed as being a cheap, nasty, deceptive imitation of the natural.[4] A manual of household taste pronounced it "inartistic and vulgar" because the authenticity and sincerity of natural materials were based on their limited potential for shapes and colors.[5] The superiority of nature lay in its rigid order. Just as Aristotle claimed that the art of the Delphi knife makers was inferior to nature because their product was multifunctional and not exclusively suited to one function, standards of "good taste" condemned the multifunctionality of artificial materials.[6] Thus in the late nineteenth century, imitation of natural materials was still the key to the invention and acceptance of new materials. Their potentially enormous range of uses was an obstacle rather than a key to their success. In fact, the early plastic materials raised the question: What are these artifacts good for?

Leo Baekeland, drawing lessons from the celluloid case, quickly realized that he should not manufacture his "Bakelite"—a synthetic material made from phenol and formaldehyde—as an imitative substitute but as an invention that would rearrange nature in new and imaginative ways. In a bestseller telling *The Story of Bakelite*, published in 1924, the journalist John Kimberly Mumford inscribed the invention of Bakelite within the big picture of a cosmogony. From the dawn of the world, nature had stored up the wastes of dead creatures from which the chemists would later derive wonderstuffs.[7] The "thousand uses" of Bakelite—for electric appliances, radios, automobiles—were no longer a weakness. They signaled its "protean adaptability." This proved to be a key to Baekeland's success, although new natural polymers—like cellophane—were still successfully launched on the market in the 1920s.

The marketing of synthetic polymers relied on two major arguments. On the one hand, the image of cheapness was reevaluated. Promoters of synthetic polymers in America presented chemical synthesis as a cornucopia of cheap products within everyone's reach. Chemistry was envisioned as a driving force toward the democratization of material goods.[8] Chemical substitutes were also presented as pillars of stability: "One plastic a day keeps depression away." They were said to provide jobs and feed the market economy thanks to the rapid obsolescence of the mass products.

On the other hand, Williams Haynes promoted chemical substitutes as a way to spare natural resources. Modern civilization, he argued, was making unprecedented demands on the world's stock of wood, iron, coal, copper, rubber, and petroleum. "The use of chemical substitutes releases land or some natural raw material for other more appropriate or necessary employment."[9] Chemically manufactured substitutes would thus contribute to the conservation and protection of nature. At the same time, in breaking the traditional alliance between one material and one specific function—considered the main characteristic of natural materials—the invention of substitutes opened up a broad field of potential innovations and came to epitomize the abstract notion of progress.

The campaign orchestrated in the 1930s to promote nylon, the new polyamid synthetic fiber 6-6 invented by William Carothers in Du Pont's laboratories, was an attempt to break with the image of synthetics as cheap substitutes for natural materials. The term *nylon* was selected after months of debate because it avoided all connotations of an artificial substitute for silk.[10]

The promoters of synthetic polymers went further in claiming the superiority of synthetic materials over those provided by nature. The argument was based on two rather antithetical characteristics of synthetics.

First, because of their invariable chemical composition, they offer uniform properties and strict quality control, whereas natural products, since they are always variable and mixed with impurities, must be submitted to repeated analyses and assays. This argument could apply to all manufactured products—to metals as opposed to wood, for instance. The second argument is more specific: synthetic polymers allow great variability in forms, uses, and tastes, because they are molded. Plasticity, which had been seen as a weakness or defect of the artificial as compared to the natural, became the most positive attribute of synthetics in midcentury.

However, only a few decades after World War II, plastics got rid of their early connotation of cheap substitutes for natural materials. When they were used by sculptors, architects, and couturiers for artistic creation they became noble materials, highly praised for their lightness, mobility, and plasticity. This changing image has to do with the processes used for manufacturing plastics. As Jeffrey Meikle points out, in the 1930s and 1940s, thermosetting plastics had encouraged the image of a static, eternally perfect future society; in the 1950s, when thermoplastic polymers, infinitely capable of being melted and reshaped, proliferated in daily use, plastics connoted disposabilility and impermanence. With curved shapes and pneumatic architecture, synthetic materials created an aesthetic of their own in which artificiality became synonymous with plastic change, contrasting with the rigidity of nature. In 1971, the French philosopher Roland Barthes devoted a few pages to plastics in his review of the mythologies of modernity. "Plastics," he wrote, "are like a wonderful molecule indefinitely changing."[11] They represent potential change, pure movement. They connote the magic of indefinite metamorphoses to such a degree that they lose their substance, their materiality, to become pure virtualities. In turn, Jean Baudrillard used plastics to describe a paradox inherent in consumerist society: the increasing mass production of items requires more and more ephemeral products: "In a world of plenty, fragility replaces rarity as the dimension of absence."[12] Thus plastics exemplify the "culture of the disposable"characteristic of the second half of the twentieth century.

Thus endowed with an "unbearable lightness of being," plastics were clearly praised as unnatural. The bright colors and shiny surfaces of vinyl and Formica were praised for their surface, their superficiality, their inauthenticity. According to Meikle, they expressed "a faith in technology's capacity for transmuting nature's imperfections so as to arrive at the dazzling perfection of the artificial."[13] "Dazzling perfection" sounds like the right phrase: the plastic age did not mean to improve on nature but to construct fake utopian worlds by accumulation of light and disposable artifacts. Thus the traditional connotion of forgery attached to artifacts turned into a positive trait.

This brief survey of a success story shows how the distinction between artifact and nature has been reconfigured by the contrast between plasticity and rigidity. The manufacture of chemical substitutes was initially justified by a very specific view of nature as a rigid economy. First, each natural material was presented as rigidly assigned to a specific function—wood for construction, or cotton for clothing, for instance. By contrast, synthetics were meant to be flexible and multifunctional. Second, nature was viewed as a strictly limited stock of resources by contrast with the endless bounty of the chemical laboratory. Nature was presented as a finite collection of products rather than as a continuous process of generation. No *natura naturans*, it was a *natura naturata*. It seems plausible that these various connotations of rigidity deeply influenced the adoption of the word *plastics* for the entire family of synthetic polymers and subsequently favored the positive connotations attached to the artificial.

## Artifacts as *Êtres de Raison* vs. Stupid Nature

Most of the synthetic polymers that became commodities in the twentieth century were designed by trial and error. Although big investments were made by chemical companies such as Du Pont or by rival nations in the case of synthetic rubber, the synthetic techniques were labor-intensive and often based on serendipity. Like many other domains, the practices of synthesis have been deeply transformed by the use of computers. To design molecules with interesting medical, magnetic, optical, or electronic properties, twentieth-century chemists, materials scientists, and pharmaceutical chemists have developed a variety of computer-assisted methods, often referred to as "rational design" by contrast with the empirical, serendipitous processes of synthesis used in the past.[14] Many algorithms are now available to design molecules, using computation, combination, and randomization. Rather than trying to survey them all, I will focus on two: computational chemistry and combinatorial chemistry.

How can one dispense with the painstaking and expensive process of synthesizing new molecules without even knowing if their properties will meet one's requirements? This is obviously a pressing question for all kinds of companies. Computational chemistry is a way of avoiding the cost of synthesis by modeling the chemical behavior on a computer from three different perspectives: thermodynamic features, electronic properties, and spatial, molecular conformation.[15] The technique was initiated in the early 1970s by Cyrus Lewenthal in the context of the Multiple Access Computer (MAC) Program at MIT based on x-ray crystallographic models. By visualizing the 3-D structure of a compound and rotating it, one can predict how a small molecule could interact with a protein. Molecular graphics are not only visualized but also manipulated. Of particular importance is the conformational analysis, which associates a relative energy to each conformation of a molecule.

The guiding principle is expressed in this advertisement from Molecular Design Ltd: "Now you can find out how well a new compound works before it does." Here is a way of producing artifacts without putting material properties to work. The creative process is no longer an interaction between physical molecules and human bodies or machines—with the pressure of money. Rather it is an interaction between an algorithm and a virtual reality.

Computational design of molecules deeply transformed the status of the artifact. First, it banished the craft dimension from the making of artifacts in the interest of rationality and efficiency. An artificial material is basically the answer to a well-defined question, whatever the question: how to bind a molecule to the receptor of a specific protein for new medicines or how to make a light, stiff, and tough material for airplanes. Modern "virtual alchemists" no longer tended the fire in dark laboratories but did not renounce the Promethean ambitions of ancient alchemists. Beyond the objective of calculating the properties and reactivity of different structures, the goal of computational methods is to "model the real world by computer in a reasonable amount of time," as Uzi Landman, director of the Georgia Tech Center, puts it.[16] They are intended to subordinate the messiness of nature to the logic of computation.

The supreme achievement would be to build up a material from nothing, using computer calculations, the most fundamental information about the atoms, and the basic principles of physics. Nanotechnologies, in particular, rest on the assumption that it is possible to control the construction of a material from the bottom up. By placing atoms and molecules at selected positions, it is possible to build structures suited to a particular design atom by atom. K. Eric Drexler, who produced a number of popular publications on nanotechnologies in the 1980s, announced prophetically that "nanotechnology would bring changes as profound as the industrial revolution."[17] Drexler depicted atoms and molecules as nanomachines. They are "universal assemblers" that could be used as machine tools by engineers in order to create molecular machines with better performance. Improving on nature is the main objective and there is no limit to the power of those handling the "universal assemblers." They handle the "engines of creation."

The revival of such archaic fantasies is not the ineluctable consequence of the rational design of molecules. Combinatorial chemistry—a computer-assisted method of discovering drugs developed in the 1990s—leads to quite different views of art and nature. It involves reacting a set of preliminary materials in all possible combinations. Instead of using the computer in order to avoid contact with physical molecules, as with computational chemistry, this method encourages the reactions, while trying to eliminate all serendipity in the process of synthesis.[18] Once a route for synthesis has been selected and optimized, in a few steps and a few months thousands of compounds are synthesized with no other purpose than that of storage. The idea is to obtain a "library" of substances.[19] Many of them are messy mixtures and prove useless when they are tested against proteins. However, they are stored since the library should contain molecules for every possible protein target, embracing the maximum diversity without redundancy. Then with the help of computer "evolutionary algorithms," a fittest structure will be selected.

This represents "rational" design because of the application of the rules of combinatorials and algorithms of selection. But it is no longer intentional. The combinatorial chemist is like the monkey randomly typing letters with the expectation that a verse of the *Iliad* will come out of these meaningless sequences of characters. It is assumed that all technological or medical questions will find an answer in a library of billions of structures designed by combining and recombining the letters provided by nature. While the ancient Greek metaphor of the letters of the alphabet is often used to describe combinatorial chemistry, a military analogy seems more appropriate in describing the second step of this technique. Thousands of the molecules stored in the library are shot at a target protein. Both the random manipulation of letters and the blind shooting differ profoundly from the traditional strategies of chemical synthesis, in which each move is carefully planned and oriented toward an end. Not surprisingly, for a number of chemists combinatorial chemistry is a despicable method of fabricating substances. Pierre Laszlo, for instance, speaks of the "moronic travesty of scientific research known as combinatorial chemistry." It is a "perversion of the latter" whose unique goal is the "proliferation of chemicals."[20]

Combinatorial chemistry is certainly a cheap and fast way of designing drugs or other interesting molecules for industrial and commercial uses. However, making and storing unnatural and improbable molecules can also be a cognitive enterprise. Just as pathology is useful for advancing physiology, designing monstrous artifacts may be a way to better understand nature.[21] To a certain extent combinatorial chemistry is an exploratory method analogous to that of eighteenth-century chemists who performed hundreds and hundreds of reactions in order to create affinity tables.[22] These tables served as instruments of prediction like the libraries of molecules. For eighteenth-century chemists and combinatorial chemists alike, knowing through making, is the most reasonable investigative strategy. The underlying assumption is that we cannot predict exactly where to find the correct solution to any problem without making all the reactions and testing all the possible structures. This means enlarging the potential of natural resources in order to be able to make use of some of them. "The Lord is subtle . . . ," too subtle for contemporary chemists. They are ready to roll the dice, provided they have gathered in their library all the possible structures in order to sort out the optimal combination in a few steps. As pointed out by Roald Hoffmann, this recent branch of chemistry has revived the old tradition of the *Ars combinatoria*, illustrated by the Catalan alchemist Ramon Lull and later by Leibniz.[23]

Combinatorial chemistry can be considered a special way of mimicking nature by simulating the blind processes of selection at work in the evolution of living organisms. It is nothing like copying natural structures because they are smart and well designed for specific purposes. Rather it is copying the nonteleological mechanisms of repetition and mass production of substances with imprecise shooting of the target that seems to be the rule in the molecular processes of replication.[24] Here we find the Bergsonian view of life as a spontaneous, aimless movement with no direction, no intention. Generating variability through combinations and recombinations and then selecting those variants that are useful is a blind and stupid process. The contrast with conventional chemical synthesis is striking. Because it involves creation without design, combinatorial chemistry is hardly an "art" if we agree that all human arts are characterized by purposes or intentions.

Both computational and combinatorial chemistry are total syntheses since they proceed from the basic units. They are both rational in the sense that they follow strict rules of design rather than the *ingenium* and the skills that are usually characteristic of art. The craft dimension of the artifact disappears. Whether it be a virtual macromolecule on the screen of a computer or a physical unnatural compound stored in the library of combinatorial chemists, the artifact is above all an être-de-raison—a creation of the mind, more precisely of human reason. Both methods face the making of artifacts as a problem of calculus. Computation and combinatorics are agents of production. However, the meaning of production is quite different in the two cases. In computational chemistry production is a demiurgic creation of virtual realities. In combinatorial chemistry production is proliferation in an attempt to exhaust all the possible combinations of elements provided by nature.

While the boundary between science and technology seems to fade away, so does the boundary between nature and art. Art is deprived of most of its traditional attributes: intentionality, ingenuity, skill, and craft. Nonetheless the boundary between nature and art is restored by the conventions governing the patenting systems. The molecules made by rational design are considered inventions rather than discoveries, hence they are patentable. They are designed as potential market goods in a close alliance between researchers and venture capitalists.

## Biomimesis: Nature Is Technology

Traditionally a material was extracted from nature, then processed for human purposes. Its structure and properties constrained the making of artifacts and determined the performance of the end product. The quality of a violin, for instance, is dependent on the quality of the wood used to make it, among other factors. By contrast the advanced materials manufactured over the past three decades are no longer preconditions of the production process. They are designed as the optimal solution to a specific problem. Given a set of desired functions or performances, let us find the properties required and then design the structure combining them. This approach guided the development of materials science and engineering in response to very specific demands raised by military and space programs in the 1960s. Rockets, nuclear reactors, and space flight created the need for materials that were not currently available.

Within a few decades of R&D on such high-performance materials, however, materials scientists and engineers realized that they had to forget about the linear scheme—structure, properties, performance—in favor of a systems approach. Structure, properties, functions, and process have to be mutually adjusted in a continuous feedback loop. They are like the four summits of a tetrahedron, holding together the creative process of any artifact.[25] It is the search for the optimization of artifacts that requires a synergy between structure, properties, functions, and process.

Moreover, these high-performance materials are generally made of several components in order to obtain the best compromise between properties—such as the lightness provided by plastics, the toughness of metals, and the resistance to high temperature of ceramics. Most of them are composite structures: they are made of a matrix reinforced by fibers. The concept of the composite emerged out of the fiber-reinforced plastics manufactured in the 1950s, but gradually composites became something different from reinforced plastics.[26] The use of long, high-modulus fibers like carbon or Kevlar® allowed chemical engineers to design new materials with never-before-seen properties. In contrast to conventional plastics, which are mass-produced, high-performance composites or more recently hybrid materials associating organic and inorganic components at the molecular level, are designed for a specific task under specific conditions. Materials by design are mapped with anisotropic structures and a specific chemical composition adjusted to endure specific efforts in the use of the end product. Thus each offers a landscape of its own. Each is unique.

At first glance, these materials as light as plastic with the toughness of steel and the stiffness or heat resistance of ceramics are a veritable paradigm of ingenuity, and most definitely unnatural. Like the chimeras invented by the ancients, they bring different species together in one body. The modern centaurs incorporate multiple species in the innner matter rather than in their external appearance.

Ironically, the search for ever more artificial materials has drawn the attention of scientists toward natural materials. Suddenly in the 1980s and 1990s, journals of materials sciences began filling up with beautiful pictures of mollusks and insects. Like the old popular books titled *The Marvels of Nature*, they enthusiastically describe the details of sea urchins and abalone shells, spider silk, penguin feathers and dolphin skin, the hedgehog spine and the porcupine quill, the beautiful colors of *Urinadae* and *Morphidae* butterflies, and even single-cell marine algae like coccolithophores.

Why did nature—thrown out the door by the triumph of plastics—return through the window? The quest for high-performance and multifunctional materials prompted this shift back to nature. Living organisms provide models of high-performance materials. In living creatures around them, and in their own bodies, scientists and engineers found inspiring models of structure, models of integration of functions, and models of processes. The spider's silk is an extremely thin and robust fiber that offers an unchallenged strength-to-weight ratio. Though mollusk shells are made out of a common raw material (calcium carbonate), they present a variety of structures—layered, tubular, porous, foamlike—with elaborated shapes and assume a variety of functions.[27] The remarkable properties of bulk materials are the result of a complex arrangement at different levels, with each level controlling the next. The hierarchy of structures with multiple levels of organization from the molecular to the macroscopic, exemplified in bones and wood, very much impressed materials scientists. This multilevel structure is viewed as crucial to the reliability of a material because the structure can respond to chemical or physical stress at different scales. It is the key to such desirable functions as growth, self-repair, and recycling.

How could those efficient, smart, and highly reliable structures be designed? The processes used by nature are no less admirable and marvelous to a chemist. Organisms synthesize these materials at ambient temperature, without high pressures. The various components are simultaneously synthesized and self-assembled, with a controlled orientation. The ingenuity of nature confounds the skill of contemporary engineers. Nature is an unrivaled master who teaches lessons to humans. To deal with this new or rediscovered master, most chemists and materials scientists have started collaborations with biologists.

Interdisciplinary collaborations may use various strategies.[28] To a number of chemists it seems hopeless to improve on nature, or even to compete with nature. As Stanford chemist Steven Boxer puts it, "We've decided that since we can't beat them [biomolecular systems], we should join them."[29] Why not start with the building blocks provided by life—whether they be proteins, bacteria, or genes—to achieve our own technological goals? An example is spider silk. After it was demonstrated that the main thread of a variety of spider was composed of two proteins, named spiderine 1 and spiderine 2, some chemists isolated the spider's gene, which encodes these two proteins, and inserted it into the mammary glands of goats.[30] Bionanotechnologies tend to blur the boundary between life and matter, in two ways: either by embedding technological devices within living organisms or by using biomaterials to manufacture artifacts such as electronic circuitry, for instance. In the former case, the key issue is to control the wet-dry interface, to ensure biocompatibility. In the latter case, biomaterials such as DNA strands have to be reengineered in order to perform an artificial task.

Instead of gradually merging biology and technology, a number of materials scientists and engineers who found in biomineral structures model solutions to their own problems adopted a mimetic strategy. For instance, Ilian Aksay from Princeton University had designed a material for a light U.S. Army shield and patented a ceramics-metal composite, when he realized that he could make a far better material in imitating the layered structure of the abalone shell.[31] Similar efforts have been made to imitate the iridescent wings of butterflies in order to design similar fabrics, and the hexagonal structures of moth eyes have inspired new antireflection structures for industrial emitting cathodes or photothermic absorbers.[32]

Models, inspiration, imitation . . . what exactly is the meaning of *mimesis* in the current term *biomimetics*? It does not invite such attempts as Hyatt's efforts to imitate natural ivory. The goal is neither to produce a faithful copy, nor to reproduce the appearance of the biological model. Biomimetism is by no means oriented toward artificial replicas of products generated by life. Does it instead involve a renewed attempt to challenge nature like the nineteeth-century apostles of chemical synthesis did? All the metaphysical debates and ambitions that inspired the legend surrounding Friedrich Wöhler's urea synthesis are gone. Contemporary materials scientists are content with picking up local models as solutions to their current technological problems, related to integration, miniaturization, and recycling.

Is biomimetics one more expression of the back-to-nature movement that characterized the fin de siècle? Beyond the arrogance of synthetic chemists does there lie a humble worshiping of nature? Such questions require us to disentangle our assumptions about nature. First, materials scientists look at nature through engineers' eyes with open anthromorphism. "We can be encouraged by the knowledge that a set of solutions have been worked out in the biological domain," writes Stephen Mann, a natural scientist who entered the field of materials science. "The challenge then is to elucidate these biological strategies, test them *in vitro*, and to apply them with suitable modification, to relevant fields of academic and technological inquiry."[33]

Biomimetics is based on the working hypothesis that nature is a designer who had to face specific problems. In fact, the proximity between nature and artifact results less from a naturalization of engineering practices than from a technicization of nature. If biology can teach us about engineering and manufacturing, it is because the living cell is now viewed as a factory crowded with numerous bionanomachines in action. A few decades of extensive use of machine metaphors by molecular biologists have popularized the notion of "molecular machinery." Contemporary biomimetics and molecular biology deny the ancient distinction between physis and technē: Biomimetics is grounded in comparative studies of human technologies and the "technologies of nature" conducted by scientists working in biomechanics. Steven Vogel, for instance, contrasts "two schools of design."[34] Julian Vincent, a chemistry professor at the University of Reading, insists on considering life as one technology among others and seeks to promote biomimesis as a case of "technology transfer":

> Over three-quarters of all inventions emerge from closely related technology. We routinely fail to take advantage of the solutions and practices of other sciences and technologies. We routinely fail to recognize the similarities between our technical problems and the solutions to similar problems in other technologies. In particular we routinely fail to tap into the four billion years worth of R&D in the natural world.[35]

The idea is to improve the Theory of Inventive Problem Solving (TRIZ from the Russian acrononym for this methodology) by including the sophisticated systems designed by nature in the database, so that inventors can more easily find useful information for their specific problems.

A second assumption is that nature is teleological. Nature's objective was the optimization of functions in an organism in order to ensure the survival and reproduction of that organism in a particular environment. Optimization has to be evaluated in terms of the best possible compromise between the necessary functions but also in terms of cost. "Since money and energy are directly equatable," Vincent writes, "it makes sense to see how natural systems apportion their energy between various functions, and how they design materials, mechanisms and structures."[36] Nature works with minimal energy, at low temperature, with cheap common raw materials. But the cost includes time. Nature has spent billions of years designing and perfecting high-performance structures capable of sustaining life, a length of time that no human, whatever his or her genius, can afford!

Despite this huge time-scale gap between nature and artifacts, it is assumed that there is a formal similarity in human and natural design strategies: given a set of functions to be achieved, nature searched for the optimal compromise between those functions at minimal energy cost. This view of nature as a collection of optimally adapted organisms is not far from the fashionable doctrine of Intelligent Design. In any case, it is a view challenged by a number of evolutionary biologists. Stephen J. Gould and Richard C. Lewontin, for instance, castigated it as "the Panglossian paradigm," a remake of the "everything is made for the best purpose" of Dr. Pangloss, Voltaire's famous hero.[37] Natural selection can be viewed as an optimizing agent only by focusing exclusively on the immediate adaptation of organisms to local conditions. This fragmented view ignores the constraints imposed by the overall architecture, by the phyletic heritage that delimits pathways of development. Every organism and a fortiori every organ in an organism is not optimally designed for its functions because nature must follow an inherited plan. Borrowing from nature local solutions to a specific set of technological problems may be misleading. Materials scientists who consider only the functions to be performed overlook other variables, notably the general constraints that determined the inner organization of each material. Biomimetism should therefore rely on a more integrative and holistic view of its models.[38]

The view of nature as a technology is fruitful as long as the analogy between nature and human technology is established at a global level, considering both of them as "technological systems." Each system—nature and human art—is a distinct entity with a coherence of its own rather than a collection of models and copies. Instead of a simple importation of notions taken from materials science into biomechanics, instead of a simple transfer of models taken from nature into technological contexts, the analogy between nature and technology emulates new promising perspectives and inventive practices of science, such as the ecological sytemic approach of the exchange between nature and humans or the soft chemistry (*chimie douce*), whose object is the study of chemical reactions at ambient temperature in open reactors, much like the chemical reactions that occur in living organisms.[39] Finally, as Vogel points out, we should evaluate the benefits of biomimetism. There is a long tradition of inspiration taken from nature in technology. The "tendency to view nature as the golden standard for the design and as a great source of technological breakthroughs" rests on a number of legends forged by inventors themselves who emphasized their debt to nature.[40]

To what extent can any recent advance in materials technologies really be attributed to biomimetism? To be sure, biomimetism has been fruitful, especially in the domain of biomaterials for drug delivery or artificial organs. It has inspired new methods of synthesis using all the available resources of chemistry and physicochemistry to self-assemble elements or to obtain the rich morphologies of many natural structures. Unlike the products of computational or combinatorial techniques, these products require a lot of craft, skill, ingenuity, imagination, and a dose of *indiscipline*. But most of them are forms of local prowess whose utility is still disputable. And biomimetics is seriously rivaled by nanotechnologists who more simply take all building blocks—molecules, macromolecules, proteins ... as devices or machines and use them as docile servants.[41]

In conclusion, the three case studies presented here attest to the complexity of the interplay between nature and artifact. First, no straightforward diachronic evolution of the representations of nature and artifact can be traced over the twentieth century. Certainly the contrast is striking between the rigidity conferred on nature during the plastic age, and the plasticity or flexibility that biomimetic strategies confer on nature. However, it would be oversimplistic to state that the cult of artificiality prevailing in midcentury prompted a fin de siècle back-to-nature movement. In the same cultural context one can find several coexisting notions of nature and artifact. Combinatorial chemistry suggests a stupid nature, while biomimetics conveys the image of nature as an unsurpassed engineer. Over the twentieth century the concepts of nature and artifact were continuously reshaped. To a certain extent the arrogance of the plastic era and the ambition of computational chemists revived the Promethean mythology attached to chemistry since the beginnings of alchemy. Similarly, the current trends in biomimetics seem to revive Aristotle's notion of art as a copy of nature. The interplay between nature and artifact sounds fairly repetitive. Like a classic play performed in modern costume, advanced technologies seem to reenact old cultural patterns in the language of modern physics and chemistry, with atomic and molecular structures replacing the four principles and substantial forms.

Does this mean that our concepts of nature and artifacts are cultural entities more or less independent from the actual practices of syntheses? To be sure, the representations of technological items are heavily constrained by cultural models. This does not mean they are culturally or socially constructed rather than shaped by the actual processes of design and manufacture. In view of the various synthetic practices examined here, the dilemma between cultural and material determinisms seems extremely reductionist. In relating both the concepts of nature and the concepts of artifact to external factors, one would overlook the creative power of their interplay. Rather than a one-way influence of culture on nature and artifact, these case studies suggest that technological choices are shaped by the kind of relation they engage in with nature. Early synthetic materials, like celluloid and Bakelite, were intended as substitutes for natural materials. Nevertheless, they became successful substitutes not through servile imitation of nature but rather by staking out their distance from nature. The image of plastics was shaped by contrast with a rigid nature. And when materials technologies were aimed at the production of artifacts totally different from those of nature, at new materials with never-before-seen properties that supposedly epitomized the domination of mankind over nature, of spirit over matter, the most successful strategy proved to be the imitation of the most modest natural materials like parts of insects and seashells. Nature and artifact are like a pair of infernal twins playing tricks on the people around them. They are mutually defined by an ambivalent relationship of connivance and rivalry.

Whatever the images attached to the notions of nature and artifact, their polarity is what defines the two terms. It thus seems impossible to escape the circle mentioned in the introduction to this paper. The circle, however, is not necessarily "vicious." Rather it illustrates the complex status of the great dichotomies that shape our culture, our perception of nature being determined not only by the art-nature duality but also by the other ancient divide between nature and society. The great divide between nature and artifact is operational at two levels. The views of nature as a clock, as a laboratory, as a computer program, or as an engineer belong to the nebulous domain of mentalities, or uncontrolled mental representations underlying technological or social practices. At the same time, they act as consciously controlled and highly sophisticated heuristic models, contributing both to the understanding of nature and to technological innovation.

## Notes

1. Maurice Merleau-Ponty, *La nature*, Cours 1956–1957, 120: "Nous ne pouvons penser la nature sans nous rendre compte que notre idée de nature est imprégnée d'artifice."

2. For a cultural history of plastics in the American context see Jeffrey L. Meikle, *American Plastic: A Cultural History* (New Brunswick, NJ: Rutgers University Press, 1995); for an overview of plastics in the British context see Susan T. I. Mossman

and Peter J. T. Morris, eds., *The Development of Plastics* (London: The Science Museum, 1994).

3. Robert Friedel, *Pioneer Plastic: The Making and Selling of Celluloid* (Madison: University of Wisconsin Press, 1983).

4. The only domain where celluloid was uniquely suited and eclipsed all rivals—the films for photography and cinematography—both caused its triumph and its defeat. While celluloid generated a new technological concept—roll film—less flammable substitutes were actively sought out (Friedel, *Pioneer Plastic*).

5. See Charles Eastlake, *Hints on Household Taste* (1872), quoted by Friedel, *Pioneer Plastic*, 88.

6. Aristotle, *Politics*, I.2.1252b.

7. John K. Mumford, *The Story of Bakelite* (New York: Robert L. Stillson, 1924).

8. Edwin E. Slosson, "Chemistry in Everyday Life," *Mentor* 10 (April 1922): 3–4, 11–12, quoted by Jeffrey L. Meikle, "Plastic, Material of a Thousand Uses," in Joseph J. Corn, *Imagining Tomorrow: History, Technology, and the American Future* (Cambridge, MA: MIT Press, 1986), 77–96.

9. Williams Haynes, *Men, Money, and Molecules* (New York: Doubleday, Donan & Company, 1936), 155. In Williams Haynes's view, the Cassandras who, considering the limits of natural resources, announced the collapse of the industrial revolution, ignored the coming of the "chemical revolution." Chemical power would displace mechanical power and restore stable relations between modern culture and nature.

10. Susannah Handley, *Nylon: The Story of a Fashion Revolution* (Baltimore: Johns Hopkins University Press, 1999). To create a popular euphoria, a gigantic two-ton model of a woman's leg was exhibited. On the chemist's crusade and the creation of the slogan "Better things, for better living ... through chemistry," see David J. Rhees, *The Chemists' Crusade: The Rise of an Industrial Science in Modern America, 1907–1922* (PhD dissertation, University of Pennsylvania, 1987); "Corporate Advertising, Public Relations and Popular Exhibits: The Case of Du Pont," in B. Schroeder-Gudehus, ed., *Industrial Society and its Museums 1890–1990* (London: Harwood Academic Publishers, 1993), 67–76.

11. Roland Barthes, *Mythologies* (Paris: Denoel-Gonthier, 1971), 171–173.

12. Jean Baudrillard, *Le système des objets* (Paris: Gallimard, [1968] 2000), 204.

13. J. L. Meikle, "Beyond Plastics: Postmodernity and the Culture of Synthesis," Working Paper no. 5, in David E. Nye, Charlotte Granly eds., *Odense American Studies International Series* (Odense: Odense University, 1993), 1–15 (quote on 12).

14. Al Globus, John Lawton, and Todd Wipke, "Automatic Molecular Design Using Evolutionary Technique" *Nanotechnology* 10 (1999): 290–299; David E. Clark ed., *Evolutionary Algorithms in Molecular Design* (Weinheim: Wiley-VCH, 2000).

15. A. J. Hopfinger "Computational Chemistry, Molecular Graphics and Drug Design," *Pharmacy International*, September 1984, 224–228.

16. See the website of the Georgia Tech Center for Computational Materials. http://www.prism.gatech.edu/-ph279sw/group.html.

17. K. Eric Drexler, *Engines of Creation* (New York, Anchor Press/Doubleday, 1986); see also "The Coming Era of Nanotechnology" in Tom Forester ed., *Materials Revolution: Superconductors, New Materials, and the Japanese Challenge* (Cambridge, MA: MIT Press, 1988); *Unbounding the Future: The Nanotechnology Revolution* (New York: Morrow, 1991); K. Eric Drexler *Nanosystems: Molecular Machinery, Manufacturing, and Computation* (New York: Wiley, 1992).

18. X.-D. Xiang et al., "A Combinatorial Approach to Materials Discovery", *Science* 268 (1995): 1738–1740.

19. The relevance of the term *library* for the storage of molecules is questioned by Roald Hoffmann in "Not a Library," *Angewandte Chemie, International Edition* (abbreviated as *Angew. Chem. Int. Ed.*), 40, no. 18 (2001): 3337–3340.

20. Pierre Laszlo, "Handling Proliferation," *Hyle* 7 no. 2 (2001): 125–140 (quote on 128).

21. In a paper titled "Unnatural Acts," Roald Hoffmann reported the case of a chemist who created a slightly different structure of DNA, with hexoses instead of pentoses as the sugar building blocks of nucleic acids. In doing "what nature chose not to do," he created "an alternative universe," which did not work but could help understand why "normal" DNA works. See Roald Hoffmann, "Unnatural Acts," *Discover* (August 1993): 21–24.

22. On affinity tables see Isabelle Stengers, "Ambiguous Affinity: The Newtonian Dream of Chemistry in the Eighteenth-Century," in Michel Serres, ed., *A History of Scientific Thought* (Oxford, Blackwell, 1995), 372–400; Lissa Roberts, "Setting the Table: the Disciplinary Development of Eighteenth-Century Chemistry as Read through the Changing Structure of Its Tables," in Peter Dear, ed., *The Literary Structure of Scientific Argument* (Philadelphia: University of Pennsylvania Press, 1991), 99–132.

23. Roald Hoffmann, "Not a Library," *Angewandte Chemie, International Edition* 40, no. 18 (2001): 3337–3340.

24. Miroslav Radman, "Fidelity and Infedility," *Nature* 413 (September 13, 2001): 115.

25. See B. Bensaude-Vincent, "The Construction of a Discipline: Materials Science in the USA," *Historical Studies in the Physical and Biological Sciences* 31, part 2 (2001): 223–248.

26. Originally the term *composite* was used in conjunction with *reinforced plastics*. The U.S. Society for Plastic Industries had a Reinforced Plastics Division, which was

renamed Reinforced Plastics and Composites Division in 1967. In France, a bimonthly magazine titled *Plastique renforcé/Verre textile* published by the professional organization bearing the same name, started in 1963 and was rechristened *Composites* with *Plastique renforcé/verre textile* as a subtitle in 1983. See Bryan Parkyn "Fibre Reinforced Composites," in Susan T. I. Mossman, and Peter J. T. Morris, eds., *The Development of Plastics* (London: The Science Museum, 1994), 105–114, and Bernadette Bensaude-Vincent, "The New Science of Materials: A Composite Field of Research," in Carsten Reinhardt, ed., *Chemical Sciences in the 20th Century Bridging Boundaries* (Wiley-VCH, 2001), 258–270.

27. H. A. Lowenstam and S. Weiner, *On Biomineralization* (Oxford: Oxford University Press, 1989).

28. *The Self-Made Tapestry: Pattern Formation in Nature* (Oxford: Oxford University Press, 1999). Philip Ball, "Natural Strategies for the Molecular Engineer," *Nanotechnology* 13 (2002): 15–28.

29. Steven Boxer, quoted in "Exploiting the Nanotechnology of Life," *Science* 254 (November 29, 1991): 1308–1309.

30. J. P. O'Brien, S. R. Fanhenstock, Y. Termonia and K. H. Gardner, "Nylons from Nature: Synthetic Analogs to Spider Silk," *Advanced Materials* 10 (1998): 1185; H. Arribart, "Du biomimétisme à l'ingéniérie génétique," in C. Sanchez, ed., *Biomimétisme et matériaux* (Observatoire français des techniques avancées, vol. 25, 2001), 139–143.

31. M. Sarikaya and I. Aksay, eds., *Biomimetics: Design and Processing of Materials* (Woodbury: AIP Press, 1995).

32. S. J. Wilson and M. C. Hutley, "The Optical Properties of Moth-Eye Anti-Reflection Surfaces," *Optical Acta* 29 (1982): 993. See also Serge Berthier, "Biomimétisme et structure des lépidoptères" in C. Sanchez, ed., *Biomimétisme et matériaux* (Observatoire français des Techniques avancées, vol. 25, 2001), 117–126.

33. Stephen Mann, "Crystallochemical Strategies" in Stephen Mann, John Webb, Robert J. P. Williams, eds., *Biomineralization: Chemical and Biological Perspectives* (Weinheim, VCH, 1989), 35–62 (quote on 35).

34. Steven Vogel, *Cat's Paws and Catapults: Mechanical Worlds of Nature and People* (New York: Norton, 1998).

35. Julian Vincent, "Structural Biomaterials and Biomimetic Strategies," in C. Sanchez, ed., *Biomimétisme et matériaux*, 313–324.

36. Julian Vincent, "Structural Biomaterials and Biomimetic Strategies," in C. Sanchez, ed., *Biomimétisme et matériaux*, 313–324 (quote on 315).

37. S. J. Gould and R. C. Lewontin, "The Spandrels of San Marco and the Panglossian Paradigm: A Critique of the Adptationist Programme," *Proceedings of the Royal Society of London*, Series B, vol. 205, issue 1161 (September 21, 1979): 581–598.

38. Advocates of evolutionary models in technology and economics develop more integrative approaches to technological products. Their holistic perspective challenges the idea that the winning technologies are always the optimal ones because of the "path dependency." See Paul David, "Understanding the Economics of QWERTY: The Necessity of History," *Economic History and the Modern Economist* (Oxford: Blackwell, 1986); John Ziman, ed., *Technological Innovation as an Evolutionary Process* (Cambridge: Cambridge University Press, 2000).

39. B. Bensaude-Vincent, H. Arribart, Y. Bouligand and C. Sanchez, "Chemists at the School of Nature," *New Journal of Chemistry* 26 (2002): 1–5.

40. Vogel, *Cats' Paws and Catapults*, 249–275. Among the most famous examples of successful copies are the Crystal Palace—designed by Joseph Paxton—whose roof allegedly copied a giant water lily; the spinneret for extruding textile fibers inspired by the organ of silkworms; barbed wire; and the Velcro invented by the Swiss engineer Georges Mestral on the model of the hooked burs that clung to his socks.

41. B. Bensaude-Vincent, "Two Cultures of Nanotechnology?", *Hyle: International Journal for Philosophy of Chemistry* 10, no. 2 (2004): 67–84.

# 14

## Concluding Comments

Roald Hoffmann

As I think about the natural and the artificial, the subject of our debate, I feel as if I am drawn into a dreamlike dance, at once ancient and mod. The dance floor? The vernal landscape of the hills around Delphi, of the temple itself. From time to time the temple grounds metamorphose into the launching pad at Cape Canaveral. People move across these spaces, across time, to their own music. It is a dream after all, a dream of the world of real things and of ideas, ideas of nature and culture. In it people set up distinctions, trying to understand the world. And then the recalcitrant world, and they themselves—the people who so neatly divvied things up—take a hard look at those distinctions and see that they do not really work. But just when they thought things were settled—lo and behold something impels them to come back, to seek out the same differences. The dance is the back-and-forth of this motion.

I am drawn into this dance by my own science, chemistry. In it I see a history, of organic and inorganic chemistry. Which was "institutionalized," giving rise to disciplines with boundaries that do seem like nature carved at its joints. So people speak comfortably of natural products, yet only with a smile of the synthesis of unnatural products.

Remaining in my discipline, I nevertheless see a driven, conscious subversion of the categories (detailed by Bernadette Bensaude-Vincent in her chapter)—the use of natural enzymes to make unnatural molecules, the insertion of new segments of genetic code instructing the synthesis of so far unknown new amino acids, probing DNA made with sugars that use hexoses and not pentoses. Or speeding up evolution in the laboratory. Cleverness in this molecular dance is choreographed in not only mimicking nature but in improving on it—what echoes of the aim of the alchemists (worked out in William R. Newman's chapter), played out today!

The dance is not just of scientists, it is of thinking and feeling people reflecting on the world and our actions. As we look at anything, either historically or in depth just at what simply is (we have done both in this book), we come to the seemingly inevitable conclusion—we must accept the lack of a deep distinction between the natural and the artificial.

My chemical colleagues like this a lot. Too often, we are typed as the producers (in ton quantities) of the unnatural. We like this particular mixing of categories, because it opens the way for all of us to have a reasonable discussion about the value of the artificial, a discussion free of "prejudice" about the inherent value of the natural. Leave that for natural food ads.

But can we stop here, in the deconstruction of difference? I do not think so, not in our time. The transformative capabilities of one species have led to a situation where half of the N or S atoms in our bodies have once seen the inside of a factory—in the Haber-Bosch process for making ammonia, or a sulfuric acid plant, on to chemical fertilizers, on to wheat, on to the bagel we ate this morning. The magnitude of our creation has made us think seriously (many of us, not all) about the ecological consequences of the bounty of creation.

So now there is a new imperative, a driving beat, for the dancers. We are back, it seems, to in fact postulating a natural world, and distinguishing the effects of human action on it. How can we preserve the world, if we do not make distinctions between what was before the explosion of human transformative capability and what came after?

Today the dancers segregate. Technocrats and greens. For a while it seems simple—which side are you on? But the fecund complexity of the real world, fed by chance, will not let order, false or real, stay put. In the dance, in thinking about the way the world really is, human intervention returns as a reality with its own value. There is no pristine landscape; the Amazon basin has been settled for thousands of years. We cannot help intervening, for self-preservation (food), or for cultural values (saving the old, both people and structures). All the choices open to us are problematic. We want simplicity, we get . . . a calculus of risks.

There may be a way to reach an equilibrium, precarious as it might be. Nature and culture, the natural and the artificial, are inseparably intertwined. Let's accept that, and not judge the man- or woman-made as automatically suspect. But let's make sure that to every act of creation or transformation there is attached an act of ethical judgment—is this particular creation of value or harm to human beings, other species, our planet? The judgment I ask for could be viewed as a human brake on our runaway, most human creativity. It is another intervention, natural in its own way. The dance goes on.

# Contributors

**Bernadette Bensaude-Vincent**, Département de Philosophie, Université de Paris X

**Horst Bredekamp**, Kunstgeschichtliches Seminar, Humboldt Universität zu Berlin

**John Hedley Brooke**, Ian Ramsey Centre, Oxford University

**Dennis Des Chene**, Department of Philosophy, Washington University at St. Louis

**Alan Gabbey**, Department of Philosophy, Barnard College

**Anthony Grafton**, Department of History, Princeton University

**Roald Hoffmann**, Department of Chemistry, Cornell University

**Thomas DaCosta Kaufmann**, Department of Art and Archeology, Princeton University

**William R. Newman**, Department of History and Philosophy of Science, Indiana University

**Jessica Riskin**, Department of History, Stanford University

**Mark Schiefsky**, Department of Classics, Harvard University

**Heinrich von Staden**, School of Historical Studies, Institute for Advanced Study

**Francis Wolff**, Département de Philosophie, Ecole normale supérieure

# Index